APPLICATIONS FOR ADVANCING ANIMAL ECOLOGY

Wildlife Management and Conservation

Paul R. Krausman, Series Editor

Applications for Advancing Animal Ecology

Michael L. Morrison
Leonard A. Brennan
Bruce G. Marcot
William M. Block
Kevin S. McKelvey

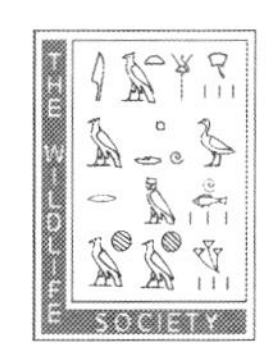

Published in Association with *THE WILDLIFE SOCIETY*

JOHNS HOPKINS UNIVERSITY PRESS | BALTIMORE

Printed in the United States of America on acid-free paper
9 8 7 6 5 4 3 2 1

Johns Hopkins University Press
2715 North Charles Street
Baltimore, Maryland 21218-4363
www.press.jhu.edu

Library of Congress Cataloging-in-Publication Data

Names: Morrison, Michael L., author. | Brennan, Leonard A. (Leonard Alfred), author. | Marcot, Bruce G., author. | Block, William M., author. | McKelvey, Kevin S., author.
Title: Applications for advancing animal ecology / Michael L. Morrison, Leonard A. Brennan, Bruce G. Marcot, William M. Block, Kevin S. McKelvey.
Description: Baltimore : Johns Hopkins University Press, 2021. | Series: Wildlife management and conservation | Includes bibliographical references and index.
Identifiers: LCCN 2020028669 | ISBN 9781421440712 (hardcover) | ISBN 9781421440729 (ebook)
Subjects: LCSH: Animal ecology.
Classification: LCC QH541 .M584 2021 | DDC 577—dc23
LC record available at https://lccn.loc.gov/2020028669

A catalog record for this book is available from the British Library.

Special discounts are available for bulk purchases of this book. For more information, please contact Special Sales at specialsales@jh.edu.

Johns Hopkins University Press uses environmentally friendly book materials, including recycled text paper that is composed of at least 30 percent post-consumer waste, whenever possible.

Contents

Preface

Our purpose in writing this book was to provide guidance on how to substantially advance an understanding of animal ecology through the application of more-rigorous sampling and analytical practices than are generally used by scientists in this discipline. Such a goal indicates our view that animal ecologists have often been failing to advance how research is conducted and how that knowledge is applied to successfully conserve wild animals. Despite improvements in sampling techniques and analytical methods, animal ecologists continue to base studies on vague and misleading terminology, carry out research that is primarily relevant to a specific time and place, and worry far more about statistical analyses and the nuances of modeling, as well as devising unnecessarily abstruse terminology, than about the actual study design that was used to collect the data. In aggregate, this has led to a failure to create investigations that substantially advance our knowledge and provide useful guidance for natural resource managers.

This book builds directly on the animal population concepts we featured in a companion book, *Foundations for Advancing Animal Ecology* (Morrison et al. 2020). Our argument in both of these volumes is that the furtherance of knowledge and the conservation of animals is best brought forward by emphasizing the study of biological populations from the standpoint of individual organisms. Although many animal ecologists understand how to design and conduct rigorous investigations, most fail to implement that understanding in their research. In *Foundations*, we showed that designing studies around ambiguous and unclear concepts, such as animal metacommunities, might describe some observed patterns of species distributions, but these are clearly not helping to advance our comprehension of the underlying causal elements of animal ecology. We therefore, in these 2 volumes, have focused on how individual animals are organized as biological populations. A misunderstanding of broader temporal-spatial relationships and anthropomorphisms about how animal species perceive their surroundings—the landscape—further inhibits how we approach research in animal ecology and make subsequent management recommendations. Although we are not the first to recognize the weaknesses in current approaches to animal ecology, our 2 books attempt to synthesize where we have been, where we are currently, and the changes that need to occur if we are to produce pertinent and information-rich basic research and applied studies of animals and their ecology.

In this volume, we emphasize study design, including observational and experimental approaches that dictate where, when, and how to gather measurements for the study of animal ecology; measurements of behavior (i.e., behavioral ecology); modeling approaches; and a summary chapter on recommended future approaches in animal ecology. It can stand alone as a guide for practicing professionals, as well as university students at all levels who have taken a general course in animal ecology, although we draw from the conceptual framework presented in Morrison et al. (2020).

Some of the topics we have included here are updated and expanded from those previously developed in *Wildlife-Habitat Relationships: Concepts and Applications* (Morrison, Marcot, and Mannan 2006). The current volume, however, draws from additional years of thought, discussion, and our own field studies, modeling projects, and data analyses, driven largely by the realization that animal ecology is in need of a new rigor and perspective to provide not just descriptions, but a better understanding of research needs and improved applications of them. Further, the measurement and analytical methods associated with the study of animal ecology have been entirely transformed over the last decade. Although our focus is on terrestrial and aquatic vertebrates, we recognize the rich literature and knowledge base on other animal taxa, particularly the mega-diverse world of invertebrates. We contend that our focal interest in organismic ecology as a basis for understanding higher levels of biological organization can equally apply to these other taxonomic groups.

Our intended audience includes upper-division undergraduates and graduate students in courses and seminars on animal ecology, wildlife ecology, wildlife management, land-use policy, and conservation biology. Although it would be helpful to have first read our companion book (*Foundations for Advancing Animal Ecology*), the approach, sampling methods, and analytical procedures we present here are accessible to anyone with a background in animal ecology and a grasp of basic statistics and data analysis. We are confident that this volume also will be an invaluable resource to researchers, professionals, and practitioners of natural resource management in public and private sectors, including those in state and federal agencies and non-governmental organizations, as well as restorationists and environmental consultants.

Our presentation of sampling methods and measurements takes a practical approach, outlining study design and experimentation and addressing key factors in the ways in which we measure and quantify animal behavior. Throughout these chapters, we describe various sampling methods and indicate how the data that were collected links back to addressing issues of animal distribution and abundance that are relevant to the biological population. We show how a failure to assess the boundaries of biological populations leads to descriptions of habitat use that are highly biased, which can create gross misunderstandings in what is driving species' occurrences in time and space and, thus, ultimately in population persistence. The issues we discuss concerning a general lack of rigor applied in common study designs have been raised by multiple authors over many decades, but that advice is seldom followed. To address this, we also include discussions of how to apply what is described in the chapters on data collection and analysis. We supply practical templates for designing studies that emphasize an initial, thorough assessment of the biotic and abiotic factors involved, plus various environmental conditions that are likely to be driving animal occurrence and abundance through time and space. We circle back to our major themes—the biological population and boundary identification—as we provide guidance on how to place any investigation in the overall context of co-occurring animal populations and the environment.

We offer this material as a road map for advancing how we study animals, which should substantially enhance communications between scientists and practitioners. Researchers must be cognizant of management needs and objectives, and managers must understand, embrace, and apply the information—including its strengths and uncertainties—needed to move forward to meet those objectives.

Numerous individuals provided support and guidance throughout this project. At Johns Hopkins University Press, we want to especially acknowledge Tiffany Gasbarrini, senior acquisitions editor for life sciences, for guiding this work. We also thank Esther

Rodriguez, editorial assistant, for her substantial efforts in pulling it together. Block appreciates the continued support of the Rocky Mountain Research Station and acknowledges insightful reviews from Jamie Sanderlin and Victoria Saab on a couple of chapters. Brennan would like to thank the founders of the C. C. Winn Endowed Chair for Quail Research in the Richard M. Kleberg Jr. Center, which supplied the space and time for him to contribute to this project, as well as Joseph Buchanan, who provided review comments for several chapters. Marcot acknowledges support from the Pacific Northwest Research Station, USDA Forest Service. McKelvey acknowledges support from the Rocky Mountain Research Station and thanks Jessie Golding, at the University of Montana, for reviews of several chapters. Joyce Van De Water provided graphic design and artwork throughout the book. Mention herein of commercial products and services does not constitute endorsement by the US government.

About the Authors

Michael L. Morrison, PhD
Professor and Caesar Kleberg Chair in Wildlife Ecology and Conservation
Department of Rangeland, Wildlife, and Fisheries Management
Texas A&M University
College Station, Texas 77843-2138

Leonard A. Brennan, PhD
Professor and C. C. Winn Endowed Chair for Quail Research
Caesar Kleberg Wildlife Research Institute and Department of Rangeland and Wildlife Sciences
Texas A&M University—Kingsville
Kingsville, Texas 78363

Bruce G. Marcot, PhD
Research Wildlife Biologist
USDA Forest Service, Pacific Northwest Research Station
620 S.W. Main Street, Suite 502
Portland, Oregon 97208

William M. Block, PhD
Emeritus Scientist
USDA Forest Service, Rocky Mountain Research Station
2500 S. Pine Knoll Drive
Flagstaff, Arizona 86001

Kevin S. McKelvey, PhD
Research Ecologist
USDA Forest Service, Rocky Mountain Research Station Forestry Sciences Laboratory
800 East Beckwith Avenue
Missoula, Montana 59801

APPLICATIONS FOR ADVANCING ANIMAL ECOLOGY

1 The Experimental Approach in Animal Ecology

Scientists these days tend to keep up a polite fiction that all science is equal.
J. R. Platt (1964:347)

Introduction

The adage that "art changes but science advances" encapsulates an important difference between these 2 pursuits. Art and science are 2 of the most important aspects of human culture. People have undertaken artistic and scientific pursuits for centuries, with the goal of improving our views and understanding of the world around us. Activities in both the arts and the sciences are influenced by our past actions, our knowledge, our perspectives, and our culture. Over time, scientists have developed sets of rules that are followed when we undertake scientific efforts intended to pursue and generate new knowledge about the world. Artists, in contrast, have far more flexibility in their pursuits, because they are constrained by fewer, if any, rules that dictate their approaches to creativity. Such is the case when the search for beauty, or art, differs from the search for truth, or science.

The study of the relationships between animal populations and their habitats is no different from any other scientific endeavor, in that the scientific method is an important tool that can be used to generate new knowledge. In the broadest sense, scientific thinking is the logical thought process one might use to discover, for example, why a car will not start by asking questions such as "Is the battery charged?" or "Is there gas in the tank?" or "Is a wire loose?" The activities of scientists who strive to understand how the universe works are called the "scientific method" (Fig. 1.1). This method appears to be easy to follow when described in its simplest form, yet its application, particularly in ecology and related fields, has generated a considerable amount of discussion and debate. These have focused on which activities are most useful or important (e.g., Platt 1964; Romesburg 1981; Peters 1991) and how to perform these activities correctly (e.g., Murphy 1990; Nudds and Morrison 1991; Drew 1994; Sells et al. 2018). Simply telling someone to "follow the scientific method" when designing and carrying out a study, although good advice, is inadequate as a set of instructions for what a person must do to conduct a successful investigation. In part, this is because there is no single "scientific method" that applies to all inquiries. Science proceeds by rigorous investigation: posing and testing hypotheses, seeking any falsifications of those hypotheses, adhering to a logical and credible interpretation of observations, and acceding to reviews and commentary by peers. There is no universally applicable set of actions, in the sense of cookbook instruc-

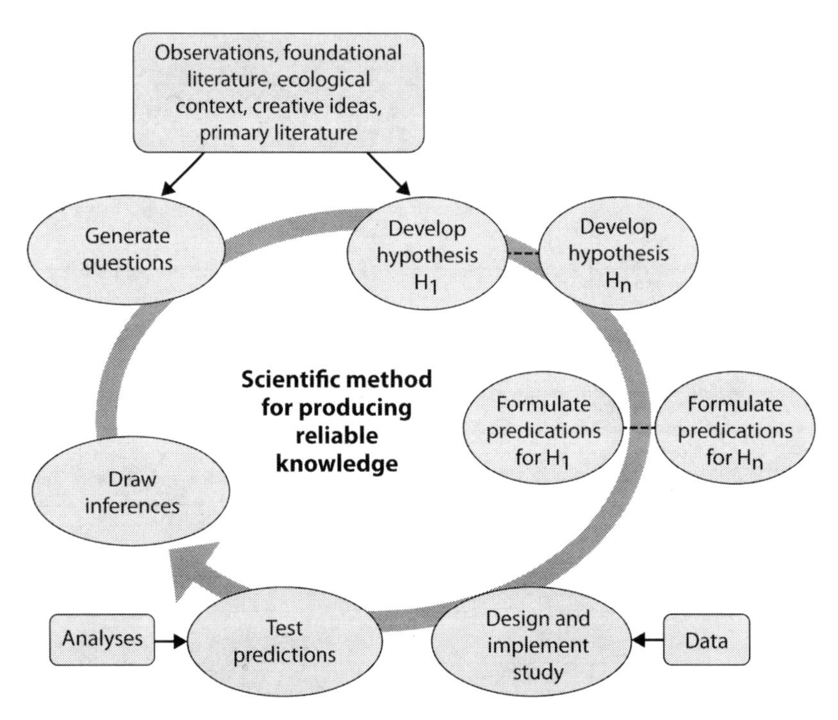

Figure 1.1. In animal ecology, a rigorous scientific method for producing reliable knowledge follows a series of logical steps to answer questions about the natural world. Each step is fundamental to the next. Inferences help inform new questions in future studies. (Adapted from Sells et al. (2018), with the permission of The Wildlife Society)

tions, particularly in the realm of ecology, where the subject matter itself is labile and resists control.

The purpose of this chapter is to build on the concepts established by Morrison et al. (2020:Chapter 1) and describe how what we designate as "the experimental approach" is used to advance the science of animal ecology. We start with a thumbnail sketch of some historic benchmarks that ultimately take us to present-day applications of this practice in animal ecology. To this end, our goal is to provide animal ecologists with an appreciation of the wide range of comparative and experimental approaches available today, as well as with how they were used to advance knowledge in the past. This chapter provides an overview of the activities associated with the scientific method and the ways in which it is related to the study of animal populations and their habitats. We also present some issues about the scientific method that have generated discussions among animal ecologists. Further, we cover the design and application of experiments in animal ecology and wildlife science, as well as some of the problems inherent in experiments under both field and laboratory situations.

We focus on the experimental approach because it is a powerful tool that has been underutilized by animal ecologists. We do not review specific statistical models, even though the design of an experiment is usually intimately linked to a statistical model of 1 kind or another. Such a review is beyond the scope of this chapter, and several excellent texts on statistics and experimental design are available (e.g., Skalski and Robson 1992; Kuehl 1994; Morrison et al. 2001; Scheiner and Gurevitch 2001; Ramsey and Schafer 2002; Guthery 2008). In addition, the statistics you apply will not repair data generated from a poorly designed study, especially after collecting a small and probably biased set of samples (see Morrison et al. 2020 for our emphasis on designing studies with reference to the biological population). We end the chapter with a discussion of when to use field and laboratory experiments and what to do when tightly controlled experiments cannot be done, because of practical constraints and uncontrollable conditions.

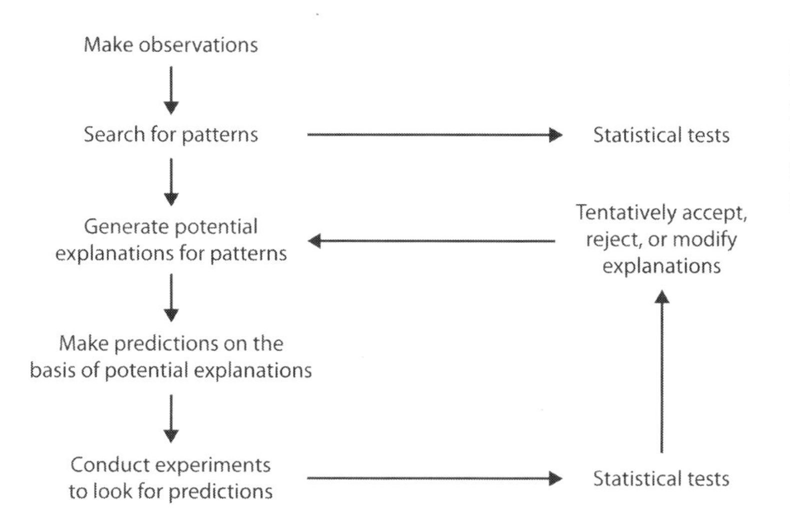

Figure 1.2. A diagram of activities in the scientific method and common places where statistical tests are employed. (Adapted from Morrison et al. (2006), with the permission of Island Press)

The Scientific Method

Carey (1994:5) defined the scientific method as "a rigorous process whereby new ideas about how some part of the natural world works are put to the test." Descriptions of the scientific method often include 3 to 6 steps or activities, depending on how they are combined: (1) making observations; (2) searching for patterns among observations; (3) generating potential explanations for patterns detected; (4) deducing predictions from the proposed explanations; (5) testing potential explanations by looking for predicted phenomena, often after some experimental manipulation; and (6) tentatively corroborating, rejecting, or modifying the proposed explanations, depending on how well their predictions were met (Fig. 1.2). In classic methodology, hypotheses are often stated as null conditions and are either falsified or fail to be falsified, but they are never "proved."

Observations are the starting point of science (Simpson 1964:88), and natural history is the underpinning of animal ecology (Herman 2002). Some philosophers argue about whether what we perceive is real, but most ecologists are willing to accept that our senses, or the sampling tools and instruments we develop, provide us with a good approximation of the existing world (Simpson 1964). That is not to say that our observations are free of error, but numerous views by different individuals from various points in space and time generally help separate observational errors from reality (Matter and Mannan 1989).

Comparisons and Experiments

Scientific studies can take a gradient of forms, ranging from literature reviews and expert judgments; to demonstrations, anecdotes, retrospective studies, and observational reports; to rigorous experimental investigations under controlled conditions (Marcot 1998). Simple natural history observations are on 1 end of this continuum, with controlled experiments on the other. In between are comparisons or quasi-experiments (where "controls"—that is, units that are not treated or otherwise manipulated—are usually absent)—the latter are sometimes called "management" experiments, because they arise from management actions—and "natural" experiments (the result of some natural disturbance, such as a fire or large storm). All experiments are comparisons in some form or another. While this may seem overly simplistic as a definition, how comparisons are structured and organized determines what kind of experiment, or even whether any experiment at all, is being conducted. In most cases, use of the term "experiment" means that some kind of no-treatment control is part of the scheme, designed to be evaluated along with treatment outcomes, and thereby detect formerly unrecognized phenomena in the natural world.

Experimentation is an excellent mechanism for increasing the amount of inference we can gain from our studies. Controlled experiments in the life sciences most often happen in laboratories, but, to a degree, they can also be conducted outdoors, given adequate space and resources. The literature on the design and execution of controlled animal experiments in the laboratory is vast and diverse (see a recent review by Liu and Fan 2018). In contrast, the literature on field-based experimental examples of controlled studies in animal ecology is quite limited, especially on wild populations. As is noted in many places throughout this book, most ecological studies are essentially anecdotal or observational, with some containing statistical analyses. Nevertheless, there are outstanding examples of manipulative field experiments, even though limited in number, that provide strong inferences for populations of wild animals (Brinkhof et al. 1997; Calsbeek and Sinervo 2002; Lecomte et al. 2004; Gruebler and Naef-Daenzer 2010). In this chapter, the bulk of attention is paid to comparative or experimental investigations conducted in the field with wild animals (e.g., Swaisgood et al. 2018).

Trial and error was probably the first, and certainly the simplest, kind of experimental comparison conducted by humans (although some might reasonably argue that trial and error is quasi-experimental). In the strictest sense, a trial-and-error approach is not a true experiment, because there is not a no-treatment control, nor are there even any kinds of non-controlled comparisons, other than the outcomes from different trials over time. Trial and error simply relates to attempts to make something work or improve how it works. If a trial is unsuccessful, then the investigator endeavors to overcome the resulting error by conducting a new trial, using a different approach. The controls in trial-and-error experiments (to use both terms loosely) are past efforts that did not work, as opposed to some kind of trial, or treatment, that did succeed. As an example, local communities in northeastern India have used such "experiments" to develop democratic approaches to managing and conserving natural forest ecosystems (Tiwari et al. 2013).

As another example, consider the invention of the wheel. The first wheels that we know of were not used for transportation, but rather for stationary tasks, such as shaping pottery. At some point, someone conceived the idea of turning these wheels upright, in order to roll them across the ground, and using them for transportation. While we know little about this transition, trial and error was no doubt fundamental to the process, which, while largely taken for granted today, is among the most revolutionary inventions in the history of humankind. There is no doubt that trial and error has played a key role in the development of other crucial advances in human culture, such as finding ways to start new fires when existing fires could not be found and cooking meat (as opposed to eating it raw), not to mention the thousands of trials (and errors) that finally led to Thomas Edison's discovery that carbonized bamboo provided a filament in lightbulbs that utilized low amounts of electricity and was reasonably durable.

Without a doubt, knowledge gained from trial and error has been fundamental to our understanding of the world around us. The problem with this process is that it is an incredibly inefficient way of acquiring new knowledge (and is sometimes dangerous—think of the first people who worked with electricity or distilled petroleum hydrocarbons). Although a trial-and-error approach has been superseded by more-efficient experimental procedures that we describe later in this chapter, it still persists in our culture, sometimes in unfortunate ways.

The construction of suspension bridges to span significant distances in the late nineteenth and early to mid-twentieth centuries provides an excellent set of examples on the potential dangers wrought by using a trial-and-error approach. The advent of the modern suspension bridge started when structural engineers streamlined the dimensions of the massive towers and cables that supported the Brooklyn Bridge (opened in 1883) as they developed their subsequent designs for the George Washington Bridge (opened in 1931) and the Golden Gate Bridge (opened in 1937). The trusses in the original Tacoma Narrows Bridge (opened in 1940) were even further reduced with respect to their mass, which quickly resulted in the collapse of that bridge, due to harmonic resonance from moderate (40 mph) winds only 4 months after it was opened to car traffic. It took about 50 years of trial-and-error design experiments in constructing ever-lighter suspension bridges, with disastrous consequences, before the error (or failure point) was finally identified (Petrowski 1995). The result was a return to deeper and heavier trusses for large suspension bridges.

What kind of linkages are there between using trial and error in the design of modern suspension bridges and using this experimental approach in animal ecology and wildlife management? It turns out that both are liable to contain real and potential pratfalls. As noted above, trial and error is an incredibly inefficient way of procuring new knowledge, because it only allows before-and-after comparisons. If a new trial works better than a previous effort, or if it results in some kind of breakthrough discovery, then great. If there are no differences, however, this is a slow and inefficient way to accumulate new knowledge. Another kind of approach, or information obtained from some other source—such as a direct comparison or an actual controlled experiment—is needed.

A classic example of such a conundrum in animal ecology involved the first attempts to restore decimated populations of wild turkeys (*Meleagris gallopavo*) in the United States. During the early twentieth century, wildlife managers repeatedly attempted to restore their populations by releasing turkeys that were produced in captive game farm flocks back into the wild, a classic trial-and-error experiment that failed time and again, because the introduced populations did not reproduce and thus became extinct. Comparisons of physiological and behavioral attributes of pen-raised and wild-caught turkeys by Leopold (1944) indicated that the wild turkeys had larger pituitary glands than the pen-raised ones, indicating potentially increased levels of the hormones that drive behaviors attributed to "wildness." The implications of these findings were that the trial-and-error approach using pen-raised stocks was a waste of time, and that such efforts might be more successful if they were focused on translocating wild birds into unoccupied habitat. The result of this change in policy was among the most successful wildlife restoration efforts in history (Dickson 1992).

Trial and error, despite being highly inefficient and, many times, ineffective, still plays a prominent role in the management of wild animal habitats and populations today. Procedures such as timber harvests, grazing regimes, prescribed fires, use of herbicides, and mechanical disturbances are applied over vast areas (see our discussion on on adaptive management later in this chapter), with only meager attempts to track or monitor the response of animal populations, if any. For example, if post-treatment differences in the abundance of the species being monitored are noted, it is virtually impossible to attribute such differences solely to management actions, when they also could be related to abiotic factors, such as weather.

Induction, Observational Studies, and Deduction

Inductive and deductive reasoning are 2 approaches to generating new knowledge. Deductive reasoning, as we will demonstrate later in this chapter, is a considerably more efficient and effective way to do so, compared with inductive reasoning. In animal ecology, however, a tradition of inductive reasoning has been dominant for many decades and, to some, it is a preferred logical approach to generalizing patterns from natural history observations (Ricklefs 2012).

Inductive reasoning covers a broad area of logical arguments that falls outside the realm of deductive proof. In mathematics, for example, proof through induction occurs when it can be shown that a recursive process is invariant. More generally, inductive arguments arise whenever it cannot be stated with certainty that a series of axioms, if correct, lead to 1 and only 1 correct solution, which is the requirement for deductive logic.

In induction, cause and effect are probabilistic, rather than deterministic. Sampling, in which an investigator collects specific observations of some phenomena and then uses these observations to make broader generalizations (usually applicable to the sampled population), therefore always leads to inferential conclusions. When engaged in inferential logic, it is important to separate processes where the degree of uncertainty can be bounded from those where the inferential uncertainties are unknown. For example, for a randomly sampled large population, the uncertainty around the arithmetic mean can be formally estimated. For large samples and populations, we can, in practice, treat the axioms associated with the calculation of a mean as essentially deterministic. In chemical reactions, where perhaps Avogadro's number of molecules are involved, uncertainty at the molecular level associated with the equilibrium state is generally ignored, but other analyses can never be treated as simply deterministic.

In Akaike's Information Criterion (AIC)–based model selection (e.g., Burnham and Anderson 2004), a single, particular explanation is more likely to be true than other tested explanations, but exactly how probable it is to be correct is unknown. Therefore, even though AIC-based understandings lend formal statistical support for a specific explanation, these analyses are intrinsically probabilistic and inferential. In animal ecology, and many other fields, it is common to develop inferential models derived from 1 place and time and extrapolate them to others—in essence, applying a second inferential step.

If, for example, we were to observe many individuals of a given species of animal in a sample of sites from a particular type of vegetation and then recognize the association between the abundance of these animals in relation to this vegetation, some level of prediction might be possible. A simple prognosis from this pattern is that we can expect to see individuals of this species when we visit different sites with the same vegetation type. A more complex, second prediction is that the number of individuals of this species will increase if we expand the extent of this vegetation type through some management activity. Several factors, such as whether the species successfully survives and breeds in the relevant vegetation type, will influence whether the second assumption is accurate. The point, however, is that predictions of this kind become possible only after patterns of association are recognized (Guthery 2008). The validity of this practice is entirely extraneous to formal uncertainty estimation and, therefore, is never fully quantifiable.

This very general and non-quantifiable level of inductive logic is probably the most common form that is used when it is applied to animal ecology. We induce process by recognizing patterns (Hanson 1965; Guthery 2008), or by reasoning from particular observations to general ideas, concepts, or theories. Humans have the ability to recognize specific configurations, partly because our brains can catalog and retain information for long periods of time, but distinguishing subtle patterns requires keen observational abilities and considerable amounts of information. Skills associated with observation almost certainly vary among individuals, but they can be enhanced for most people through practice and knowledge.

Scientists usually search for patterns in nature in a more formal manner than casual observers. Scientific observations often are made systematically, or they can come from random samples, with either practice allowing quantification of the uncertainties associated with generalization. If a relationship between a species and a particular vegetation type is suspected, then a series of surveys for that species might be conducted throughout the year, and then over several years, in a random sample of areas within the vegetation type, in order to determine if the relationship is likely to be both real and persistent. Such a study could be enhanced by also sampling outside of this vegetation type, to reaffirm the relationship with it. For example, decades of work initially suggested that northern spotted owls (*Strix occidentalis caurina*) are obligates in old-growth coniferous forests. But it wasn't until people started sampling slickrock canyons like Grand Canyon and those in Zion, Bryce, and Canyonlands National Parks, that they discovered thriving populations of a related subspecies, Mexican spotted owls (*Strix occidentalis lucida*). Statistics or other numerical analyses can be employed to help detect patterns or correlations among variables that otherwise might be difficult to observe (Fig. 1.2).

It can even be argued that science should be defined by being able to provide testable and correct predictions—that is, observation, description, and explanation alone do not constitute science if there is no accurate predictability. To this end, natural history is a necessary underpinning for animal ecology, by providing vital observations and descriptions. Statistical analyses may supply some degree of explanatory power, but demonstrating correct predictions is the hallmark of scientific inquiry and knowledge. If we accept this premise, then consider how much some aspects of animal ecology (such as evolutionary ecology) fall short of effective predictability. The development of evolutionary adaptations in animals can be observed and described, and perhaps even explained, but only in a post hoc manner. We still lack a predictive science of evolution, although the field of genomics seems to provide a possible pathway to predictability in evolutionary studies (e.g., Bridgham 2016).

Further, it is important to understand that a correlation between 2 variables does not imply a cause-and-effect relationship between them. For example, a positive correspondence between the abundances of 2 species—say, mosquitoes and deer ticks (*Ixodes scapularis*)—in an area does not indicate that an increase in the amount of mosquitoes caused a rise in the number of deer ticks, or vice versa. The 2 species simply could be responding to the effects of an increase in rainfall on their respective resources, and therefore appear in greater quantities independently of each other. Conversely, if the abundances of 2 species are positively correlated and 1 species (e.g., flycatchers) depends on another for some needed resource (e.g., woodpeckers, to excavate the flycatchers' nest cavities), then there could be a mechanism relating the covariance of the 2 species. The correlative nature of patterns alone, however, reveals little about these mechanisms, which often have hidden causes underlying and inducing them.

Once patterns among observations have been identified (and perhaps even confirmed) on some level, various individuals might be stimulated to wonder what processes or mechanisms might have created the patterns—that is, they begin to ask why. This speculation can result in the development of potential explanations for the observed patterns. Such speculative explanations are called "research hypotheses" (Romesburg 1981:295), and, as such, they articulate a mechanism or process that might be responsible for the pattern of interest.

The long-standing tradition of naturalists collecting myriad observations (and specimens) of animals and then using insights from these factors to derive generalized theories or inferences is a classic example of inductive reasoning. Darwin's (1859) theory of natural selection is the most profound example of inductive reasoning in the life sciences. Using insights derived from numerous specific observations during his residence in England and those made during a 5-year voyage on the *Beagle*, along with many other lines of evidence, he generalized that species have evolved in response to factors related to a phenomenon he called "natural selection." Despite the cultural opposition of some people to evolution that continues to this day, natural selection is 1 of the few theoretical constructs in the life sciences—and, hence, animal ecology—that has not been falsified or disproven. In a much more limited but still highly influential context, Grinnell (1917) induced the general niche relationship patterns of California thrashers (*Toxostoma redivivum*), made over many months and years from observing and collecting specimens of these birds in California. Thus inductive reasoning has given us 2 extremely important and useful theoretical concepts that are fundamental to animal ecology—natural selection and niche theory—that remain so to this day. But, because conclusions associated with inductive logic are intrinsically probabilistic, failures are also to be expected. Nevertheless, we are optimistic that certain conclusions in animal ecology will change with additional and more rigorous research.

We can achieve higher levels of certainty if we use induction as a step in a larger process. Induction is sometimes used to describe the way in which we generate research hypotheses (Platt 1964; Popper 1981; Medawar 1984), because it is through induction that we identify how to recognize patterns. The intellectual activities associated with hypothesis generation, although they involve pattern recognition, probably require more information, creativity, and insightfulness than pattern recognitions based on induction alone. Neither process, however, is well understood or completely under our control (Matter and Mannan 1989). Thus the word "method" in the term "scientific method" is somewhat misleading if it is interpreted to mean that all people will be able to perform all steps or activities in science equally, if given enough instruction. We emphasize that anyone can improve their chances for participation in the creative aspects of science by being well informed in their area(s) of interest and understanding how the scientific method works.

Potential explanations of patterns or phenomena in nature are constrained by what is already known about the world. They cannot, for example, contradict the laws of physics or many generalized theories, such as evolution based on natural selection. Yet even with these constraints, it is possible (and even likely) that more than 1 research hypothesis can be proposed for a given pattern or phenomenon.

The inductive approach to science is a useful way in which to conduct a study when there are few or no theoretical concepts or coherent causal hypotheses available. As with technical stock market analyses, repeated patterns can be observed and used to predict future events, even with no understanding of why the patterns occur. Indeed, if a pattern is strong and stable, there may be little practical need to seek causes. We have happily used observational models of the effects of gravity to solve myriad problems ever since the time of Isaac Newton, but we still have no clear understanding of what gravity is or what causes it. Probably the vast majority of humanity, including most scientists, don't view this as a problem. A difficulty with this approach is that, like trial and error, it can be an extremely inefficient way to gain knowledge, because the cataloging of patterns is often based on meeting some kind of specific objective, rather than testing a more generalized theory or hypothesis. At its worst, the empty search for patterns can give the appearance of active knowledge acquisition while providing little or no actual information. As noted throughout this book, these issues have been a major shortcoming of animal-habitat studies over the decades, and this is something that we are striving to correct.

For many decades, the introductions in legions of papers in ecological journals contained some kind of statement—usually toward the end of the introduction—to the effect that "our objectives were to estimate the breeding habitat characteristics and home range size of species (fill in the blank) at the (fill in the blank)." Whereas this kind of approach is commendable on 1 level, because it gives focus and structure to a research paper (and was almost assuredly strongly correlated with the proposal that funded such a study), scientists can argue that the widespread use of pattern demonstrations without formal hypothesis testing has severely hampered progress in animal ecology. Nearly 40 years ago, Romesburg (1981) noted that ecology journals are filled with huge masses of unrelated facts, and, as a result, we have scant evidence of reliable knowledge in our field, a situation that still remains today (Sells et al. 2018). In a way, this is a reflection of our culture. Animal ecologists in the United States and elsewhere have long valued natural history, but this attitude is clearly changing (and has been doing so for a long time), not for the better. Moreover, to many people, the process of advancing animal ecology within science is not appreciated, so we are stuck in a paradigm of unrealistic political and societal expectations regarding the means to produce robust knowledge about the ecology of wild animal populations and the habitats that support them.

Why Experiments Are Critical for Advancing Knowledge

The primary purpose of an experiment is to test an idea or hypothesis about how the natural world works. The goal of conducting an experiment is to evaluate the validity of a research hypothesis, often using deductive reasoning, as noted in the section on "Deduction Revisited" below. A secondary purpose is to corroborate or describe the existence or nature of a pattern (i.e., "fact-finding"; see Kneller 1978:116). The design of fact-finding experiments in animal ecology is the same as or similar to that used to test research hypotheses, but in the former, there is no explanation or mechanism being tested, and no predictions are being made, because the purpose is to find out what happens when, through manipulation, some natural event is mimicked. Whereas these fact-finding ventures are usually described as experiments, because they require active manipulation, they are more properly viewed as a branch of observational knowledge acquisition. For the remainder of this chapter we assume that experiments are designed to determine the cause that then results in some type of effect.

Kinds of Experiments

There are many kinds of experiments. What they all have in common is that they contain approaches allowing the separation of potential causes, where some potential causes can be eliminated and, therefore, increase the inference strength associated with the remaining causes. The most common logic is the following: given that everything else is constant, if we change 1 thing, the observed results are due to that single element and nothing else.

This logic can be seen in experimental processes associated with product design. Assume that you would like to build a faster microprocessor. You could change the structure of the memory cache, instruction pipeline, branch heuristics, and so forth of the microprocessor you are designing. You have groups working on each of these issues and would like to test their designs. If you built a chip that incorporated all of these design changes and tested it, you would have no idea which of these alterations were responsible for performance increases. You would have an experimental observation, but it would not allow an expansion of causal knowledge. The importance of determining cause is obvious in this case: 1 revision might actually make the chip slower, with this performance degradation masked by another, very positive change. Logically, you should make chips that only contain 1 alteration, in order to isolate the associated performance differences. If you thought that the changes might interact, you could additionally build chips that include all possible combinations and evaluate them both individually and in groups. In animal ecology, by far the most common experiment takes the form of some sort of treatment where untreated areas, termed "controls," provide the isolating mechanism. Unlike chip design, where between-chip variation is low and would be overwhelmed by the effects of design changes, experiments involving animals are expected to be fraught with large and often uncontrollable variability in all aspects of the experiment. Therefore, experimental design approaches that separate the treatment effects from the "noise" created by all of the uncontrolled changes will be needed.

Experimental Design Considerations

The application of treatments and selection of experimental units are associated with 2 concepts, randomization and replication (Fisher 1947), which help control variability and are integral to the design of an experiment. Randomization is 1 way to protect the experiment from unknown sources of bias (Skalski and Robson 1992; Morrison et al. 2001). When experimental units are selected, there inevitably will be variation between or among them, and these differences could influence the effects of the treatment. Therefore, a designation of which experimental units receive treatment should be made randomly (or randomly with some constraints; see Cox 1958a), so that "a treatment is not consistently favored or impaired by extraneous or unexpected sources of variation" (Skalski and Robson 1992:16). For example, Franzblau and Collins (1980) supplied food at randomly selected territories within their study area, so differences among the territories would not uniformly bias the effects of their treatment. If, instead, they had arbitrarily provided food to territories on the northern half of their study area, some environmental factor in that section could have compromised their experimental results by either ameliorating or enhancing the effects adding food had on territory size.

The responses of variable(s) within experimental units to treatments will usually not be identical, no matter how carefully a test is controlled. Therefore, the responses of several experimental units must be assessed to increase the precision of the estimate of the variable of interest and thus help determine if the predictions of the experiment were met. This approach is called "replication," and each set of treatment and control units is termed a "replicate." The number of replicates needed in an experiment (see Chapter 2) is positively related to the amount of change in it that will mask the effects of the treatment. The greater the amount of variation or error, the larger the number of replicates that will be needed. A host of factors can influence experimental error (see Skalski and Robson 1992 for a review), but it can be estimated from other, similar experiments or from preliminary surveys. Error in an experiment can be reduced through an appropriate design. For example, if experimental units are to be selected along an environmental gradient that will influence the effects of the treatment, 1 way to reduce error is to group experimental units along the gradient and assess the effects of the treatment on groups. This technique, called "blocking," is an attempt to reduce experimental error by assigning as much of it as possible to differences between groups, or blocks (Skalski and Robson 1992). All of this points to the need to have adequate natural history knowledge prior to conducting an experiment. Doing so without such knowledge can result in murky findings, because relevant sources of noise were not adequately addressed. Robust experiments using randomization and replication also require a third element, control. "Control" refers not just to including components that are not subject to specific treatments, but also to ensuring that at least other anticipated major influences within a system are held constant, so they do not confound outcomes.

By now, you should be able to recognize that many aspects of ecological inquiry do not and cannot conform to the strict criteria and standards of randomization, replication, and, especially, control. This is particularly true with studies of populations and landscapes that essentially constitute irreproducible, unique configurations and conditions. Such situations have samples sizes of 1. This is the challenge and the bane of much of animal ecology: trying to understand, explain, and predict causality using the experimental approach. We attempt to induce general principles, as well as deduce and predict the dynamics of ecological systems that, themselves, exist in flux and in contexts of uncontrollable influences from beyond a study area's boundaries (Wiens 2016).

Deduction Revisited

Deductive logic demands the formal elimination of all potential causes except 1, which can be identified unambiguously. The exercise of formal deductive logic is not possible in field experimentation since, if for no other reason, such experiments require induction from samples to populations. As noted above, we also cannot logically and physically eliminate all potential but incorrect causes. There is too much variation in nature, and we don't thoroughly understand the systems we study. We can, however, apply concepts from deductive logic to guide our approach to problems and our reasoning behind the choice of experimental designs.

Romesburg (1981), along with others (such as Guthery 2008), advocated using deductive reasoning whenever possible in animal ecology research. In this form of reasoning, some kind of generalized theory or hypothesis is posed. Then the research investigation is designed around the collection of data, in order to test whether the proposed theory or hypothesis is supported (or fails to be supported). This approach to science has been called "hypothetico-deductive reasoning," or, more recently, a "research hypothesis," as noted by Guthery (2008:18–22).

A research hypothesis is a statement about what a scientist expects to find regarding certain phenomena in nature, based on data collected in an experimental context. For example, rather than saying that "our objective was to measure the biomass of native grass species 5 months after the application of a prescribed fire in 4 pastures on Rancho Cuchomondo" (which is induction with a limited potential inference), an investigator using a deductive approach could state, "Our research hypothesis was that the biomass of native grasses 5 months after the application of a prescribed fire in 4 pastures on Rancho Cuchomondo would be 3 times greater on loamy soils, compared with sandy soils, due to differences in soil fertility." Note that even in this simplified example, the use of a hypothetico-deductive approach to test a research hypothesis, rather than an inductive approach to meet a vague research objective, will result in a much stronger inference from what is essentially the same set of data.

Posing a research hypothesis also allows an investigator to develop answers to both "what" (in this case, the biomass of grasses) and "why" (soil fertility) questions about the study topic. In addition it forces a researcher to think about what kinds of linkages, patterns, and mechanisms might be responsible for the phenomenon under study during the design period, before the study is actually conducted (Guthery 2008:18–22). Furthermore, even if the collected data fail to support the research hypothesis being posed, the investigator can argue that scientific progress has been made. For example, if the data in the bunchgrass biomass example above showed that the post-fire biomass of native grasses on loamy soils was only twice as great as the biomass of these grasses on sandy soils, more inference was obtained from the data than would have been possible from the inductive approach.

In ornithology, ecologists have struggled with understanding how the effects of parental care influence the post-fledgling survival of offspring. Such care is a major expense in terms of time and energy, but it can potentially impact fitness. The first-year survival of offspring is also confounded by changes in environmental conditions. Using coots (*Fulica atra*), Brinkhoff et al. (1997) employed an experimental approach to test the "date hypothesis" versus the "parental care hypothesis." Their results provided support for the latter. Using a similar experimental approach, Gruebler and Naef-Daenzer (2010) found similar support for the parental care hypothesis as the primary factor influencing the post-fledging survival of barn swallows (*Hirundo rustica*).

The model for a hypothetical-deductive test forms the conceptual foundation for designing a hypothetico-deductive experiment, but there is no simple formula or methodology that can be used as a recipe book to provide experimental design details. Each experiment is an exercise in creativity, because the investigator must be able to identify an element that plays an important role in the research hypothesis being tested, manipulate it without changing other important elements, and logically deduce (i.e., predict) what will happen after this manipulation. Consideration of several rules and concepts about experimental design should help strengthen experimental results. Below, we briefly review several factors that should be thought about during the design of experiments. Our coverage focuses on field experiments, because they are more common in animal ecology than those conducted in laboratories.

An important rule associated with experimentation in any branch of science is that an experiment must be repeatable. This means that other scientists must be able to successfully duplicate the experiment, if they wish. The methods, therefore, must not include any subjective assessments of important variables, and they must be carefully and accurately described in a written report or publication. Data from replications must be collected in the same manner as the original study. Field experiments in ecology, however, rarely are repeated. But that is another matter, the implications of which are discussed a little later.

Another important consideration when designing a hypothetico-deductive experiment is that the experimental units involved (e.g., individual animals, plants, or plots of land) must accurately represent the domain to which the hypothesis is supposed to apply. In other words, inferences from an experiment will usually be pertinent only for the locale from which samples were taken (Morrison et al. 2001). We should not formally and statistically measure the strength of an inference associated with extrapolation beyond the system that was studied. But, because of demands to make decisions on imperfect information, we can, and do, infer the findings from 1 study's place and time to other places and times. The sampling frame (Skalski and Robson 1992), or the pool from which samples are drawn, should match the overall statistical population of interest, though in practice this is not often possible. Further, if the hypothesis being tested is about a process that operates within a given vegetation type, the plots used in the experiment ideally should be selected at random from the entire distribution of that particular vegetation type.

In reality, experimental units used in field experiments often are constrained by a lack of access to study areas, a dearth of travel funds or personnel, or any other typical limitation that we face as animal ecologists. Another shortcoming is that we often rely on agencies to implement the treatments in managerial-type or similar forms of experiments. Treatments may be applied inconsistently (e.g., by using different methods or timing), thereby adding more noise and variation to the system. Also, many experiments rely on convenience sampling, since they are conducted on a university's or agency's experimental forest or rangeland. Whether those areas are representative of the larger landscape may be questionable. Moreover, in some situations, experimental units are purposely chosen to be similar to each other, as a way to reduce variation. Thus, for several reasons, experimental units might come from a subset of the target population, and inferences drawn from the experiment are then severely limited (see Skalski and Robson 1992).

Manipulation of an element that is central to the hypothesis being tested is a crucial component of an hypothetico-deductive experiment. In experiments, these manipulations are usually called "treatments." Ideally, important biotic and abiotic factors not being manipulated are controlled by the experimenter. Thus changes in the variable of interest after the manipulation can be attributed to the treatment. In field experiments, these other factors usually are not under direct control, but when they become altered during the course of an experiment, their effects must be assessed. This is done by collecting information about the variable of interest, before and after treatment, both on areas where the treatment was applied and on the controls. As mentioned earlier, controls are experimental units that are not manipulated but are assessed in the same manner as the units that are manipulated or treated (Hairston 1989:24–26). There are no set criteria to determine the period of time over which data on the variable of interest should be collected, but Hairston (1989:24) noted that the lack of adequate baseline (i.e., pretreatment) data was a common failing among ecological experiments (see the examples below). Measurements after treatment should extend at least through the period during which the variable of interest is expected to change.

An assessment of such changes before and after treatment on the affected areas, as well as on the controls, provides a measure of how much the variable of interest would have altered in the absence of the manipulation, if other key influences can be accounted for and factored out. This is the classic before-[and]-after-control-impact (BACI) study design (Chevalier et al. 2019). For example, Franzblau and Collins (1980) hypothesized that food availability is a primary element determining the size of territories for birds. They predicted that, if their hypothesis was true, adding food to the territories of rufous-sided (or eastern) towhees (*Piplo erythrophthalmus*) should result in a decrease in the size of these territories. The first steps in their experiment were to identify a sample of the territories for this species and measure the size of each territory. As noted above, inferences from an experiment are restricted to the universe, or sampling frame, from which the samples are drawn. In this experiment, the region from which territories were selected (ideally, done in a random fashion) represented the area to which the results of the experiment would apply. Food was then added to the treatment, or experimental, territories, and others were left untreated, as controls. The size of each territory was remeasured after the addition of food, and alterations in the size of the treatment territories were assessed relative to those in the controls. The value of the controls becomes apparent if, for example, the size of all territories had decreased because of some variation in the environment other than the addition of food, or because of some effect caused by the observer. If this had happened, there would have been no way to evaluate changes in the treatment territories without being able to compare them with those in the controls, and even then, such inferences would be tenuous.

A way to get around this potential experimental conundrum is to design a controlled experiment so that pre-treatment data are part of the overall study parameters. Taking a comparative approach by contrasting the differences in the responses of animal populations between or among certain management actions, is certainly better than simple trial and error, but inferences can still be limited in such instances, even when controls are present, as noted in the above rufous-sided towhee territory experiment. Unless some kind of pre-treatment or baseline assessment is considered for the estimated metrics of habitat, populations, or both, it can be impossible to know if the end-point differences observed in response to the management actions are real.

For example, consider 2 pastures of 2,000 hectares each, where the goal is to increase the abundance of a focal species by deferring cattle grazing from 1 pasture and maintaining grazing on the other. After 2 years of grazing deferment, the densities of the focal species (six-lined race-runners, *Aspidoscelis sexlineata*) on the no-treatment site are observed to be twice that of those on the grazed pasture. You could be tempted to infer that the grazing deferment resulted in the desired management outcome, but making such a determination would be potentially misleading, because there was no prior information available on the abundance of race-runners in either pasture. Pre-treatment data, for example, might have shown that race-runners were twice as abundant on the pasture selected for grazing deferment before the cattle were removed, so no difference in the post-grazing response of race-runners could be inferred. In any case, without such pre-treatment data, it is impossible to know if the difference in the post-treatment abundance of the focal species of the experiment is real, or if it is simply an artifact related to some other aspect of the pasture or to previous conditions.

Examples of management comparisons or experiments that use pre-treatment data are rare in the literature on animal ecology. This is because collecting such data adds at least 1 or 2 years to a research project, thus potentially increasing overall project costs by as much as 25 to 50 percent. This point alone makes acceptance of a research project by funding agencies even more difficult than usual. Collecting pre-treatment data also means that the experimental design of an investigation must be clearly mapped out in relation to the management treatment(s) that will be applied.

A study by Provencher et al. (2002), which examined the extent to which habitat management at Eglin Air Force Base in the Florida Panhandle for endangered red-cockaded woodpeckers (*Picoides borealis*) impacted populations of other species of terrestrial vertebrates, is a unique example of such a management experiment. These woodpeckers need open (<50% canopy cover), parklike stands of southern longleaf pine (*Pinus palustris*) forests with little hardwood understory. Such a forest structure was maintained historically by frequent (every 2–3 years) fires started by lightning. Today, the open forest structure required by this endangered species is maintained by applications of prescribed fire, mechanical tree removal, herbicides, or some combination of such treatments (usually tree removal and herbicides). The study's authors designed a field experiment that compared the responses of multiple vertebrate species with 3 hardwood reduction treatments (chainsaw tree removal, herbicide application, and a combination of the 2) with no-treatment controls and pre-treatment data (Fig. 1.3). After 4 years of such treatments, their research showed that only the bird populations of Carolina chickadees (*Poecile carolinensis*), northern cardinals (*Cardinalis cardinalis*), and tufted titmice (*Baeolophus bicolor*) were impacted negatively. The treatment effects on more than a dozen other species of birds were either neutral or positive. Yet making such an inference would have been impossible in the absence of pre-treatment data on the abundance of these birds. In other words, having pre-treatment data allowed Provencher and his coauthors to make strong inferences based on the end-point measurements in their bird abundance data, or response variables.

Further Thoughts on the Hypothetico-Deductive Method

Deduction is the derivation of a conclusion by reasoning or logic, where the conclusion follows from the initial premises. When using the hypothetico-deductive method, scientists devise tests that put their hypotheses at risk—that is, they perform experiments or propose novel observations that potentially refute, cast doubt on, or outright falsify their candidates for explanations. Romesburg (1981) noted that the causal mechanism or process described in a research hypothesis might be difficult to observe directly. Thus, "a research hypothesis must be tested indirectly because it embodies a process, and experiments can only give facts entailed in the process" (Romesburg 1981:295). A general model of a hypothetico-deductive test might be, "If my hypothesis holds (i.e., if the mechanistic process I envision is actually ongoing), and if I manipulate a crucial element involved in the process while simultaneously controlling other critical elements, certain things must happen, or, conversely, certain things cannot happen."

The core elements of the general model of a hypothetico-deductive test are predictions—observable events or patterns—derived from what should happen after a particular manipulation, given that the tentative explanation being tested is valid. Predictions in a hypothetico-deductive test, then, are logically deduced from the premise of the hypothesis, and they should be specific to the hypothesis being tested. If the predictions generated in such a test could come about by processes other than the 1 being tested (e.g., by a competing hypothesis), then it would constitute, at best, only a weak test (see the section below on "Inferential-Theoretic Approaches").

According to Popper (1981), the best tests of research hypotheses are falsification tests—those in which all elements are controlled except the element crucial to the proposed explanation, and those that involve attempts to find results that are logically prohibited from occurring if the hypothesized explanation is true. If experiments produce logically prohibited results, the hypothesis is falsified and should be modified or discarded. Failure to find the prohibited results does not prove the truth of hypothesis, but it does increase confidence in the supposition

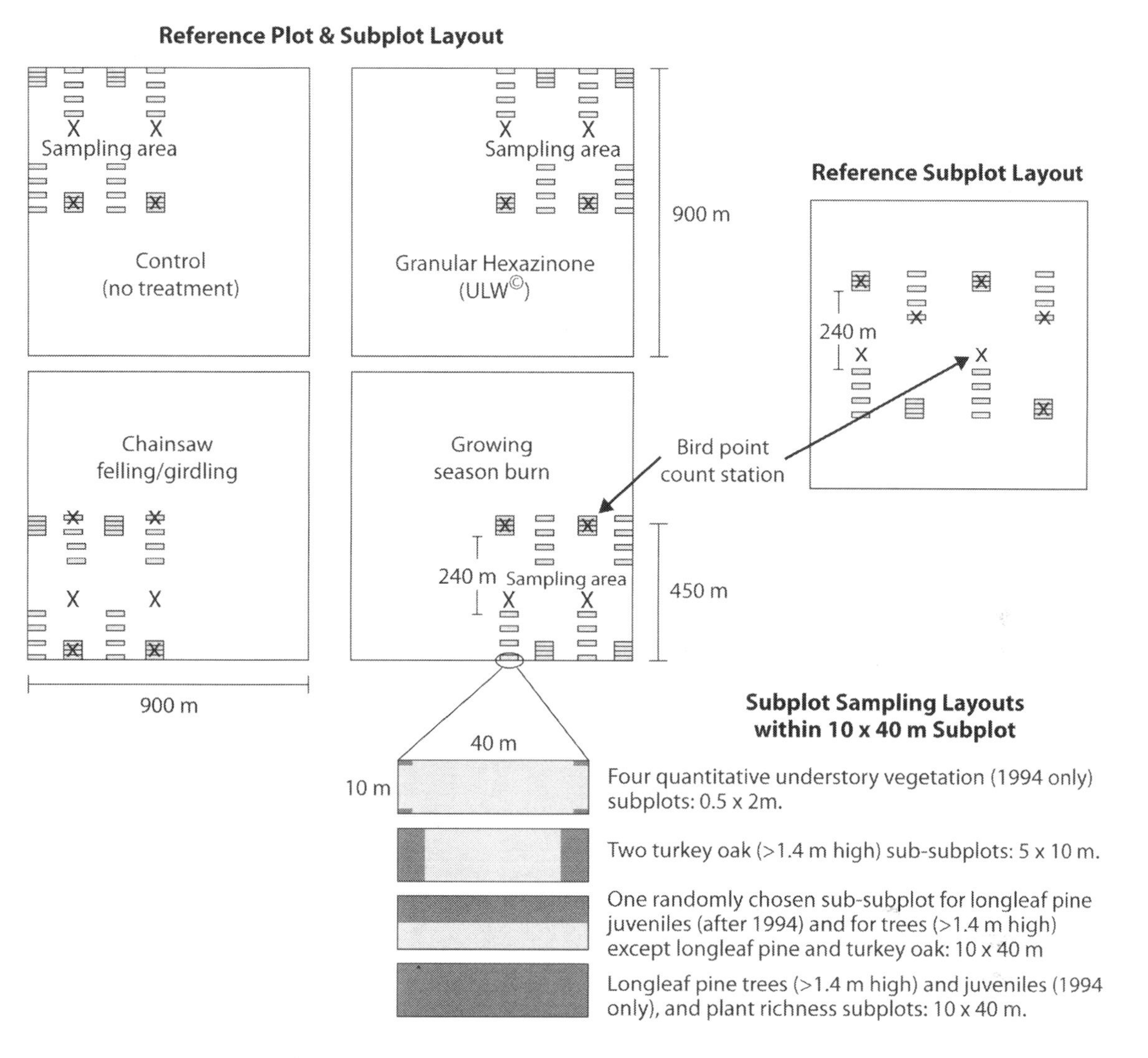

Figure 1.3. A sample layout of 8-hectare experimental longleaf pine (*Pinus palustris*) restoration plots. It shows 1 of 6 reference plots and sampling areas in 1 of 6 blocks in a randomized complete block design, consisting of 4 whole-plot treatments, at Eglin Air Force Base in Florida. Midstory hardwood reduction treatments were applied to the 81-hectare plots. Subplot centers are separated by 10 or 50 meters. The image depicts 1 subplot and 4 layers of subplots. (Adapted from Provencher et al. (2002), with the permission of The Wildlife Society)

(Simpson 1964). In formal deductive logic, you are not seeking to reject hypotheses. The logic itself formally proves that a statement is true. For example, if A = B and B = C, therefore A = C (a conclusion that is true), this is a formal deductive statement. We only seek to eliminate hypotheses through falsification when we are working in an inductive reasoning environment. By eliminating possible explanations, we increase the inferential strength of those that remain. Clearly, if you can unambiguously prove that something is true, you don't have to bother with this approach. Nonetheless, a high level of confidence in the validity of a given hypothesis should happen only after it has been tested repeatedly, in a variety of ways (e.g., replication in different times and locations, preferably over a relatively long period, such as multiple generations of the species under investigation).

Tests of the kind Popper (1981) described are possible only if an investigator can envision events or patterns that will refute the hypothesis under consideration. If the hypotheses are vague or general, they sometimes can be interpreted to explain nearly any related observation in nature or any outcome of a pertinent manipulation. Hypotheses of this kind are considered to "unfalsifiable"—that is, not amenable to testing. An example is a statement that "a giant white gorilla lives on Mount Fuji." This is unfalsifiable, because there are no primate tracks or any other evidence that such an animal resides on that mountain. Such hypotheses sometimes can be improved and made meaningful by more explicitly describing the mechanism or process being tested, or by making them more specific. For example, rewording the unfalsifiable hypothesis above to say that "there are no giant white gorillas on Mount Fuji" results in a falsifiable hypothesis. You only need 1 reliable sighting of such an animal on Mount Fuji to reject the hypothesis. Identifying the critical elements to manipulate in an experiment and avoiding making what Popper (1981) called "risky predictions" often are easier steps to take if the details of the proposed process are explicitly described.

Predictions that come from hypotheses about ecological phenomena often involve defining and documenting linkages between animals and habitats, as noted by Morrison et al. (2020). These responses are not likely to be identical among individuals—and especially among populations—no matter how carefully an experimental test is controlled, at least partly because each individual in a population has a unique genome and environmental history. For example, providing the same amount of an important nutrient to individual plants, even under highly controlled conditions, will not result in the same amount of growth in each plant. Statistical tests traditionally have been used to assess whether the predictions in a hypothetico-deductive test are met. In the above example, a scientist might have hypothesized that the nutrient in question plays an important function in the growth of plants and articulated a mechanism through which that nutrient aids growth. A prediction in this hypothetico-deductive test might be that plants receiving the nutrient will grow more rapidly than those to which it is not applied. A statistical test could be used to compare the average growth of plants in the 2 groups. The null statistical hypothesis in this example typically would be that there is no difference between the mean growth rates of the 2 groups. Rejecting or failing to reject the null statistical hypothesis determines whether the predictions of the test have been met. Thus assessments of patterns through statistics can play an important role in the validation of research hypotheses when the statistical test is embedded in the broader framework of a hypothetico-deductive test. Differences between research hypotheses and statistical hypotheses, and recent concerns about the use and misuse of statistical tests in animal ecology, are outlined below.

Statistical Hypotheses

It is fundamental to understand that a research hypothesis and a statistical null hypothesis are not the same thing (Johnson 1999; Guthery 2008). Conducting some kind of statistical test to identify whether there is no probability of difference—that is, a "null hypothesis"—between or among means of variables is not the same thing as posing a research hypothesis. Statistical tests, or other forms of analyses and estimation, can be used to evaluate a research hypothesis, but it is the data themselves, and the interpretation of these data by the investigator, that determine whether the data support—or fail to support—a research hypothesis. In other words, it is the creativity of the human mind, not the application of some kind of statistical test, that results in advancing natural resource sciences, including animal ecology (Guthery 2008).

The traditional approach used in animal ecology to make inferences about patterns is with statistical tests, in which a statistical null hypothesis either is rejected or not rejected, based on the value of a computed test statistic and its associated *P*-value. (A

P-value is the probability that taking another sample of the same size and replicating the collection method would result in the same mean and variation in values. It is not the probability that the calculated mean and variation values are "true.") Concerns about the utility of statistical testing with *P*-values are not new (e.g., Berkson 1938, 1942; Cox 1958a, 1958b), but more-recent challenges to it (e.g., Cherry 1998; Johnson 1999; Anderson et al. 2000; Krausman 2017) suggest that a shift in the way data are analyzed in ecology may be emerging (Guthery et al. 2001; also see Cumming 2014), albeit slowly. At a minimum, students and professional biologists should be aware that concerns exist about null hypothesis statistical testing, as well as that alternatives are available.

Problems identified with null hypothesis testing include but are not limited to the following: (1) it is used when it is not required; (2) the meaning of *P*-values is often misinterpreted; (3) the assumptions of statistical tests are ignored or misunderstood; (4) many null hypotheses are known to be false before the test is conducted, or, if rejected, provide little information; (5) the decision to reject or not reject a null hypothesis is based on a fixed cutoff value (usually $\alpha = 0.05$) that is subjectively selected; and (6) statistical significance may or may not be associated with biological significance (Cherry 1998; Johnson 1999; Anderson et al. 2000). Designing experimental studies to identify biologically significant relationships between animals and their habitats should be the apex of what we strive to do as animal ecologists.

A vast majority of hypothesis tests reported in the ecological literature fail to address statistical power (the probability of correctly rejecting a null hypothesis that is false) and effect size (the smallest change that has biological meaning in the context of a study; see Steidl et al. 1997). Low power and low effect size typically result from a limited amount of replication in field ecology studies (Lemoine et al. 2016). These problems are a mixture of a misunderstanding (or misuse) of the traditional procedures and the weaknesses inherent in the procedures. Recommendations for correcting the problems, therefore, include calls for a wiser use of traditional approaches and employment of alternative approaches.

Cherry (1998) recommended that investigators decide whether they are interested in assessing the presence or absence of effects (i.e., what is assessed in most null hypothesis tests) or in estimating the size of effects. He suggested that the focus of many studies in animal ecology is estimation and recommended that greater emphasis be placed on calculating and presenting point estimates, standard errors, and confidence intervals for the characteristics of interest. Similarly, Johnson (1999) and Anderson et al. (2001) cautioned against presenting only *P*-values from statistical tests (i.e., "naked *P*-values") without including estimates of effect size and measures of precision. Presenting confidence intervals in addition to or instead of *P*-values provides information about the precision with which a parameter has been estimated, which, in turn, allows the reader to evaluate whether a lack of statistical significance is most likely due to lack of an effect or small sample sizes (Johnson 2002). Confidence intervals also encourage biologists to consider effect size in their interpretation of results (Johnson 2002; also see below). In some situations, the use of confidence intervals may be more meaningful than statistical tests and could reduce tests of "silly null hypotheses" (Robinson and Wainer 2002)—that is, those known to be false before the tests are conducted. For example, Cherry (1998) noted that the null hypothesis tested in many studies of habitat use-availability is that an animal uses different environments in proportion to their availability on the landscape. In most cases, this null is known to be false a priori. He recommended that this kind of question might best be answered with estimates and confidence intervals, avoiding null statistical hypothesis tests altogether. Rejecting or failing to reject a null statistical hypothesis has clear implications about how we view the existence of whatever pattern is under investigation. If the test is part of an hypothetico-deductive experiment, then the outcome either lends support to the research hypothesis under consideration or falsifies it. All a statistical rejection of a null hypothesis can show is that the null is unlikely, rather than that it is actually false. It is important, therefore, to understand the kinds of errors that can be made when deciding the outcome of a statistical test. A "Type I error" (i.e., a false positive) is made if the null hypothesis is rejected when it is true, and a "Type II error" (i.e., a false negative) occurs if the null hypothesis is not rejected when it is false. The probability of a Type I error, denoted by alpha (α), in a statistical test is set by the investigator, and convention has established that α often is 0.05—that is, $P < 0.05$. The probability of a Type II error in a test is denoted by beta (β). "Statistical power" $(1 - \beta)$ is the probability of correctly rejecting a null hypothesis that is false.

Committing a Type II error would, in the typical design of a statistical test, indicate that no pattern exists when, in fact, 1 does. This kind of error could be problematic in animal ecology. For example, in a study about patterns, a Type II error could lead a biologist to conclude that there was no decline in the abundance of a species after some human-induced manipulation when there actually was a decrease. For species of significant conservation concern, it would make sense to set Type II error rates low (such as 0.1, instead of 0.05) to avoid determining that either no effect or a decline occurred when the opposite might have happened. An analysis of statistical power during the design of a study can help avoid Type II errors, as power increases with a greater sample size (n) and a larger effect size. Effect size is broadly defined as "the difference between the null hypothesis and a specific alternative" (Steidl and Thomas 2001:17) and could be identified as the smallest change that has biological meaning in the context of a study (Steidl et al. 1997). The effect size of a statistical test is a statement of the magnitude of some causal phenomenon. Effect size is typically presented as Cohen's d statistic, which is calculated as

$$(m_2 - m_1) / SD_{pooled}, \text{ where}$$
$$SD_{pooled} = \text{sqrt}[(SD_1^2 + SD_2^2) / 2]$$

where m_i = the mean of the ith sample and SD_i = standard deviation of the ith sample. Values of Cohen's d can be generally interpreted as a small effect size ($d \approx 0.2$), moderate effect size ($d \approx 0.5$), and large effect size ($d \approx 0.8$).

Identifying what is a meaningful biological effect is perhaps the most challenging aspect of power analyses (Steidl and Thomas 2001), but it also is among the most beneficial components (Cherry 1998). Because statistical significance is related to sample size, even very small differences between, say, the mean growth rates of plants in 2 groups (see the example above) can be statistically different, given a large enough sample. A more important question is whether the difference is biologically meaningful.

The use of power analyses encourages animal ecologists to think about meaningful biological effects before beginning a study. Our conclusion about null hypothesis statistical testing is that these analytical techniques clearly suffer from problems, some of which are inherent, and others that are a product of misuse (but see Robinson and Wainer 2002). Greater reliance on confidence intervals, tolerance intervals, and the identification of meaningful biological effects (i.e., a wiser use of traditional methods and expert knowledge about the species being studied), as well as employment of the alternative methods outlined above, may reduce these problems. Nonetheless, we emphasize, as did Guthery et al. (2001) and Guthery (2008), that statistical tests are primarily a tool to identify or elucidate patterns. They are not an end in themselves and are most useful when they are part of a broader experimental approach designed to test an explanation about a particular pattern found in nature.

Information-Theoretic Approaches: Research Hypotheses or Parameter Estimations?

Chamberlain (1897) advocated working with multiple research hypotheses, as he felt that a single hypothesis is more likely to become accepted without being adequately tested. We do not think that the scientific method requires multiple explanations, but it is essential that research hypotheses be viewed as potential candidate explanations until they are tested. Prior to testing their hypotheses, scientists must remain objective about them, no matter how convincing such hypotheses may sound. If more than 1 potential explanation for a given pattern exists, tests should help determine which is most likely to be valid. Deciding which of several competing hypotheses to test first depends, in part, on which best fits existing theories and empirical information. Also, perhaps most importantly, hypotheses that are not testable are basically just belief systems and, therefore, are not part of a rigorous scientific inquiry.

Anderson et al. (2000) described an alternative to statistical null hypothesis testing, which they called the "information-theoretic method." This method is based on a formal relationship, found by Akaike (1973), between "Kullback-Leibler information [a dominant paradigm in information and coding theory] and maximum likelihood [the dominant paradigm in statistics]" (Anderson et al. 2000:917). In this approach, what would be statistical hypotheses (both null and alternative) in traditional tests are developed a priori as models about relationships within the system of interest. Models are then ranked based on Akaike's Information Criterion (AIC), which describes how well an empirical set of data fits each model, while minimizing the number of estimated parameters (Anderson et al. 2000). Probably the most fundamental problem with this approach is that it provides no information on how good a model actually is. Rather, it simply supplies information content relative to other models. There is absolutely no logical claim that can be made for the top model being the correct model. Further, this approach only indicates correlational relationships. It does not and cannot disclose causality in a system. Professional biologists and students who have used Program MARK (White and Burnham 1999) are familiar with this approach. For example, in this program, AIC values could be used to evaluate a set of models that incorporates the influence of sex, age, and time on the survival rates of marked animals.

Anderson et al. (2000) argued that the information-theoretic approach is superior to null statistical hypothesis testing, because no null model that is based on a fixed α level is rejected (or not rejected). Rather, a set of models is evaluated and ranked based on best inference. Guthery et al. (2001) generally favored the use of the information-theoretic method but cautioned that because it was, by definition, a parametric estimation approach, assumptions about probability distributions are required before any inferences that are drawn from it can be useful. Furthermore, these authors noted that their approach offered no protection against trivial hypotheses. Models developed using this approach could be just as silly, and inconsequential, as many null models in statistical tests. As noted earlier, the information-theoretic approach cannot reveal causality in a system, only correlations. Guthery et al. (2005) also warned that investigators should not get caught up in substituting the use of model selection analyses for past ritualistic uses of null hypothesis statistical testing. Although rare, it is possible to publish scientific papers in animal ecology based on complex quantitative analyses without statistics, model selection, or *P*-values (for examples of 2 contemporary animal ecology papers with astatistical analyses, see Guthery et al. 2005; Rader et al. 2011).

A second alternative to null statistical hypothesis testing is Bayesian statistics. This approach can be used to predict the outcome of environmental changes in terms of their likelihood or their odds. It relies, in part, on the ability of biologists to describe a priori likelihood functions for ecological outcomes (i.e., prior probabilities), given specific environmental conditions, such as the relationship between the environmental states and sizes of a given population. This approach has been controversial, but it is becoming more accepted and used by wildlife ecologists (Link and Barker 2010). It may have value when standard experimental designs or statistical tests are constrained by real-world problems (e.g., small sample sizes or inadequate replicates or controls). Advantages and disadvantages of Bayesian statistics are outlined in Chapter 5, but our focus here is on the use of existing knowledge and expert opinion to estimate prior probabilities, as well as on the influence of errors in prior probabilities on predicted outcomes. A growing area of research uses graphical models to hypothesize and test causal relationships of variables within a system, such as in the field of climate science (Ebert-Uphoff and Deng 2012). Nonetheless, statistical methods of causal discovery are not often employed in animal ecology (although they were applied in Marcot 2019) beyond the use of structural equation and Bayesian network modeling (Eisenhauer et al. 2015; Marcot and Penman 2019).

The Scientific Method Revisited

Issues Associated with the Scientific Method

Many discussions about the scientific method have focused on formulating and testing hypotheses. Yet what constitutes a research hypothesis is a subject that sometimes causes confusion among students and professional biologists alike, because the term is used in a variety of ways. Romesburg (1981) recognized this problem and clearly differentiated between "research hypotheses" (i.e., those that include a mechanistic process and explain a pattern) and "statistical hypotheses," which he defined as conjectures about classes of facts (i.e., patterns). Unfortunately, the distinction between these 2 kinds of hypotheses continues to be overlooked (Guthery et al. 2001; Guthery 2008). Carey (1994:9) noted, "In the jargon of the scientist, just about any claim that may require testing before it is accepted will be called a hypothesis." Thus a query about whether a species is more abundant in 1 mountain range than in another might be framed as a hypothesis. But Carey (1994) was clear in arguing that statements of this kind are not research hypotheses, because they do not include an explanation. We suggest that questions which address patterns without an accompanying rationale are best framed as statistical hypotheses. Tentative interpretations of how some part of the world works (i.e., research hypotheses) should be assessed with the hypothetico-deductive method. Calsbeek and Sinervo (2002) provided an exceptionally fine example of an experiment to test the "ideal despotic distribution theory," which predicts that the "quality" of habitat used by a territorial animal should vary, based on the availability of resources. The authors, using side-blotched lizards (*Uta stansburiana*), manipulated the quality of lizard territories by adding rocks (to increase an opportunity for thermal regulation by providing more heat) to some territories and removing rocks (to decrease such an opportunity), much like food was manipulated in the rufous-sided towhee example earlier in this chapter. Lizard progeny on the higher-quality territories had significantly higher rates of growth and survival than those on the lower-quality territories. Such a result would be virtually impossible to unravel using only direct observations in the absence of any experimental manipulations.

Challenges to the Scientific Method

The philosophical and technical issues described in this chapter are subjects that might be considered internal to our field, because they primarily involve scientists working in animal ecology who are debating among themselves about how they should do their jobs. The issue explored in this section involves public perceptions about the products of science and about scientists themselves. It thus has implications which go beyond improving the details about how science in animal ecology is designed and conducted.

A basic tenet of science is that we live in a material universe that can be at least partly understood (Feder 1990). How the universe works is the same for everyone (at least in our view), but our individual understanding and perceptions of what is happening differ among people, because of societal, cultural or even tribal values. Simpson (1964:vii) described this situation by noting that everyone lives in 2 worlds, 1 public and 1 private: "The public world is the objective, material, outer world that exists around us regardless of what we know or think about it. The private world is just what we do know and think about that public world; it is the world as it seems to us, as we perceive and conceive it." The private world of an individual can be influenced by his or her gender, race, culture, and individual experiences, and it can include misconceptions and erroneous ideas. Science is the best way we know of to align and match our private worlds with the public world.

Challenges to the content of science (i.e., our current knowledge about how the universe works), and even to the scientific process itself, are founded on the idea that individual biases—due to, for example, a person's gender or sociopolitical or economic background—severely constrain our ability to objectively perceive and understand reality (e.g., Harding 1986). Individuals who espouse this idea, sometimes called "social relativists" or "postmodernists," claim that what we know about the universe is flawed, because of the inevitable biases of those who did the work. There can be little doubt that politics and other cultural influences have played a significant role in who has participated in science. For example, until recently, males almost completely dominated the field of animal ecology. Furthermore, even the questions asked in scientific endeavors are likely to be influenced by societal context. Thus what we know about the universe is, to some extent, culturally driven. We disagree, however, with the notion that bodies of accepted knowledge are flawed (i.e., do not match reality) because of these influences. If the scientific process is performed properly, the answer that is derived should be the same, no matter who does the work. This is not to say that the process itself is perfect. Mistakes in the history of modern science are common. But the process tends to be self-correcting, so that even if personal biases, or outright fraud (Goodstein 1995), cause an individual scientist to misinterpret or misrepresent the outcome of an experiment, other scientists doing similar work should soon expose the error. The challenges of social relativists and postmodernists are of concern, not because they cause scientists to seriously question either their practices or the ideas in which they have developed great confidence, but instead because the general public may interpret such challenges as legitimate, become disillusioned with answers provided by science, and turn to less-reliable sources of information to assist them in making decisions (Gross and Levitt 1994; Gross et al. 1996).

Application of the Experimental Approach

In the past, animal ecologists did not often participate in empirical experimental studies, partly because the questions they asked were generally about patterns of natural phenomena, not the processes that caused them. Frequently studied species were those that were of interest from a management perspective, and queries addressed patterns of relative abundance, habitat use, survival, and productivity. Animal ecology is focused on describing these patterns, because the information provided generally is sufficient to develop successful management strategies. Studies that document patterns in nature often require intensive and extensive surveys of animals or painstaking observations of animal behavior (see Morrison et al. 2020:Chapter 3), but (usually) not experiments.

In the past, animal ecologists were not alone with respect to ignoring experiments. Hairston (1989) noted that ecologists used them infrequently in the 1950s and 1960s, despite the pleas of advocates of the experimental approach. Of note is that some of the early field and laboratory experiments performed by ecologists related directly to animal-habitat relationships. For example, Harris (1952) attempted to test the idea of habitat selection experimentally by presenting individual prairie deer mice (*Peromyscus maniculatus bairdi*) and woodland deer mice (*P. m. gracilis*) with a choice of artificial woods or artificial fields. He found that the test animals preferred the artificial habitat that most closely resembled the natural environment of their own subspecies and concluded that these mice were reacting to visual cues provided

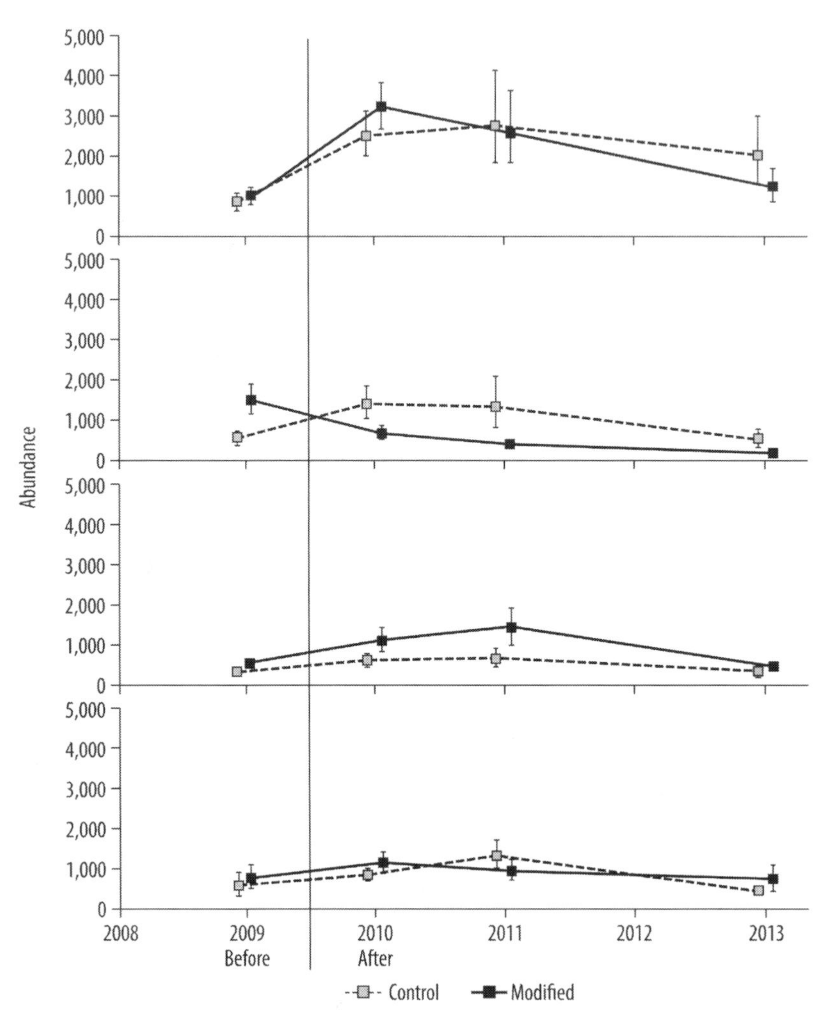

Figure 1.4. A graph of abundance estimates for a mark-resight analysis of crested auklet (*Aethia cristatella*) surface activity for each of 4 plots and plot halves (control shown as *gray squares* and modified shown as *black squares*) before and after vegetation removal at Gareoli Island, Alaska, in 2009–2013. Abundance was a function of plot half and year for both abundance and the recapture rate, using a zero-truncated mark-resight model in program MARK. Data are shown as means and 95 percent confidence intervals. (Adapted from Major et al. (2017), with the permission of The Wildlife Society)

by the artificial vegetation. Wecker (1963) later attempted to determine whether early experience (i.e., learning) played a role in habitat recognition and selection in deer mice. He found that behaviors associated with habitat selection were primarily controlled by what Mayr (1974) called "closed genetic programs"—that is, they are not greatly affected by early experience. Similar experiments conducted by Klopfer (1963), however, revealed that habitat selection by chipping sparrows (*Spizella passerina*) could be modified to some extent by early experience.

Experiments are a potentially powerful tool for increasing our understanding of ecological processes and phenomena, and their use in ecological research, including animal ecology, has increased dramatically in recent years (Lecomte et al. 2004; Klaus and Noss 2016; Major et al. 2017; also see Fig. 1.4). Although somewhat controversial, experimental removal of predators to test their potential role as factors limiting population performance for species of conservation concern appeared in the literature more than 2 decades ago (Tapper et al. 1996; Côté and Sutherland 1997). Such removal experiments are among the most common applications of the experimental approach in animal ecology for both birds (Conover and Roberts 2017) and mammals (Lieury et al. 2015; Boertje et al. 2017; also see Fig. 1.5). Removing predators, as is often done in wildlife management, is not the same as predator control (i.e., killing does not necessarily control anything you are seeking to inves-

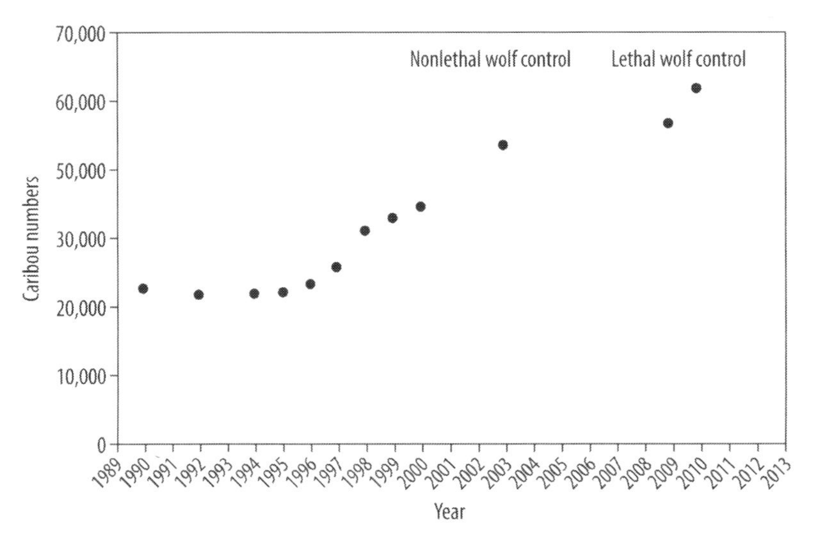

Figure 1.5. A graph of the growth of the Fortymile caribou (*Rangifer tarandus*) herd in Alaska, based on photo censuses from 1990 to 2010 and timelines of 2 forms of wolf (*Canis lupus*) control. (Adapted from Boertje et al. (2017), with the permission of The Wildlife Society)

tigate in your study). This is another example of the need for experiments before we start implementing a specific management practice.

An important consideration related to the selection of replicates or experimental units is that, to maximize experimental efficiency, they should be independent of each other. Independent experimental units "are dispersed in time and space so that the response [to a treatment] of any one unit has no influence on the response of any other unit" (Smith 1996:17). Selecting independent replicates in field experiments is sometimes difficult, because of interactions between animals, as well as the movements of animals, water, or other materials between experimental units. For example, if the movements or behavior of rufous-sided towhees in territories that received supplemental food had influenced the size of control territories, or vice versa, the replicates used by Franzblau and Collins (1980) would not have been independent. Another way that independence sometimes is violated in experiments is by sampling repeatedly from 1 site (e.g., an area treated with prescribed fire) and considering each sample as a replicate. The samples in this situation are not independent of each other, because they all come from a single application of the treatment. Analyzing samples of this kind as if they were independent replicates is 1 form of pseudoreplication (Hurlbert 1984). Use of a crossover design, where experimental treatments and controls are switched after a certain period of time, can be a useful tactic in overcoming these kinds of problems (Tapper et al. 1996; also see Fig. 1.6). Namely, you need to know which part of the biological population you are working on. Otherwise, you have to assume that the study responses would apply everywhere, when, in fact, they may not.

Field Experiments

Field experiments are powerful tools, in part because researchers can bring about specific conditions in study plots that, without manipulation, either might never happen or take decades to occur in nature. Ecologists often favor field experiments over laboratory experiments (Hairston 1989), because the former are conducted in natural settings, and, hence, the results generally are considered to be more believable or applicable to field management. The price of increased realism, however, is a loss of control over most environmental variables. This loss may preclude testing some research hypotheses, because the critical elements may be difficult or impossible to manipulate in field situations. It also is usually a sampling problem, where the areas sampled are too small to be representative of the system being studied.

The dependence of field experiments on controls

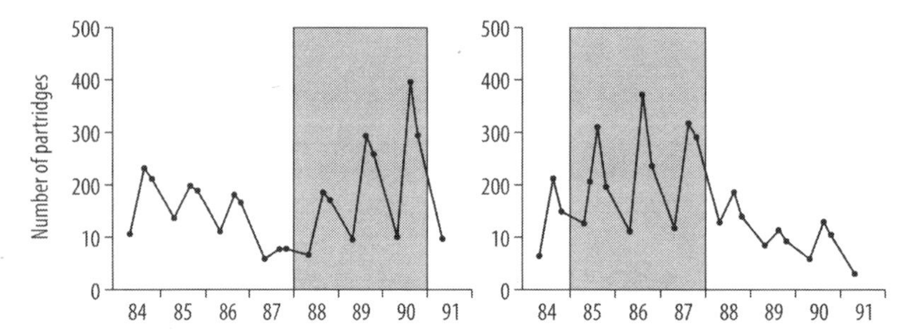

Figure 1.6. Graphs of yearly and seasonal changes in the population of grey partridges (*Perdix perdix*) at Milston (*left*) and Collingbourne (*right*) in England. *Shaded blocks* show years when breeding-season nest-predator reductions took place, first on Collingbourne and then on Milston, indicating an increase in partridge abundance in response to predator reduction. (Adapted from Tapper et al. (1996), with the permission of the British Ecological Society)

and the spatial scale at which many such experiments must be conducted make them vulnerable to several practical problems (Hilborn and Ludwig 1993; Morrison et al. 2001). Finding sites or defining populations to serve as controls may be difficult, especially if the experiment concerns endangered species or their habitats, limited environments, or mechanisms that operate at broad spatial scales. Furthermore, as the spatial scale rises, experimental error is also likely to go up, because of an increase in heterogeneity within and among experimental units. Enlarging the sample size to deal with this error may be impractical or impossible because of the cost and availability of personnel and equipment. The small sample sizes and unique site conditions often associated with field experiments also make randomization and a proper interspersion of treatments difficult to achieve. Thus, while field experiments can be relatively easy to design, they can be difficult to execute. This difficulty usually increases as the spatial scale of the experiment expands. Unfortunately, mechanisms that operate at broad spatial scales sometimes are among the most important for understanding the causes of ecological patterns of animal populations and the habitat that supports them (Lawton 1996b).

Another problem inherent in field experiments (and, to some degree, in all ecological experiments) is that patterns and phenomena in ecology often have multiple, interacting causes (Peters 1991). Experimental designs do exist that examine the effects and interactions of more than 1 treatment at a time, but they require more replicates than simpler experiments. Thus the practical problems outlined above are exacerbated.

Field experiments rarely are repeated (Hairston 1989). This is not surprising, given the costs and difficulties associated with designing and executing them, as well as the fact that many take years to complete. Furthermore, most scientists probably would prefer to test a new idea than repeat someone else's work. Confidence in a research hypothesis, however, should come only after it has been tested in a variety of ways (Kneller 1978:117). A lack of repetition in field experiments may mean that an unknown number of hypotheses in ecology are accepted prematurely.

The difficulty of implementing field experiments and the cost and time needed for their execution may also discourage those that go beyond crude levels of explanation. For example, an experiment could be done to test whether the application of prescribed fire causes an increase in the abundance of an endangered species. This experiment might involve comparing the abundance of the species in the treatment and control plots before and after burning. The results may support the idea that a causal relationship exists between fire and the abundance of the species, but additional experiments would be needed if biologists were interested in knowing, for example, whether fire increases the abundance of the species by increasing the amount of available food; or by improving the structure of vegetation for cover, relative to the needs of the species; or both. Such factors could be included as covariates for an initial assessment of their role in the system being studied. Biologists also would want to know whether the increase in abundance was due to a functional or a numeric response, and whether that abundance is positively related to survival and productivity. The frequent need to explain patterns in nature at multiple levels forms part of the concerns that Peters (1991) expressed about the utility of ecological experiments (also see Gavin 1991). Small study plots make it difficult to prevent immigration, which can mask treatment effects, such as abundance as a response variable.

The level of reliable knowledge needed for management purposes depends on the situation(s) managers are concerned about. If, in the example above, prescribed fire was consistently effective in increasing the abundance of an endangered species, and if patterns of land ownership and other social constraints permitted burning, then the level of explanation provided by the hypothetical experiment might be sufficient. If, in contrast, burning was not feasible throughout the range of the species (e.g., potential damage to human-built structures), it would be important to know what features of the habitat were changed by fire, so alternative manipulations (e.g., cutting, grazing) could be devised to produce the same effect.

One class of field experiments pertains to studying changes over time. The archetypal version of this is the BACI framework (Chevalier et al. 2019). Another is the application of adaptive management studies that compare outcomes with expectations (see the "How to Proceed" section below for additional information).

Laboratory Experiments

The design of ecological experiments in laboratories is conceptually no different from those in field situations. The primary dissimilarity between these 2 types of experiments is the degree of control over experimental conditions. In a laboratory, many important variables can be controlled. This level of regulation makes designing experiments simpler conceptually (although not necessarily easier), because the investigator can include and manipulate the environmental elements needed to test the idea under study and can exclude or control almost everything else. The questions that can be asked in laboratory experiments are thus more precise than those in field experiments (Hairston 1989). Control of nearly all variables in an experiment also facilitates the replication of trials within an experiment, as well as duplication of the experiment itself, because conditions do not change over time and presumably can be re-created by other scientists.

Significant advances in our understanding the natural world have been made in laboratory experiments in many scientific fields, including animal ecology. Yet there is reluctance by some (perhaps many) ecologists to accept the findings of laboratory experiments in their own field of study (Mertz and McCauley 1980), primarily because "its artificiality may simply swamp processes of ecological relevance" (Peters 1991:137).

Lawton (1996a) listed and discussed the primary arguments against experiments in an ecotron, which is a set of highly controlled environmental chambers designed to replicate miniature terrestrial ecosystems (sometimes called "bottle communities"). Not surprisingly, the arguments he mentioned that opposed bottle experiments are the same as those leveled against many other laboratory experiments in ecology. They focused on the lack of applicability of the experiments to actual wildland environments, as well as included concerns about the choice of species composition and the lack of important biological processes, such as major perturbations, immigration, emigration, seasons, appropriate scale, and the like. A major apprehension was that manipulating a single factor in a laboratory setting that does not include the full array of ecological interactions might have produced results that could have either unrealistically emphasized or diminished the importance of that factor (Peters 1991). Because of these concerns, the results of laboratory experiments in ecology have been most widely accepted when they have addressed questions about physiological processes, bioenergetics, or animal behaviors (Hairston 1989). Yet laboratory experiments can also reveal unexpected findings. Using captive northern bobwhites (*Colinus virginianus*), Larson et al. (2012) found that seeds from non-native guineagrass (*Urochloa maxima*) and native switchgrass (*Panicum virgatum*) were consumed equally by volume, even though field data from telemetry studies generally showed that bobwhites tended to avoid areas with non-native grasses. Apparently, the similarity in protein content between switchgrass and guineagrass (16% and 18%, respectively) was a factor that resulted in equal consumption of seeds from these 2 types of grasses. But a question still remains—did bobwhites eat these seeds in the wild, or was there something else going on, such as a lack of other foods? While the unexpected results are indeed intriguing, this example illustrates that there can be problems with inferences from laboratory studies in animal ecology.

Improvements in the design of laboratory settings may help reduce some of the concerns about the applicability of results that come from them. The "big bottle" experiments of Lawton (1996b) represented an effort to make an experimental setting for the study of miniature terrestrial ecosystems more realistic, despite the concerns that were expressed. As Lawton (1996a:668) noted, "Many ecologists seem to understand the need for simple experiments in pots and greenhouses; and yet a minority of colleagues become highly critical and concerned when an attempt is made to make artificial, controlled environments more rather than less realistic!"

A specific concern about artificiality in laboratory experiments is that the settings (e.g., cages, aquaria) can constrain or prevent behaviors that animals would perform in natural situations (Peters 1991). Many important ecological interactions, such as competition and predator-prey relationships, directly involve animal behaviors. Clearly, laboratory settings must be designed to allow animals to respond naturally to experimental conditions if the results are to be more acceptable to ecologists (see Matter et al. 1989; Glickman and Caldwell 1994).

Population-habitat relationships obviously involve habitat selection behaviors. Animals seeking a place to live can respond to a set of environmental conditions in at least 2 ways—stay and try to establish themselves, or leave. Experimental enclosures with no exits potentially are adequate settings for assessing the preference of animals for a given environmental condition. For example, if fish are placed in a tank of water with a gradient of temperatures, most generally stay in the part of the tank where the temperature is most acceptable to them (Warren 1971:186). Enclosures, however, are inadequate for assessing whether conditions, as a whole, inside the enclosure represent a suitable environment for an animal, because that animal cannot leave.

Enclosures with exits, that thus allow animals to depart, have been used to study a variety of topics, including social behavior and population size (Butler 1980; Gerlach 1996, 1998), habitat relationships (Wilzbach 1985), and emigration and population regulation (McMahon and Tash 1988; Nelson et al. 2002). Permitting animals to leave an enclosure may represent a significant improvement in the design of ecological experiments in controlled settings, because the animals' responses to the conditions inside the enclosure indicate whether these are suitable (Matter et al. 1989; although see Wolff et al. 1996). Enclosure experiments are rare in terrestrial animal ecology, but they have been widely used for aquatic and fisheries research (Englund and Cooper 2003; Matsuzaki et al. 2009).

Simulation Experiments

There are many situations in animal ecology that simply do not lend themselves to long-term experimental investigations. Tracking the demographic responses of just 1 vertebrate species to changing environmental conditions, even in the absence of deliberate manipulations, often takes legions of personnel and huge amounts of funding. Past and ongoing studies of northern spotted owls in the Pacific Northwest, which have cost many millions of dollars, are an example of such outlays.

Simulation modeling and analyses offers a mechanism for conducting "What if?" kinds of experiments to examine the potential performance of a population in relation to variation(s) in habitat and environmental factors. Rader et al. (2011) conducted such a simulation series to test whether nest-predator reduction could potentially result in northern bobwhite population increases over a 10-year timeframe in south Texas by mitigating for potential losses from excess heat, drought, and loss of habitat. The demographic parameters for their model were based on a long-term telemetry study of northern bobwhite survival and nest success in that part of the state. These data were supplemented by an infrared camera study to document the top nest predators in this region. The initial model output was tested against known estimates of northern bobwhite productivity and density from the peer-reviewed literature and was found to correspond well with these empirical estimates. The authors then ran a series of different management scenarios to examine whether the removal of the top 3 most abundant nest predators would result in a population increase. In general, these simulation experiments revealed only a weak potential for northern bobwhite populations to respond to nest-predator reduction, and there was little evidence that predator reduction could compensate for decreases due to drought, heat, or habitat loss. The take-home inferences from these simulation experiments were that (1) efforts to reduce northern bobwhite nest predators were most likely a waste of time and money, and (2) management actions to conserve or improve nesting habitat should probably receive higher priority than predator reduction. It is notable that previous researchers came to the same conclusion about not needing to reduce nest predators with regard to northern bobwhite population management, based on natural history observations (Lehmann 1984) and manipulative field experiments (Guthery and Beasom 1977), 2 completely different, but complimentary, lines of evidence.

DeMaso et al. (2011) developed a similar simulation analysis to evaluate model sensitivity to variability in northern bobwhite demographic parameters. In a follow-up study, DeMaso et al. (2014) observed that long-term (100-year) population projections in areas with 5 percent brush cover were substantially different, compared with similar analyses on areas with 11 percent and 32 percent brush cover. Projected population abundance on the sites with the 2 higher percentages of brush cover was twice that compared with the least brush cover site. Population persistence on the 11 and 32 percent brush cover sites was 0.94 and 1.0, respectively, whereas persistence on the 5 percent brush cover site was only 0.54. The unexpected inference from the simulation experiments conducted by these researchers was that even extremely minor, non–statistically significant differences in demographic parameters can result in vastly different population trajectories over time (Fig. 1.7). Without having simulated population trajectories over 100 years, the potential relationship between adequate brush cover and long-term population persistence may never have been revealed.

Although simulation experiments can be run to formulate and test hypotheses, White et al. (2014) cautioned that tests of statistical significance are often inappropriately applied to simulation results. They argued that simulations can provide a large number of sample sizes of replicate model runs, thereby artificially inflating statistical power, resulting in highly significant results—tiny P-values—regardless of ef-

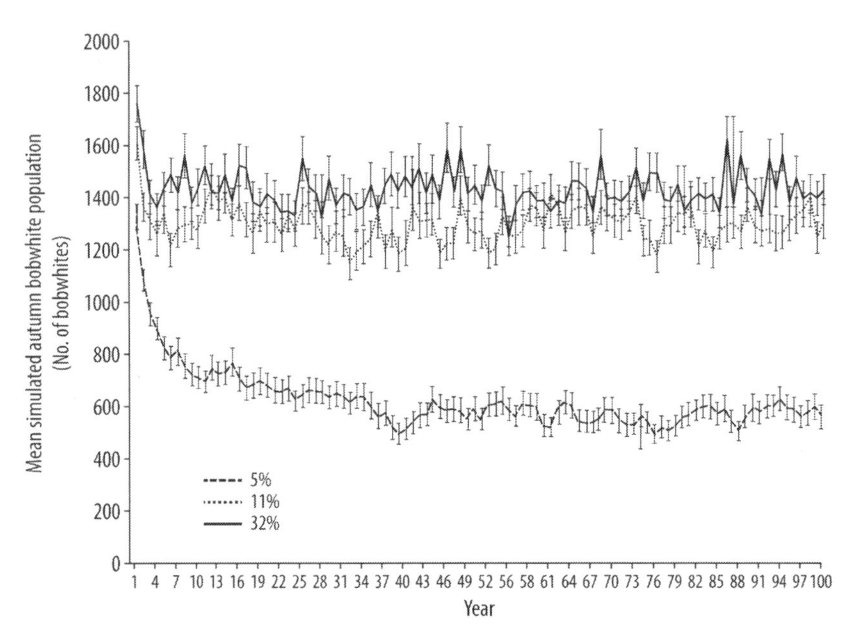

Figure 1.7. A graph of simulated long-term means (plus or minus standard errors) for autumnal northern bobwhite (*Colinus virginianus*) populations in areas with 5 percent, 11 percent, and 32 percent brush canopy coverage in Brooks County, Texas. (Adapted from DeMaso et al. (2014), with the permission of the Society for Range Management)

fect size. They also argued that tests of null hypotheses with no differences, such as between simulations of a treatment and a control, are irrelevant, because these differences are known a priori. Their concerns should be respected, but they should not be tempered to avoid any statistical comparisons of the outcomes of simulation experiments, such as in the above-mentioned study by Rader et al. (2011), which avoided using any statistical tests.

Study Duration

Although long-term studies in animal ecology are becoming more common, long-term experiments remain rare. Wiens (1984) lamented the lack of such studies in ornithology, where attention had been (and, in our opinion, continues to be) focused on simple queries that yielded answers in days, weeks, or months, rather than complex questions that required a decade or more to sort out. Funding cycles, resource agency priorities, and academic demands (e.g., graduate students should conduct research projects that can be completed in 2 to 4 years) are but a few of the factors that work against developing a culture of long-term studies in animal ecology.

Research on acorn woodpeckers (*Melanerpes formicivorus*) at the Hastings Reservation in California (Koenig and Mumme 1987) and on Florida scrub-jays (*Aphelocoma coerulescens*) at the Archbold Biological Station in Florida (Woolfenden and Fitzpatrick 1985) are 2 examples of how such investigations can be based on a sequence of dissertation studies that maintain long-term data but are complimentary, because they test and evaluate different hypotheses in an ongoing manner. In other cases, some species of animals, such as northern bobwhites, have been investigated for nearly a century, but at numerous locations, during different time intervals, throughout their distribution range. In the case of this well-studied game bird, a massive set of literature—at least 2,000 scientific journal articles—has allowed scientists to develop what has been defined as "pillars of knowledge," or well-understood linkages between populations, habitat, and weather (i.e., *r*-selection, successional affiliation, adaptive plasticity, and weather influences), that hold up across the entire geographic distribution of this species (Guthery and Brennan 2007). Other game species, such as white-tailed deer (*Odocoileus virginianus*), have also benefited from this type of approach, as a critical mass of information has emerged from numerous studies that have been conducted in enough different places over long timeframes (Hewitt 2011). Unfortunately, this is an extremely inefficient way to conduct animal ecology research that links demographic factors of populations with elements of habitat.

Short-term data merely produce short-term insights. How extensive a period of time a long-term study should cover is a function of numerous factors, some of which (especially funding) are beyond the influence of an investigator. Wiens (1984) noted that the duration of a study should be determined by logic and analytical needs, not by a lack of funding or by boredom.

How to Proceed

Romesburg (1981) felt that many of the ideas on which the science of wildlife management has been founded are hypotheses that have not been tested. He argued convincingly that animal ecologists do a good job of making observations, searching for patterns, and formulating hypotheses, but attempts to apply the hypothetico-deductive method to verify hypotheses are rare. He encouraged wildlife biologists to participate more actively in the search for causal mechanisms through experimentation. In contrast, Peters (1991) suggested that the search for causal mechanisms in ecology was overemphasized, and that more effort should be placed in a quest for predictive relationships, in order to help us solve (or at least identify) ecological problems.

Given the difficulties inherent in both laboratory and field experiments, as well as the somewhat contradictory advice in the literature about both the scientific method and its application, a legitimate question for those studying wildlife-habitat relationships is, How to proceed with investigations? The answer, in our opinion, depends on the kind of examination that is undertaken and the resources available to the researcher. In animal ecology, the questions at the heart of many studies are species driven. That is, investigations are done primarily because information about some aspect relevant to the life history of a species is needed for management purposes. Often even basic data about these species, such as the types of vegetation they use, are lacking, so research that documents patterns of this kind is needed. On the other hand, if an investigation concerns identifying or verifying patterns in nature, then hypothetico-deductive tests are neither necessary nor applicable.

Studies of patterns are crucial in both science and management, and they should not be viewed as second-class activities (Weiner 1995; Lawton 1996b; Wiens 2016). In science, patterns stimulate the development of explanations and, therefore, are a vital part of the scientific method. Much of what we know about animal ecology is based on observational studies (Herman 2002; see also the many books by E. O. Wilson). One might argue that observational studies are the underpinnings of the focused experimental work that allows us to explore mechanisms. Management decisions frequently need to be made quickly and, almost always, without all of the necessary information. Predictions derived from patterns alone may be sufficient for a management action, given that a particular risk is acceptable to the decision maker. For example, if we know that the abundance of a species is declining and the habitat on which it depends is also becoming reduced, some form of habitat conservation probably should be initiated, even though a causal relationship between the 2 patterns has not been definitively established. In this example, the decision maker may wish to invoke a precautionary principle and be more risk averse in issues pertaining to threats to the species.

The important role that patterns play in science and management requires that they be accurately identified and articulated. Thus, studies of specific configurations, conducted at several spatial scales, over relatively long periods of time, can help ensure that the true natures of the patterns under investigation are revealed. Also, "attention should be paid to details such as using unbiased sampling techniques, collecting adequate samples, and employing appropriate statistics" (Nudds and Morrison 1991:759).

Following these suggestions will help ensure that the detected patterns are real and are accurately described. Inaccurate descriptions can lead to inappropriate management actions and will hinder the development of realistic explanations (Kodric-Brown and Brown 1993).

When the purpose of an investigation is to discover the nature of a process or mechanism underlying a pattern in nature, a hypothetico-deductive test is needed. The species or the ecological system to be studied in a such a test might be chosen on the basis of how amendable the species or system is to control and manipulation. Neither laboratory nor field experiments are perfect for answering questions about ecological processes, yet both have value. Their relative benefits can perhaps be assessed by thinking of them as tools designed for a specific function. Like most tools, each will do "some things well, some things badly, and other things not at all" (Lawton 1996a). We suggest that, whenever possible, investigators make use of both field and laboratory experiments to gain the highest level of confidence in their hypotheses.

Murphy (1990) advocated applying the hypothetico-deductive method to single, large-scale manipulations, such as the creation of refuge boundaries, and treating them as experiments. He proposed that we could make predictions about what would happen after the manipulation is completed, wait to see what actually happens, and then make inferences about the processes that caused the outcomes. Sinclair (1991) echoed these ideas, and Nudds and Morrison (1991:758–759) noted, "Hypothetico-deductive research is not characterized by whether it is experimental, because hypotheses can be tested with data not collected by experiment."

There are almost certainly some situations where hypotheses can be conclusively falsified by a single observation, or by a few observations not derived from experiments, but the level of confidence in accepting or rejecting an explanation under investigation is dependent on the number of alternative rationales that could also account for observed outcomes. Ecosystems and the ecological processes that drive them are sufficiently complex for alternative explanations for observed phenomena to usually be abundant (Peters 1991). Hence large-scale manipulations—which generally are not replicated or randomly applied, and during which there often is no control or no assessment of critical environmental variables—are not likely to be exclusionary tests of underlying processes. Drew (1994:597) concluded, "The assertion that any question can be made scientifically rigorous by forcing it into the hypothetico-deductive method is false. Demographic and environmental indicators can be useful management tools, but when attempts are made to raise management targets to the level of scientific hypotheses—and exclusionary hypotheses at that—we only confuse the issue and give managers a false sense of assuredness."

The process of manipulating the environment, monitoring what happens, and changing the treatment if we do not get the results we expect is called "adaptive management," also referred to in the literature as "adaptive resource management" and "evidence-based knowledge" (McConnachie and Cowling 2013; also see Morrison et al. 2020:Chapter 6). This approach is potentially very useful, provided that an opportunity for modifying management practices actually exists. Such an approach could supply a wealth of information about what happens when we change the environment, under specific conditions, at a particular point in time. It also may offer hints about processes that cause the results we observe. Caution should be used, however, if adaptive management exercises are treated as hypothetico-deductive tests, because most are not designed as conventional experiments and thus cannot be analyzed as such (Guthery 2008).

Therefore, a legitimate concern of many animal ecologists is that environmental manipulations, particularly those dealing with large areas or occurring over long periods of time, may never provide much information about cause-effect relationships. Determining causal relationships is central to deciding what management can, and cannot, control. We submit that activities that can cause significant changes in natural environments but are relatively rare—such as building a telescope on a forested mountaintop that is habitat for an endangered squirrel species (Rushton et al. 2006)—are not likely to afford an opportunity to conduct hypothetico-deductive experiments. Inferences from studies of the effects of these 1-time events will be limited, due to a lack of replication and randomization. Nonetheless, impact studies of this kind are common and are not without value. Morrison et al. (2001) used the term "quasi-experiment" to categorize investigations in which either randomization or replication—or both—were sacrificed. BACI (mentioned earlier in this chapter) refers to the design of 1 common kind of quasi-experiment (Stewart-Oaten 1986). This design relies on measuring an ecological variable of interest (e.g., the occurrence or abundance of an endangered species) on the site where a disturbance is to take place, as well as on a similar site nearby, for a long-enough period before the disturbance to assess how the variable changes over time on both sites. Many endangered species, however, are rare and therefore may not be represented by enough individuals to assess abundance. Hence a study may simply consider occupancy. The variable is then measured over a span of time after the disturbance, and the relative difference in the variable between the disturbed site and the nearby site, both before and after the disturbance, is evaluated (Stewart-Oaten 1986; Morrison et al. 2001). Inferences beyond the disturbed site usually are not an objective in this kind of study. Therefore, random sampling from a larger target population is not as crucial as it is in an experiment (Stewart-Oaten 1986). Diller et al. (2016) used a classic BACI design to test whether the experimental removal of barred owls (Strix varia) influenced the demography of northern spotted owls, which it did, in a positive manner.

Complications that might be caused by differences between the disturbed site and a nearby site—Morrison et al. (2001) called these "reference sites"—potentially are reduced by tracking the variable of interest long enough to understand how it behaves on the 2 sites. Bayesian statistics (see above, and Chapter 5) may also provide an effective, alternative means by which to analyze the results of manipulations that lack classical experimental designs. Bayesian statistics and quasi-experiments are not substitutes for careful experimentation, but they may be useful when the latter is not possible (Hilborn and Ludwig 1993). Such approaches, however, are not to be confused with trial and error (discussed earlier in this chapter).

The methodologies people use to understand how the world works vary, according to circumstances and subject matter. Some subjects and situations may not lend themselves to experimentation. In these cases, understandings may need to be based on relatively weak inferences, derived either from patterns or creative analyses of manipulations, without replication or randomization. Properly designed experiments embedded in hypothetico-deductive tests are perhaps the best tool available for comprehending ecological processes and, when feasible, should be favored.

Strong Inferences

In his classic article, Platt (1964) made the analogy that implementing the scientific method is like climbing a tree. At the first fork, the results from an experiment form the basis for deciding to move to a branch on the left or a branch on the right. The next experiment gives us the basis for the following choice, and so on. It is crucial that individuals participating in the study of population-habitat relationships in animal ecology have a thorough understanding of all activities involved in the scientific method. This does not mean that each person will perform all of the activities in every investigation he or she undertakes. Such an understanding does, however, help ensure that the approach taken in a study will be appropriate for the kind(s) of question(s) being asked. The benefit of the scientific method is its power in helping humans understand the world in which they live. Thus all scientific activities—observing, recognizing patterns, and formulating and testing hypotheses—should be carried out so as to move us toward a better understanding of the natural world, doing so by generating reliable knowledge and using an experimental approach whenever and wherever possible.

Summary

There are numerous examples throughout this chapter that link back to the major themes outlined in Morrison et al. (2020) and are reiterated throughout this book. In order to effectively apply the experimental approach in animal ecology, we must start with an initial focus on individual organisms and how they gather together into collections, such as breeding pairs, leks, herds, populations, and so on. We need to know how a species is distributed in space before we can decide what to measure about that species. Taking the perspective of the organism of interest, while striving to keep our human perspective free of bias, is essential for laying the foundation upon which to base a controlled, manipulative experiment to test a hypothesis and generate reliable knowledge for management and conservation activities.

LITERATURE CITED

Akaike, H. 1973. Information theory as an extension of the maximum likelihood principle. Pp. 267–281 *in* B. N. Petrov and F. Csáki, eds. Second International Symposium on Information Theory. Akadémiai Kiadó, Budapest, Hungary.

Anderson, D. R., K. P. Burnham, and W. L. Thompson. 2000. Null hypothesis testing: Problems, prevalence, and an alternative. Journal of Wildlife Management 64:912–923.

Anderson, D. R., W. Link, D. J. Johnson, and K. Burnham. 2001. Suggestions for presenting the results of data analysis. Journal of Wildlife Management 65:373–378.

Berkson, J. 1938. Some difficulties of interpretation encountered in the application of the chi-squared test. Journal of the American Statistical Association 33:526–536.

Berkson, J. 1942. Tests of significance considered as evidence. Journal of the American Statistical Association 37:325–335.

Boertje, R. D., C. L. Gardner, M. M. Ellis, T. W. Bentzen, and J. A. Gross. 2017. Demography of an increasing caribou herd with restricted wolf control. Journal of Wildlife Management 81:429–448.

Bridgham, J. T. 2016. Predicting the basis of convergent evolution. Science 354:289.

Brinkhof, M. W. G., A. J. Cave, and A. C. Perdeck. 1997. The seasonal decline in the first-year survival of juvenile coots: An experimental approach. Journal of Animal Ecology 66:73–82.

Burnham, K. P., and D. R. Anderson. 2004. Multimodel inference: Understanding AIC and BIC in model selection. Sociological Methods and Research 33:261–304.

Butler, R. G. 1980. Population size, social behavior, and dispersal in house mice: A quantitative investigation. Animal Behavior 28:78–85.

Calsbeek, R., and B. Sinervo. 2002. An experimental test of the ideal despotic distribution. Journal of Animal Ecology 71:513–532.

Carey, S. S. 1994. A beginner's guide to scientific method. Belmont, CA.

Chamberlain, T. C. 1897. Studies for students: The method of multiple working hypotheses. Journal of Geology 5:837–848.

Cherry, S. 1998. Statistical tests in publications of The Wildlife Society. Wildlife Society Bulletin 26:947–953.

Chevalier, M., J. C. Russell, and J. Knape. 2019. New measures for evaluation of environmental perturbations using Before-After-Control-Impact analyses. Ecological Applications 29(2):e01838.

Conover, M. R., and A. J. Roberts. 2017. Predators, predator removal, and sage-grouse: A review. Journal of Wildlife Management 81:7–15.

Côté, I. M., and W. J. Sutherland. 1997. The effectiveness of removing predators to protect bird populations. Conservation Biology 11:395–405.

Cox, D. R. 1958a. Planning experiments. John Wiley & Sons, New York.

Cox, D. R. 1958b. Some problems connected with statistical inference. Annals of Mathematical Statistics 29:357–372.

Cumming, G. 2014. The new statistics: Why and how. Psychological Science 25(1):7–29.

Darwin, C. 1859. On the origin of species by means of natural selection, or the preservation of favoured races in the struggle for life. John Murray, London.

DeMaso, S. J., W. E. Grant, F. Hernández, L. A. Brennan, N. J. Silvy, X. B. Wu, and F. C. Bryant. 2011. A population model to simulate northern bobwhite population

dynamics in southern Texas. Journal of Wildlife Management 75:319–322.
DeMaso, S. J., F. Hernández, L. A. Brennan, N. J. Silvy, W. E. Grant, X. B. Wu, and F. C. Bryant. 2014. Short- and long-term influence of brush canopy cover in northern bobwhite demography in southern Texas. Rangeland Ecology and Management 67:99–106.
Dickson, J. G., ed. 1992. The wild turkey: Biology and management. Stackpole, Harrisburg, PA.
Diller, L. V., K. A. Hamm, D. A. Early, D. W. Lamphear, K. W. Dugger, C. B. Yackulic, C. J. Schwartz, P. C. Carlson, and T. L. McDonald. 2016. Demographic response of northern spotted owls to barred owl removal. Journal of Wildlife Management 80:691–707.
Drew, G. S. 1994. The scientific method revisited. Conservation Biology 8:596–597.
Ebert-Uphoff, I., and Y. Deng. 2012. Causal discovery for climate research using graphical models. Journal of Climate 2517:5648–5665.
Eisenhauer, N., M. A. Bowker, J. B. Grace, and J. R. Powell. 2015. From patterns to causal understanding: Structural equation modeling (SEM) in soil ecology. Pedobiologia 58:65–72.
Englund, G., and S. D. Cooper. 2003. Scale effects and extrapolation in ecological experiments. Advances in Ecological Research 33:161–213.
Feder, K. L. 1990. Frauds, myths, and mysteries: Science and pseudoscience in archaeology. Mayfield, Mountain View, CA.
Fisher, R. A. 1947. The design of experiments, 8th edition. Oliver & Boyd, London.
Franzblau, M. A., and J. P. Collins. 1980. Test of a hypothesis of territory regulation in an insectivorous bird by experimentally increasing prey abundance. Oecologia 46:164–170.
Gavin, T. A. 1991. Why ask why? The importance of evolutionary biology in wildlife science. Journal of Wildlife Management 55:760–766.
Gerlach, G. 1996. Emigration mechanisms in feral house mice—a laboratory investigation of the influence of social structure, population density, and aggression. Behavioral Ecology and Sociobiology 39:159–70.
Gerlach, G. 1998. Impact of social ties on dispersal, reproduction and dominance in feral house mice (*Mus musculus domesticus*). Ethology 68:684–694.
Glickman, S. E., and G. S. Caldwell. 1994. Studying behavior in artificial environments: The problem of "salient elements." Pp. 197–216 *in* E. F. Gibbons Jr., E. J. Wyers, E. Walters, and E. W. Menzel Jr., eds. Naturalistic environments in captivity for animal behavior research. State University of New York Press, Albany.
Goodstein, D. 1995. Conduct and misconduct in science. Annals of the New York Academy of Sciences 775:31–38.
Grinnell, J. 1917. The niche-relationships of the California thrasher. Auk 34:427–433.
Gross, P. R., and N. Levitt. 1994. Higher superstition. Johns Hopkins University Press, Baltimore.
Gross, P. R., N. Levitt, and M. W. Lewis, eds. 1996. The flight from science and reason. New York Academy of Sciences, Albany.
Gruebler, M. U., and B. Naef-Daenzer. 2010. Survival benefits of post-fledgling care: Experimental approach to a critical part of avian reproductive strategies. Journal of Animal Ecology 69:334–341.
Guthery, F. S. 2008. A primer on natural resource science. Texas A&M University Press, College Station.
Guthery, F. S., and S. L. Beasom. 1977. Responses of game and non-game wildlife to predator control in south Texas. Journal of Range Management 52:144–149.
Guthery, F. S., and L. A. Brennan. 2007. The science of quail management and the management of quail science. Pp. 407–420 *in* L. A. Brennan, ed. Texas Quails: Ecology and Management. Texas A&M University Press, College Station.
Guthery, F. S., J. J. Lusk, and M. J. Peterson. 2001. The fall of the null hypothesis: Liabilities and opportunities. Journal of Wildlife Management 65:379–384.
Guthery, F. S., A. R. Rybak, S. D. Fuhlendorf, T. L. Hiller, S. G. Smith, W. H. Puckett Jr., and R. A. Baker. 2005. Aspects of thermal ecology of bobwhites in north Texas. Wildlife Monographs 159:1–36.
Hairston, N. G., Sr. 1989. Ecological experiments: Purpose, design, and execution. Cambridge University Press, Cambridge.
Hanson, N. R. 1965. Patterns of discovery. Cambridge University Press, Cambridge.
Harding, S. 1986. The science question in feminism. Cornell University Press, Ithaca, NY.
Harris, V. T. 1952. An experimental study of habitat selection by prairie and forest races of the deer mouse, *Peroyscus maniculatus*. Contributions from the Laboratory of Vertebrate Biology No. 56. University of Michigan, Ann Arbor.
Herman, S. G. 2002. Wildlife biology and natural history: Time for a reunion? Journal of Wildlife Management 66:933–946.
Hewitt, D. G., ed. 2011. Biology and management of white-tailed deer. CRC Press / Taylor & Francis Group, Boca Raton, FL.
Hilborn, R., and D. Ludwig. 1993. The limits of applied ecological research. Ecological Applications 3:550–552.
Hurlbert, S. H. 1984. Pseudoreplication and the design of

ecological field experiments. Ecological Monographs 54:187–211.

Johnson, D. H. 1999. The insignificance of statistical significance testing. Journal of Wildlife Management 66:763–772.

Johnson, D. H. 2002. The role of statistical hypothesis testing in wildlife science. Journal of Wildlife Management 66:272–276.

Klaus, J. M., and R. F. Noss. 2016. Specialist and generalist amphibians respond to wetland restoration treatments. Journal of Wildlife Management 80:1106–1119.

Klopfer, P. H. 1963. Behavioral aspects of habitat selection: The role of early experience. Wilson Bulletin 75:15–22.

Kneller, G. F. 1978. Science as a human endeavor. Columbia University Press, New York.

Kodric-Brown, A., and J. H. Brown. 1993. Incomplete data sets in community ecology and biogeography: A cautionary tale. Ecological Applications 3:736–742.

Koenig, W. D., and R. L. Mumme. 1987. Population ecology of the cooperatively breeding acorn woodpecker. Monographs in Population Biology 24. Princeton University Press, Princeton, NJ.

Krausman, P. R. 2017. P-values and reality. Journal of Wildlife Management 81:562–563.

Kuehl, R. O. 1994. Statistical principles of research design and analysis. Duxbury Press, Belmont, CA.

Larson, J. A., T. E. Fulbright, L. A. Brennan, F. Hernández, and F. C. Bryant. 2012. Selection of seeds of an exotic and three native grasses by northern bobwhites (*Colinus virginianus*) Southwestern Naturalist 57:319–322.

Lawton, J. H. 1996a. The ecotron facility at Sillwood Park: The value of "big bottle" experiments. Ecology 77:665–669.

Lawton, J. H. 1996b. Patterns in ecology. Oikos 75:145–47.

Lecomte, J., K. Boudjemadi, F. Sarrazin, K. Cally, and J. Colbert. 2004. Connectivity and homogenisation of population sizes: An experimental approach in *Lacerta vivipara*. Journal of Animal Ecology 73:179–189.

Lehmann, V. 1984. Bobwhites on the Rio Grande Plains of Texas. Texas A&M University Press, College Station.

Lemoine, N. P., A. Hoffman, A. J. Felton, L. Baur, F. Chaves, J. Gray, Q. Yu, and M. D. Smith. 2016. Underappreciated problems of low replication in ecological field studies. Ecology 97:2554–2561.

Leopold, A. S. 1944. The nature of heritable wildness in turkeys. Condor 46:133–197.

Lieury, N., S. Ruette, A. Devillard, F. Drouyer, B. Baudox, and A. Million. 2015. Compensatory immigration challenges predator control: An experimental evidence-based approach improves management. Journal of Wildlife Management 79:425–434.

Link, W. A., and R. J. Barker. 2010. Bayesian inference with ecological applications. Academic Press / Elsevier, London.

Liu, E. and J. Fan, eds. 2018. Fundamentals of animal laboratory science. CRC Press / Taylor & Francis Group, Boca Raton, FL.

Major, H. L., R. T. Buxton, C. R. Schacter, M. D. Conners, and I. I. Jones. 2017. Habitat modification as a means of restoring crested auklet colonies. Journal of Wildlife Management 81:112–121.

Marcot, B. G. 1998. Selecting appropriate statistical procedures and asking the right questions: A synthesis. Pp. 129–142 *in* V. Sit and B. Taylor, eds. Statistical methods for adaptive management studies. British Columbia Ministry of Forests Research Program, Victoria. http://www.for.gov.bc.ca/hfd/pubs/docs/lmh/lmh42.htm.

Marcot, B. G. 2019. Causal modeling and the role of expert knowledge. Pp. 298–310 *in* L. A. Brennan, A. N. Tri, and B. G. Marcot, eds. Quantitative analyses in wildlife science. Johns Hopkins University Press, Baltimore.

Marcot, B. G., and T. D. Penman. 2019. Advances in Bayesian network modelling: Integration of modelling technologies. Environmental Modelling & Software 111:386–393.

Matsuzaki, S. S., U. Nisikawa, N. Takamura, and I. Washanti. 2009. Contrasting impacts of invasive engineers on freshwater ecosystems: An experiment and meta-analysis. Oecologia 158:673–686.

Matter, W. J., and R. W. Mannan. 1989. More on gaining reliable knowledge: A comment. Journal of Wildlife Management 53:1172–1176.

Matter, W. J., R. W. Mannan, E. W. Bianchi, T. E. McMahon, J. H. Menke, and J. C. Tash. 1989. A laboratory approach for studying emigration. Ecology 70:1543–1546.

Mayr, E. 1974. Behavior programs and evolutionary strategies. American Scientist 62:650–659.

McConnachie, M. M., and R. M. Cowling. 2013. On the accuracy of conservation managers' beliefs and if they learn from evidence-based knowledge: A preliminary investigation. Journal of Environmental Management 128:7–14.

McMahon, T. E., and J. C. Tash. 1988. Experimental analysis of the role of emigration in population regulation of desert pupfish. Ecology 69:1871–1883.

Medawar, P. 1984. Pluto's republic. Oxford University Press, New York.

Mertz, D. B., and D. E. McCauley. 1980. The domain of laboratory ecology. Pp. 229–244 *in* E. Saarinen, ed. Conceptual issues in ecology. D. Reidl, Dordrecht, Netherlands.

Morrison, M. L., W. M. Block, M. D. Strickland, and W. L.

Kendall. 2001. Wildlife study design. Springer, New York.

Morrison, M. L., L. A. Brennan, B. G. Marcot, W. M. Block, and K. S. McKelvey. 2020. Foundations for Advancing Animal Ecology. Johns Hopkins University Press, Baltimore.

Morrison, M. L., B. G. Marcot, and R. W. Mannan. 2006. Wildlife-habitat relationships: Concepts and applications, 3rd edition. Island Press, Washington, DC.

Murphy, D. D. 1990. Conservation biology and scientific method. Conservation Biology 4:203–204.

Nelson, A. R., C. L. Johnson, W. J. Matter, and R. W. Mannan. 2002. Canadian Journal of Zoology 80:2056–2060.

Nudds, T. D., and M. L. Morrison. 1991. Ten years after "reliable knowledge": Are we gaining? Journal of Wildlife Management 55:757–760.

Peters, R. H. 1991. A critique for ecology. Cambridge University Press, Cambridge.

Petrowski, H. 1995. Engineers of dreams: Great bridge builders and the spanning of America. Knopf / Random House, New York.

Platt, J. R. 1964. Strong inference. Science 146:347–353.

Popper, K. 1981. The myth of inductive hypothesis generation. Pp. 92–99 *in* R. D. Tweney, M. E. Doherty, and C. R. Mynatt, eds. On scientific thinking. Columbia University Press, New York, USA.

Provencher, L., N. M. Gobris, L. A. Brennan, D. R. Gordon and J. L. Hardesty. 2002. Breeding bird response to midstory hardwood reduction in Florida sandhill longleaf pine forests. Journal of Wildlife Management 66:641–661.

Rader, M. L., L. A. Brennan, K. A. Brazil, F. Hernández, and N. J. Silvy. 2011. Simulating northern bobwhite population responses to nest predation, nesting habitat, and weather in south Texas. Journal of Wildlife Management 75:61–70.

Ramsey, F. L., and D. W. Schafer. 2002. The statistical sleuth. Duxbury Press, Pacific Grove, CA.

Ricklefs, R. E. 2012. Naturalists, natural history, and the nature of biological diversity. American Naturalist 179:423–435.

Robinson, D. H., and H. Wainer. 2002. On the past and future of null hypothesis significance testing. Journal of Wildlife Management 66:263–271.

Romesburg, H. C. 1981. Wildlife science: Gaining reliable knowledge. Journal of Wildlife Management 45:293–313.

Rushton, S. P., D. J. A. Wood, P. W. W. Lurz, and J. L. Koprowski. 2006. Modelling the population dynamics of the Mt. Graham red squirrel: Can we predict its future in a changing environment with multiple threats? Biological Conservation 131:121–131.

Scheiner, S. M., and J. Gurevitch, eds. 2001. Design and analysis of ecological experiments, 2nd edition. Oxford University Press, New York.

Sells, S. N., S. B. Bassing, K. J. Barker, S. C. Forshee, A. C. Keever, J. W. Goerz, and M. S. Mitchell. 2018. Increased scientific rigor will improve reliability of research and effectiveness of management. Journal of Wildlife Management 82:485–494.

Simpson, G. G. 1964. This view of life: The world of an evolutionist. Harcourt, Brace & World, New York.

Sinclair, A. R. E. 1991. Science and the practice of wildlife management. Journal of Wildlife Management 55:767–773.

Skalski, J. R., and D. S. Robson. 1992. Techniques for wildlife investigations: Design and analysis of capture data. Academic Press, New York.

Smith, R. L. 1996. Ecology and field biology, 5th edition. Harper Collins, New York.

Steidl, R. J., J. P. Hayes, and E. Schauber. 1997. Statistical power analysis in wildlife research. Journal of Wildlife Management 61:270–279.

Steidl, R. J., and L. Thomas. 2001. Power analysis and experimental design. Pp. 14–36 *in* S. M. Scheiner and J. Gurevitch, eds. Design and analysis of ecological experiments, 2nd edition. Oxford University Press, New York.

Stewart-Oaten, A., W. W. Murdoch, and K. R. Parker. 1986. Environmental impact assessment: "Psuedoreplication" in time? Ecology 67:929–940.

Swaisgood, R. R., L. A. Nordstrom, J. G. Schuetz, J. T. Boylan, J. J. Fournier, and B. Shemai. 2018. A management experiment evaluating nest-site selection by beach-nesting birds. Journal of Wildlife Management 82:192–201.

Tapper, S. C., G. R. Potts, and M. H. Brockless. 1996. The effect of experimental reduction in predation pressure on the breeding success and population density of grey partridges *Perdix perdix*. Journal of Applied Ecology 33.965–978.

Tiwari, B. K., H. Tynsong, M. M. Lynrah, E. Lapasam, S. Deb, and D. Sharma. 2013. Institutional arrangement and typology of community forests of Meghalaya, Mizoram and Nagaland of North-East India. Journal of Forestry Research 24(1):179–186.

Warren, C. E. 1971. Biology and water pollution control. W. B. Saunders, Philadelphia.

Wecker, S. C. 1963. The role of early experience in habitat selection by the prairie deer mouse, *Peromyscus maniculatus bairdi*. Ecological Monographs 33:307–325.

Weiner, J. 1995. On the practice of ecology. Journal of Ecology 83:153–158.

White, G. C., and K. P. Burnham. 1999. Program MARK:

Survival estimation from populations of marked animals. Bird Study 46, Supplement:120–138.
White, J. W., A. Rassweiler, J. F. Samhouri, A. C. Stier, and C. White. 2014. Ecologists should not use statistical significance tests to interpret simulation model results. Oikos 123(4):385–388.
Wiens, J. A. 1984. The place of long-term studies in ornithology. Auk 101:102–103.
Wiens, J. A. 2016. Ecological challenges and conservation conundrums. John Wiley & Sons, New York.
Wilzbach, M. A. 1985. Relative roles of food abundance and cover in determining the habitat distribution of stream-dwelling cutthroat trout (*Salmo clarki*). Canadian Journal of Fisheries and Aquatic Science 42:1668–1672.
Wolff, J. O., E. M. Schauber, and W. D. Edge. 1996. Can dispersal barriers really be used to depict emigrating animals? Canadian Journal of Zoology 74:1826–1830.
Woolfenden, G. E., and J. W. Fitzpatrick. 1985. The Florida scrub jay: Demography of a cooperatively breeding bird. Monographs in Population Biology 20. Princeton University Press, Princeton, NJ.

2 Measurement of Habitats and Populations for Habitat Classification

Introduction

As we established in Morrison et al. (2020), animal ecology involves describing places inhabited by a species in the context of its population(s). Thus it is insufficient to go to a location, find a species, and characterize its habitat without also understanding how that habitat relates to aspects of the species' population, such as density, position within its geographic range, demographic parameters, and the like.

Animal ecology is focused on documenting patterns and then designing studies to understand the processes that shape those patterns (Pickett et al. 1994). This and the following chapter can perhaps best be described as an exercise in pattern seeking. That is, in most studies, we are searching for ways to describe patterns of habitat use and how they influence or relate to the status of a population, rather than testing specific hypotheses about the causes of the relationships we observe in nature. Pattern seeking is a necessary early step in the analysis of an animal's habitat and falls into the practice of natural history observation that has marked the history of animal ecology (Block and Brennan 1993; Fig. 2.1). This practice can most readily be contrasted with various descriptive and experimental methodologies, as developed in Chapter 1. As such, much of what is described here and in Morrison et al. (2020:Chapter 4) represent a first step in a chain of questions that ultimately leads to deeper understandings of population processes and thereby increases our ability to efficiently manage and conserve species.

In this chapter, we review 2 major aspects of measuring animal habitat: what to measure and how to measure it. Implicit in our discussion is a careful evaluation of the "niche gestalt" of an animal—that is, what that animal perceives as important to its reproduction and survival. As discussed elsewhere in this volume and in Morrison et al. (2020), we must try to see through the eyes of that animal and understand its sensory perceptions. We also review common sampling/estimation methods to assess population parameters and then relate them to how habitat is classified. The key here is to identify parameters that are likely to respond to the occurrence of and changes in environmental conditions and other limiting factors. Measuring occupancy, abundance, or density may be misleading, as they can mask underlying changes in demography that influence the stability or trajectory of the population (Van Horne 1983; Gundersen et al. 2008; Boyce et al. 2015). Hence, investigators may need to consider multiple population parameters to describe their system. In addition, we discuss common sampling methods and analyses

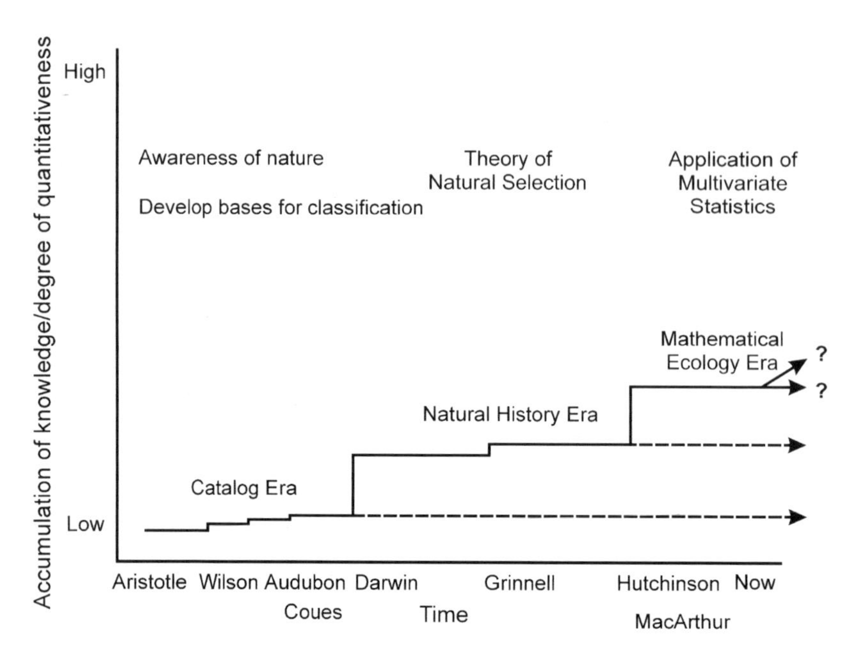

Figure 2.1. A graph of the historical development of the habitat concept in ornithology. (Reproduced from Block and Brennan (1993:Fig. 2), courtesy of *Current Ornithology*)

employed in measuring wildlife habitat. These techniques include the use of organism-centered plots versus various randomization procedures, how vegetation and other environmental features have been quantified, and calculations made of habitat diversity and heterogeneity, all of which vary by spatial scale. That is, variables used to describe a specific breeding site will probably differ from those used to depict a forested stand, a landscape, or a region.

What Habitat Variables to Measure

When searching for patterns, it is important to recognize that knowledge is a path containing stages. At each step along the path, the important variables to measure will change, because the questions have shifted. At the earliest stages of knowledge, the most important question might be, "Where can I find the organism?" Much of the early study of wildlife was devoted to this specific query (see Morrison et al. 2020:Chapter 2). So the first patterns might simply be those of occurrence. Once you know where to find an organism, other questions naturally follow. Are there associations that will allow you to find these organisms reliably in the future, or in other areas? Are there associations that will dependably help you to locate high densities of these organisms? Much of what is described as the study of "wildlife-habitat relationships" (WHR) is associated with these 2 questions. After they have been answered adequately, others revolve around more fundamental and causal relationships, such as "What life-history traits are facilitated by the use of specific habitats?" and "Which habitats are most critical for reproduction and survival?" Beyond these simple population descriptors, we might also be interested in the effects of disturbances, stressors, and other impacts on wildlife populations. Importantly, the variables measured, the precision of those measurements, and, most likely, the necessary study designs will all change. An investigation targeted simply at finding organisms will probably measure the wrong variables and have an inappropriate study design to meaningfully ask questions about life-history characteristics.

Green (1979:10), commenting specifically about ecological responses to disturbances and stressors, offered 4 criteria for variable selection, with a focus on impact studies: (1) spatial and temporal variability in biotic and environmental variables, (2) the feasibility of sampling with precision at a reasonable cost,

(3) relevance to the impact effects and the sensitivity of responses to them, and (4) some economic or aesthetic value. These criteria also apply broadly to experiments, quasi-experiments, and descriptive studies. In designing an investigation, a severe weakness in any single aspect, or a series of weaknesses across several of them, will place strong limitations on the scientific credibility of the research and the applicability of its results to management situations. Further, investigators cannot measure everything, everywhere. Collecting data just for the sake of doing so is of limited value unless the data convey reliable information (Anderson 2001). Identifying the key variables to measure is crucial for a well-designed study that can address its research objectives (Morrison et al. 2008). This requires structuring your data collection by using the results of previous investigations to help guide the selection of variables and, by understanding the ecology of the relevant species, to identify factors that enhance or limit their populations (Herman 2002).

Incorporating the variability inherent in the system of interest is crucial in designing a study. It should be formulated to capture the range of variation and its distribution, as the availability and amount of resources rarely occur in classic Gaussian (i.e., normal) distributions. This includes natural, stochastic, and systematic changes, and measurement and sampling errors. An important component of variability is temporality. Researchers must keep in mind that various elements in landscapes are always changing. Therefore, studies should be of sufficient duration and include adequate spatial replication to include the shifts in conditions that a species experiences. Also, a species' ecology is not constant throughout its annual and life-stage cycles, and species-habitat relationships may vary by spatial scale. Resources such as food and cover vary over the course of the year, so animals may move or alter their behaviors in response to these fluctuations. Many birds migrate either short or long distances to take advantage of these types of changes. Ungulates, such as black-tailed deer (*Odocoileus hemionus*) or elk (*Cervus canadensis* [*C. elaphus canadensis*]), move to winter ranges, which more often offer forage free of snow cover. Studies must be clear regarding the applicability of the work—whether it pertains to a species on its breeding or wintering grounds, during migration, or throughout its full annual cycle. As noted by Johnson (1981), habitat selection operates at a hierarchy of spatial scales. Therefore, a study should be designed to measure variables at the appropriate spatial scale and during the right time period in relation to the research hypotheses being addressed (Guthery 2008:18–22).

There is a cost-benefit analysis that researchers conduct, either formally or informally, when choosing variables for measurement. An investigator must explicitly identify the precision necessary to reach project goals, and then match all sampling to this. Project goals are often stated in terms of being able to measure an effect of a certain size, and the sample size must be adequately large to have the power to validate the effect, should it occur. Therefore, because all measurements require time and money, the feasibility of the study crucially depends on carefully choosing the variables to be measured. A classic waste-of-time syndrome that seems to have become inherent in habitat research is a study design where the objective is to measure nearly every potential habitat component possible over the duration of the study, and then run correlations to try and identify variables that are important in describing the distribution of the animal of interest. Doing so, however, results in a tradeoff between measuring lots of variables at only a few plots (at the expense of including more plots) and having an adequate sample size to ensure meeting the objectives for precision, effect size, and so forth. The fear of missing something important can drive such decisions and is therefore somewhat understandable. Given that most investigations are operating with limited time and scarce budgets, however, extensive thought should go into the selection of variables before field studies are started. Newer remote-sensing capabilities, such as LiDAR (discussed later in this chapter), provide opportunities to be more efficient in the selection of habitat variables and the placement of sample plots.

As noted above by Green (1979), the variables that are measured should have relevance and sensitivity to the question at hand. This requires researchers to first have a basic understanding of the species under investigation and be able to hypothesize the factors that limit their distribution and population. Part of this understanding is gleaned from lessons learned in previous studies, to help narrow the set of potential habitat variables to those demonstrably related to habitat use and to avoid duplicating prior mistakes. If questions remain regarding the need to include certain variables, then preliminary analyses of pilot data should be used to reduce or refine the variables being collected in the field. This provides more time to increase the sample size and, concomitantly, the precision of the remaining variables.

Green (1979) also stated that the selected variables could have some economic or aesthetic value. This criterion may or may not be a factor, depending on the purpose of the study. Recent policy directions in the listing, designation of critical habitat, and recovery of species under the US Endangered Species Act require an analysis of the economic effects of proposed actions. Thus investigations conducted prior to initiating treatments can elucidate the costs and benefits of implementing a proposed management treatment. Variables of economic interest might also be included in a study if the intended audience for the results has specific concerns. For example, the relationship between bird diversity or abundance and timber output would have more relevance for foresters than would that between birds and shrub cover.

Conceptual Frameworks

Characterizing and measuring habitats and populations has been the focus of naturalists and ecologists for millennia (Morrison et at. 2020:Chapter 2). Indeed, early hunter-gatherers needed to know where and when to find animals for food, clothing, and shelter. They learned by trial and error and passed that information on to succeeding generations. For centuries afterward, naturalists observed and described places where they found organisms. Some indigenous peoples related the occurrence, abundance, and distribution of animals with cultural and economic values to place names, identifying specific locations with words that conjured what was found there or with names that honored ancestors. This was done, for instance, by the native speakers of Tlingit and Haida in southeast Alaska (Thornton 2012).

At about the turn of the twentieth century, ecologists moved toward quantitative descriptions by measuring attributes of the habitats where they found species. Initially, these descriptions were univariate estimates of habitat variables. Research publications presented this information in large tables and often tested the differences among locations, conditions, or species using univariate statistics. Building on the concept of a niche being a multidimensional space, ecologists explored ways to apply multivariate statistics to a set of variables in a single analysis, to better characterize species' habitats (James 1971; Capen 1981; Golley 1993). For the past 30 or 40 years, the science of wildlife habitat ecology has pretty much stagnated, as efforts have obsessed over the newest and best way to analyze data and the not-so-much-better ways to understand animal ecology. This certainly is not to belittle the tremendous advances made in various forms of analyses, as they are furthering our goal to understand animal ecology. Instead, we want to emphasize that analysis is a tool, not our ultimate goal in predictions.

Morrison et al. (2020) outlined a theoretical framework by identifying the appropriate scales of study as they relate to the population(s) under investigation. Historically, the lack of such a framework has been a major impediment to sound resource management and conservation planning (e.g., Huston 2002). For example, there are thousands of models predicting habitat information for terrestrial and aquatic vertebrates that are based on regression, correlation, several multivariate methods, Bayesian analysis, and model selection, as well as on a habitat suitability index (HSI), habitat evaluation procedures (HEP), and Gap Analysis techniques (Scott et al. 1993; also see Chapter 5). By and large, the extent of these efforts was usually restricted to a specific place and time, and many of them are untested or not validated independently. The results of such models are then qualitatively compared with similar, previously conducted studies to show how they concur with or differ from past findings. Given the constrained scope of these investigations, they have only a limited ability to predict changes in distribution or habitat use (O'Connor 2002). Resource managers are left with a set of often-conflicting independent studies, without useful guidance on how to synthesize and apply them to real-world management situations.

A relatively new analytical approach involves taking the results of independent studies and conducting a meta-analysis to seek consistent patterns (Franklin et al. 2004). Meta-analyses are particularly appealing when (1) moderate to large amounts of empirical observations are available, (2) the results vary across studies, (3) the expected magnitude of the effect is relatively weak, and (4) the sample sizes of individual investigations are limited (Arnqvist and Wooster 1995). As promising as it sounds, however, meta-analysis is not a panacea. Issues with missing data, data quality, and data exclusion, and the combination of dissimilar datasets, may reduce the utility of meta-analyses and the validity of conclusions drawn from them (Gurevitch et al. 2001).

Regardless of the scale, variables should represent factors that limit the abundance and distribution of species (O'Connor 2002). O'Connor noted that various factors can achieve this, but the influence of any single relevant element is not additive to the influence of others, because different combinations may interact in disparate ways for each situation. Clearly, a paradigm shift is needed in the way we conceptualize and study the factors driving the distribution and abundance of animals if we are to advance our understanding of what is happening and implement meaningful management and conservation actions (e.g., Huston 2002; O'Connor 2002). A clever advance along this line of inquiry is the concept of "slack" in the variability of habitat relationships for a species. Northern bobwhites (*Colinus virginianus*) constitute a species with a huge geographic distribution (the eastern two-thirds of the continental United States and most of Mexico) and can occupy a wide variety of forest and rangeland vegetation configurations. Guthery (1999) argued that across the northern bobwhite's geographic distribution, many different arrangements of patches of woody and herbaceous cover can meet the habitat needs of this species. In cases where herbaceous cover was limited, this slack in available habitat could be overcome if sufficient woody cover was present. Therefore that site could be occupied by northern bobwhites. The same applies with respect to insufficient woody cover that is compensated for by more-abundant herbaceous cover.

Many early studies described aspects of the places where animals were found. More-recent studies have addressed hypotheses that could explain the persistence of animals, including vegetation structure, floristics, interspecific competition, disturbance, predator activity, food availability, and abiotic conditions. Most of these factors are interrelated (e.g., habitat and predator activity), and most investigations conducted to examine such components have shown them to operate in certain locations and at certain times. Given the interactions among multiple processes and limiting factors, research should be scaled appropriately to capture the full range, distribution, and degree of variation for the animals under investigation. Trying to isolate a specific variable that drives a response appears to be misdirected in most applications. Although we are not suggesting that experimental and observational studies of specific factors are unwarranted, we propose that a different overall framework is needed to improve our ability to predict animal persistence across space and time. Simply measuring selected environmental variables and certain aspects of the occurrence of organisms, and then fitting them into some type of statistical correlation model, whether or not these components are statistically significant, can still fail to correctly predict future events and outcomes. Cooke and Carroll (2017) noted this when developing models to predict the spread of mountain pine beetles (*Dendroctonus ponderosae*) in forests in eastern Canada under climate change conditions.

Animals are influenced by multiple factors, and they can control their exposure to only some of them. Most of these vary in time and space, and their effects on species will differ accordingly. This results in a continual shifting of limiting factors, thereby complicating our ability to predict where, when, and how specific elements influence populations and distributions. Many of these are non-limiting, have little or no effect on the rate of the process of interest, or work in concert with other factors so their separate, individual effects may seem minor. Consequently, a focus on these elements may provide a very circumscribed (if any) influence on habitat selection or population processes. Typically, process rates will be correlated only with variation in a limiting resource. Hence it is important to correctly identify such factors to understand their influence on the processes under investigation. Most ecological research, however, is conducted using a mixture of measurements that were made under both limiting and non-limiting conditions, and this produces datasets with a great deal of variance, which prevent identifying a species' actual response to a specific factor. The high level of variance that is often found in correlations between ecological processes and presumed causal factors may not be a sampling error involving random noise, but, rather, a mechanistic consequence of shifts between limiting resources or the influence of other restrictive factors, such as mortality or dispersal (Wiens 2016).

Modeling Approaches

Details on the types, merit, and limitations of ecological models are discussed in detail in Chapter 5, so our discussion here provides a general overview on how the modeling process relates to decisions about what to measure. Models are effectively descriptions of complex, real-world systems. They can be conceptual, verbal, graphical, or numerical and can represent hypotheses that can be tested with data, generally using a statistical model. Animal ecologists use models for many reasons. For instance, one might be designed to examine how populations alter as a result of human-induced environmental change (e.g., wind development projects, forestry, agriculture) by making simulated projections into the future and defining boundaries of populations in the most realistic manner available. Animal ecologists with long-term (20 years or more) datasets may want to develop models that can characterize historical patterns and use them to project future configurations. Here, however, we will concentrate on conceptual and statistical models.

The purpose of a conceptual model is to formalize a process, so it can be broken down into a group of testable hypotheses. It focuses on causal relationships and seeks to describe a process in a coherent and logical manner. While all conceptual models are, to some extent, based on data and data-derived understandings, there is no general formalized relationship between a modeled process and specific data streams. Computer models are a special group within conceptual models that seek to numerically simulate processes. These are often referred to as "mechanistic models." A useful way to think about such computerized models is to view them as animated extended hypotheses. They ask, "If things operate in ways similar to the way the model is structured, what behaviors are to be expected?" Mechanistic models can be extremely complex, and they are common in applied physics fields, such as meteorology and climate science. Predictions of hurricane trajectories, tomorrow's weather, El Niño-Southern Oscillation (ENSO) cycles, and long-term climate projections are largely dependent on mechanistic models. Because they tend to be complex and, therefore, are prone to error propagation, mechanistic models generally are most useful in disciplines such as physics, where processes can be described more accurately.

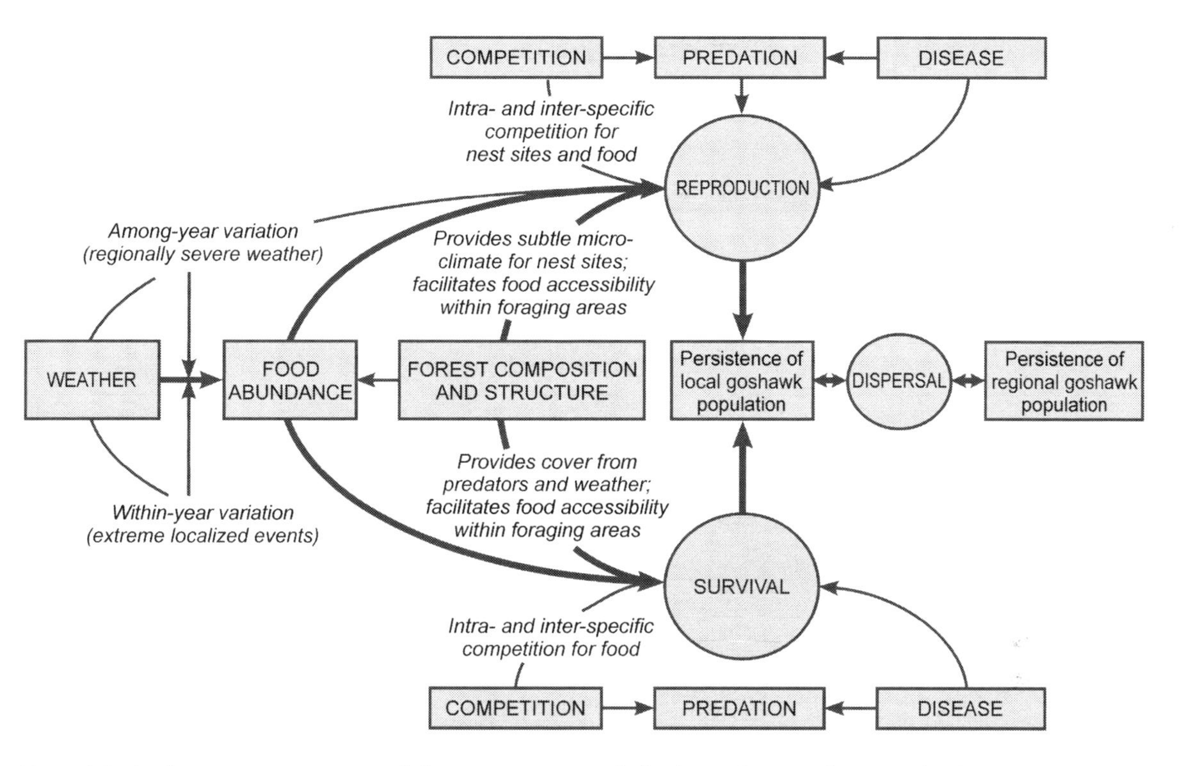

Figure 2.2. A schematic representation of the various pathways by which physical and biotic factors interact to influence vital rates and the population being studied. *Thicker lines* indicate pathways with relatively stronger effects. (Reproduced from Reynolds et al. (2006:Fig. 2), courtesy of the Cooper Ornithological Society)

In wildlife biology, conceptual models are useful for framing factors that influence the distribution and population of a species of interest. For example, Reynolds et al. (2006) used a conceptual model to depict how various biotic and abiotic factors interact to influence northern goshawk (*Accipiter gentilis*) populations (Fig. 2.2). This type of model can provide the basis for a series of research studies to test predictions arising from the model, which can iteratively identify factors to explain the variation in population parameters (Brennan and Block 2018).

Unlike conceptual models, statistical models are formally based on and limited by data (Brennan et al. 2019). Whereas conceptual models seek to explain a process predicated on causal mechanisms, statistical models are more limited in scope. They only ask if things are related (e.g., general linear models, logistic regression), are different (e.g., analysis of variance [ANOVA], t-tests), and are grouped (e.g., cluster analyses, ordination). None of these tests directly implies causality. Causal understanding arises from a study's design aspects, not from the tests themselves. For example, the results of an ANOVA test of whether 2 treatments had different effects on a population can only be related to the treatment if the statistical design removed all other possible causes for the observed difference. As such, statistical tests only have meaning within the contexts in which they are applied.

Fitting a statistical model to data entails estimating model parameters and then evaluating the strength of the model in predicting real-world situations (White 2001; Wester 2019). A statistical model is only as good as the data used to estimate the parameters and the selection of the situation in which the data were gathered. Models based on limited spatial or temporal data are likely to be constrained in their application, because of variability in time and space

(Van Horne 2002). Even the most elegant statistical model cannot include all biologically important causal relationships for a given location. The reasons for this are twofold. First, researchers rarely know the suite of factors that influence habitat use a priori, nor do they understand the relative strength of those relationships. Second, even if a model highlights the strengths of relationships, these associations are correlative and may not be the variables that actually influence organisms' choices. This signals the importance of variable selection, which underpins a model's validity.

In both conceptual and statistical models, there is a tradeoff between general and complex models. General models use few terms, which may increase their applicability to a wide range of systems. They may describe broad patterns, but they are not as useful for making specific predictions, because they are less likely to be directly related to processes than complex models. Whether they are simple or complex, models cannot be universally validated, because they are, by nature, local—that is, built on local data or (if conceptual) on local understandings. Therefore, they are likely to apply only within tight and often unknown boundary conditions, or they may have failed to incorporate processes and therefore rely on correlations that may be fleeting or spurious (Van Horne 2002). The term "validating" in a modeling context is technically incorrect, because it implies, without any particular element of support, that the model is valid. Therefore, rather than stating that you are validating some type of model, it would be more appropriate to use language such as "testing the validity (or veracity) of a model." Therefore, knowing the scope of a model, both in terms of its conceptualization and testing, is crucial to its proper application.

Application of a Conceptual Framework

Understanding how an organism responds under varying environmental conditions across its range is critical for the restoration, conservation, recovery, and management of species (Wiens 2016). Any framework that is developed must be readily accessible to the resource personnel who ultimately make on-the-ground management decisions. Habitat variables, especially when sampled over short timeframes, usually provide little information about key processes, such as birth and death rates, that are needed to make sound management decisions (e.g., Tyre et al. 2001). Below we introduce and review many of the methods that have been employed to assess the use of resources by wildlife, including relatively traditional as well as more-recent methodologies. Each has fundamental strengths and weaknesses, with its merit depending on the questions being asked in a particular study.

Spatial Scale

Here we will be concerned with spatial scale as it relates to the measurement of environmental features that influence the distribution, abundance, and viability of animals and populations. Johnson's (1980) seminal paper on habitat selection identified a hierarchy of scales at which a species selects habitat. He defined "first-order selection" as the physical or geographic range of the species. Within this geographic range, individuals or social groups use home ranges, which constitute "second-order selection." The use of a specific site within the home range defines "third-order selection." "Fourth-order selection" involves the procurement of tangible resources (e.g., food, dens) from that site. A correct determination of scale is the cornerstone of habitat analysis and model development, and, in turn, species conservation. That is, we should match the relative size or extent of the measured variables, data collection methods, and analysis with the scale we wish to apply to our results. For example, there is probably little reason to spend time collecting microhabitat variables if a researcher is only interested in describing the distribution of an animal across general vegetation types. Green (1979) asserted that it is not the degree to which a model meets perfection, but, rather, its adequacy for fulfilling a prescribed purpose that renders it valid. There is no reason to use models having a great deal of precision if the biological change to be detected and managed is highly variable.

The scales at which wildlife-habitat relationships can be examined fall along a continuum that is not unlike the ways an animal selects and uses "proper" habitat. Animals choose different aspects of the environment, and they do so at different scales (Mayor et al. 2009). For example, a bird species might elicit a settling response, based on a particular forest type, and then select varying specific conditions within that forest type to display, forage, and nest. The conditions for each function, however, may differ in topography and in their vegetation structure and composition. The manner in which an animal perceives its environment, and the way we relate these perceptions to an organized method of study, have the utmost importance in model development. We can examine habitat use at its broadest, or biogeographic, scale, passing through successively finer-scale evaluations until we reach the level of the individual. Whereas large spatial scales do not necessarily imply coarse spatial resolutions, as a practical matter, spatial scale and spatial resolution are almost always coupled. As we move from relatively broad-scale, or macrohabitat, variables to relatively fine-scale, or microhabitat, variables (Fig. 2.3), changes occur in both the scale of our measurements and the methodologies we must employ. As noted below, mixing scales (scale mismatch) into a single analysis has been a common problem in habitat analysis (and ecology

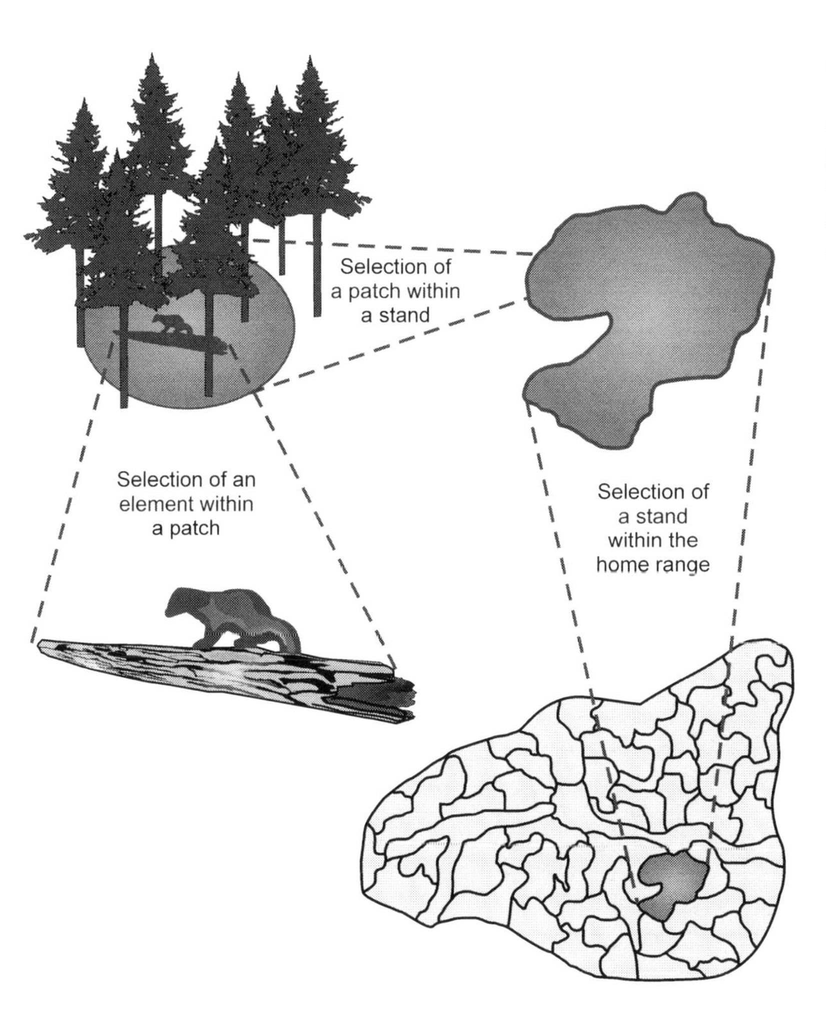

Figure 2.3. An illustration of the hierarchical nature of habitat selection, using fishers (*Martes pennanti*) as an example. (Reproduced from Weir and Harestad (2003:Fig. 1), with the permission of The Wildlife Society)

in general), a problem that is beginning to be resolved through the use of hierarchical modeling.

As discussed above, it is important to note that these fine-scale models almost always vary between locations and time periods, and between or among populations. The magnitude of these variations determines the generality of the model, where "generality" refers to the ability of a model developed for a single time and place to be applicable at other times and places. Much of the wildlife-habitat literature has been criticized because of its time and place specificity (e.g., Irwin and Cook 1985). These critiques reflect a misunderstanding of the relationship between the precision of how the variables that were measured and the scale of application that was possible. We should add that most researchers conducting habitat-relationships studies seldom acknowledge either the generality or the specificity of their data and models and indicate the locations and conditions under which managers should or should not use them. Hence the decision to develop relatively extensive (i.e., broad-scale) models or more-intensive (i.e., fine-scale) models should be based on the objectives and design of a particular study. The more extensive approach typically cannot tell us such things as how an animal reacts to changes in litter depth, the local density of trees when delineated by species, or the occurrence of a predator in a specific patch of vegetation, although this methodology is probably necessary in the management of localized populations of animals. Once the degree of specificity is determined, a researcher can then determine the type, resolution, and geographic extent of data collection that is required.

Van Horne (1983) outlined a hierarchical approach to viewing wildlife-habitat relationships (Fig. 2.4). Her Level 1 ("level" here refers to a categorization of studies, not biological levels) was applied to intensive, site-specific studies of individual species. Level 2 used more generalizable variables and probably would allow the application (or relatively easy adaptation) of a model to other locations. Level 3 was the most extensive approach, which developed relationships for a host of species. Her scheme could easily be divided into many more categories, but it serves to indicate how habitat relationships can be studied, depicted, modeled, and applied along a continuum of resolution. Figure 2.5 illustrates the major

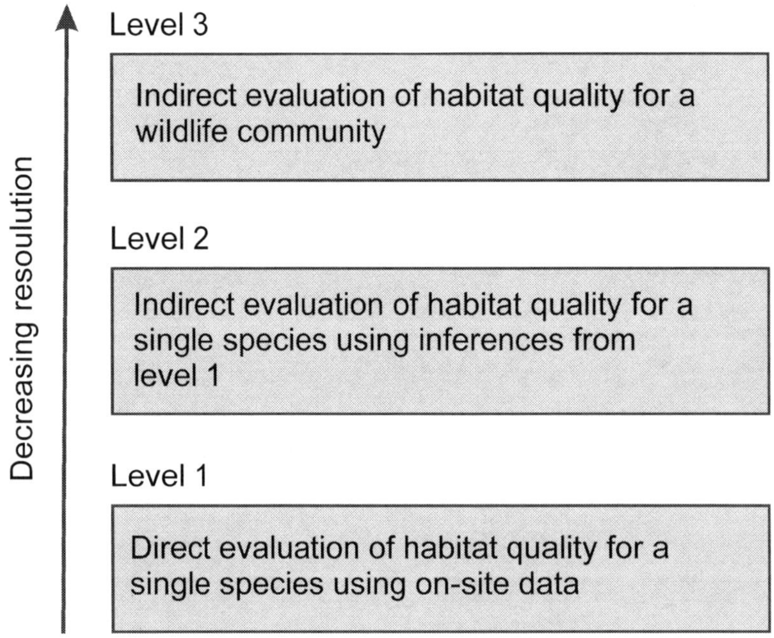

Decreasing resoulution

Level 3

Indirect evaluation of habitat quality for a wildlife community

Level 2

Indirect evaluation of habitat quality for a single species using inferences from level 1

Level 1

Direct evaluation of habitat quality for a single species using on-site data

Figure 2.4. A schematic of hierarchical descriptions of habitat quality assessments. (Reproduced from Van Horne (1983:Fig. 1), with the permission of The Wildlife Society)

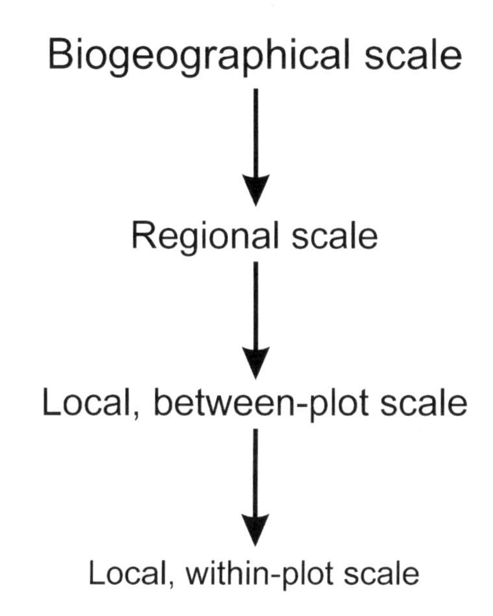

Figure 2.5. An outline of the spatial scales at which ecological field studies are conducted.

spatial scales at which ecological field studies are typically conducted.

It is understandable that wildlife managers are often frustrated by the failure of most models to be effective in their specific location. Yet we have a dilemma. Models based on broad measurements of vegetation can seldom be applied to local situations (e.g., a small management unit or refuge), whereas models developed at a fine scale can rarely be adequately generalized to be used for other locations (Block et al. 1994). The patterns of habitat use that emerge from our studies are sensitive to the scale of the comparison, because varying relationships may exist in different subsets of the samples being compared. As shown by Wiens and Rotenberry (1981; see also Wiens 1989:56–57), a species will exhibit different patterns of habitat use, depending on which portion of its distribution across the landscape is sampled by the observer (Fig. 2.6). In Figure 2.6, sampling across the entire habitat gradient that is depicted would result in yet another averaged pattern of habitat use. None of these findings would be wrong per se, but their "correctness" would be based on the scale of the question being asked.

How can this dilemma be solved? In many situations it is possible to adapt a model developed for use in a particular location to the conditions that exist at another site. This probably will be more successful with process models of how organisms behave and how they use and select resources, than with brute-force statistical correlation models that can be overfit to a particular study location. In other words, models that include fundamental, mechanistic explanations of the activities and responses of animals to environmental conditions are more likely to be applicable across time and space, relative to those that are based solely on statistical correlations (with the caveat that we usually should not mix scales in a typical statistical analysis; examples of how this may be done are given below).

Therefore, as summarized by Wiens (1989:239), studies across broad geographic areas are likely to overlook important details that account for the dynamics of local populations. Far from being idiosyncratic noise, the variations within or among local populations may contain important mechanistic information about the factors causing the organisms' responses. These differences tend to disappear at broader scales, because of the consequences of averaging, unless the study is designed to determine such variation by, for example, stratified subsample or blocking designs.

Mayor et al. (2009) argued that scale-dependent habitat studies limited our ability to quantify organisms' selections across study areas and populations. Hale et al. (2019) concluded that many studies of fish habitat are too small in scale to adequately describe such habitats. Clearly, a new approach—which encompasses multiple scales, including large spatial scales, relating habitat to measures of population or fitness—is needed to understand the factors that limit a population. Rettie and Messier (2000) used a cross-scale approach to characterize habitat use by woodland caribou (*Rangifer tarandus caribou*). They found that habitat selection was primarily driven by the need to avoid predation by wolves (*Canis lupus*), and, secondarily, by forage availability.

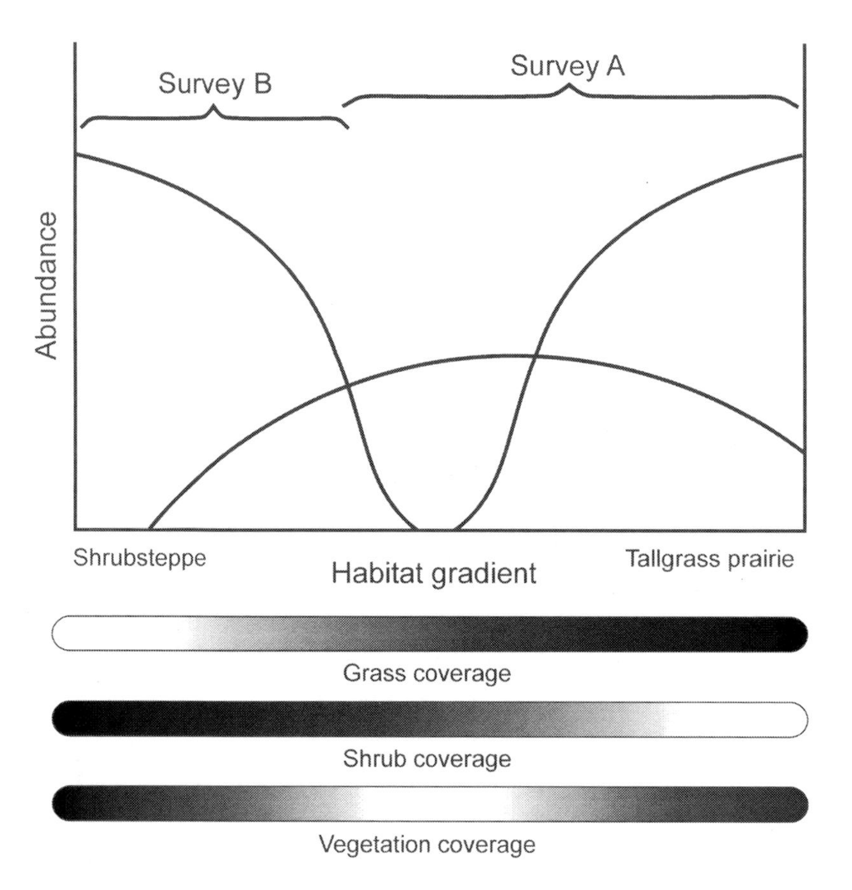

Figure 2.6. A representation of the effects of scale on a study of patterns of habitat association. In this schematic diagram, species had characteristic distributions on a gradient from shrubsteppe through short- and mixed-grass prairies to tallgrass prairie. Grass coverage, shrub coverage, and overall vegetation height changed on the gradient shown below the graph. In survey A, a large portion of the gradient was sampled but extreme shrubsteppe sites were omitted, whereas in survey B, only shrubsteppe and a few grass-shrub sites were studied. The species involved exhibited different patterns of habitat association in the 2 surveys. A particular species, for example, exhibited a strong negative association with grass coverage and a positive association with shrub coverage in survey A, but it failed to show either association in survey B. (Reproduced from Wiens and Rotenberry (1981:Fig. 3), with the permission of John Wiley & Sons)

We cannot specify, within the scope of this volume, the detail required to develop proper habitat relationships for every situation. What we are trying to convey here is the thought process that researchers must use on a study-specific basis. Indeed, studies of animal ecology are limited by the available resources for conducting research. Scientists must recognize the limitations of their investigations and carefully acknowledge the spatial and temporal applicability of their results. Much of this can be addressed with a research design that includes variations in conditions and limiting factors across space and time, at scales appropriate to the intended use of the results. Studies with adequate spatial replication, or that are conducted over a sufficiently long timeframe, are more the exception than the rule. In the following sections, we will present and discuss several examples of the types of variables collected by researchers attempting to develop predictive relationships. The list of potential studies is long, and our examples should not be taken as indicating only the best or even the most common techniques.

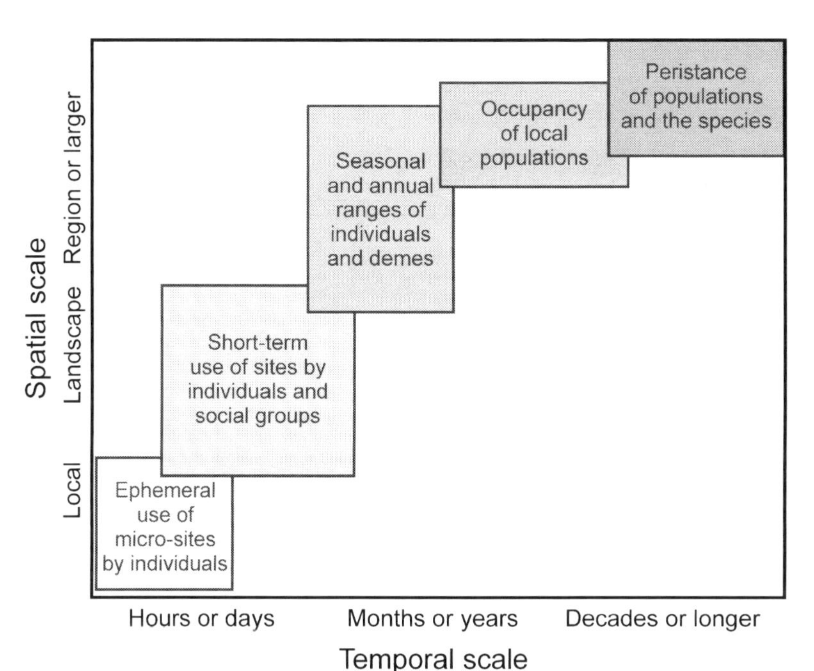

Figure 2.7. A diagram of the link between spatial and temporal scales of habitat selection. The diagonal *text blocks* reflect animal mobility; the *axis units* reflect the various requirements appropriate for a long-lived, wide-ranging species. (Reproduced from Mayor et al. (2009:Fig. 2), with the permission of Taylor & Francis)

Temporal Scale

Many investigations focus on spatial scale, but temporal scale may be even more important to consider (Mayor et al. 2009). Food, for example, can differ more through time than it can by space. Even so, neither spatial nor temporal variation occurs independently, so animal ecology should consider the interaction between both (Fig. 2.7). Habitat requirements during 1 phase of an animal's annual or life-history cycle may vary from those during other parts of the year or the life cycle. Dispersal, migration, wintering behavior, courtship, nesting/bedding, and the like may all require different conditions. Hence a set of conditions used for nesting may not reflect those needed for dispersal. In Chapter 3, we elaborate on the importance of considering temporal scale in studies of animal ecology.

Population Measurements

It is beyond the scope of this book to review all of the methods available for estimating population parameters. We provide a summary developed by Sanderlin et al. (2019) in Table 2.1, although many excellent works have been published on this subject (e.g., Seber 1965; Cooperrider et al. 1986; Ralph et al. 1993; Bookhout 1994; Heyer et al. 1994; Buckland et al. 2001; Williams et al. 2002; Amstrup et al. 2005).

Although we are not covering this topic in detail, estimates of the population numbers must be reliable. In other words, are our assessments of absolute (or even relative) abundance a fair reflection or index of the number of animals actually present? Are they adjusted by incorporating detection probabilities? Are estimates among multiple observers unbiased and comparable? Can appraisals derived by an observer (or a group of observers) in 1 study area be used to validate a model developed by other people at a different location? Throughout this book we discuss the varying effects of observer errors, which are encountered because of hidden biases, and other factors that work against the development of habitat models. Here, however, we are interested in the errors associated with estimating numbers of animals. We remind readers that there are many problems

Table 2.1. **Animal population state variables and vital rates for monitoring, analysis methods, analysis programs, and sampling considerations**

Monitoring state variable(s) and vital rates	Analysis method	Analysis program(s)	Sampling considerations
Abundance, density, λ (i.e., finite rate of increase)	Closed (Otis et al. 1978) and open (Jolly-Seber model [Jolly 1965; Seber 1965]) models (or both, *sensu* robust design models [Pollock 1982]); distance-based methods (Buckland et al. 2001); spatially explicit capture-recapture (SECR) methods (Royle and Young 2008); matrix models[2] (Caswell 2001)	MARK[1]; CAPTURE (Otis et al. 1978; Rexstad and Burnham 1991); distance sampling (Buckland et al. 2001); SPACECAP (Gopalaswamy et al. 2012)	• A lot of effort to capture, mark, and recapture animals (CMR) is required, although physical marks are not necessary (e.g., non-invasive genetic CMR [Waits 2004], camera traps). • Population closure and tag loss (or, with genetic CMR, genotyping error [Taberlet et al. 1999]) should be minimized or incorporated into population models (Wright et al. 2009). Population closure is minimized by limiting the number of repeat sample days and avoiding sampling during migration or high rates of mortality/recruitment. • Multiple sampling methods (Besbeas et al. 2002; Abadi et al. 2010) may improve abundance or density estimates. • For distance-based methods, use an accurate estimation of distance/angles from observer to animal. Animals located on the line/point have a detection probability of 1, and animals are not influenced by observation (see Buckland et al. 2001 for specific considerations). • For density methods, the effective trapping area may not be known and SECR methods are preferable.
Survival, nest success	Known-fate analyses Kaplan-Meier method [Kaplan and Meier 1958; Pollock et al. 1989a, b]); band-recovery analyses (Brownie et al. 1978, 1985; Barker 1997); Cormack-Jolly-Seber model (CJS) (Cormack 1964; Jolly 1965; Seber 1965)	MARK[1]	• Effort for nest survival can be allocated toward either increased time searching for nests or more visits. • For known-fate analyses, survival probability should be restricted to days on which the fate could be observed, and disturbance to nests and surrounding habitat should be minimized during visits. • Accurate estimation of death times and well-specified origin times are needed with telemetry studies. • Censoring is random and independent of survival. • Sampling periods are instantaneous, and recaptured animals are released immediately. • Emigration from the sampled area is permanent.
Reproduction, recruitment	Jolly-Seber model (Jolly 1965; Seber 1965); reverse-time models (Pollock et al. 1974; Nichols et al. 2000; Nichols 2016)	MARK[1]	• Sampling periods are instantaneous, and recaptured animals are released immediately. • Emigration from the sampled area is permanent.
Movement/ dispersal	Multistrata (multistate) capture-recapture model (Arnason 1972, 1973; Brownie et al. 1993; Schwarz and Arnason1996); least-cost models (Adriaensen et al. 2003); geneflow via population substructure (Hedrick 2005); genetic assignment tests (Waser and Strobeck 1998)	MARK[1]; GIS; STRUCTURE (Pritchard et al. 2000); GENELAND (Guillot et al. 2005)	• Simultaneous sampling of multiple sites is preferable. • CMR sampling is often costly, and telemetry may provide more information. • Indirect methods (e.g., genetic monitoring) may increase sample sizes and the power to detect changes.

Table 2.1. continued

Monitoring state variable(s) and vital rates	Analysis method	Analysis program(s)	Sampling considerations
Occupancy, local extinction, and colonization probability	Occupancy models (MacKenzie et al. 2006)	PRESENCE (Hines 2006); MARK[1]; R package **Unmarked** (Fiske and Chandler 2011)	• Repeat surveys (visits and sampling points) are required during a period of closure. • Sampling unit and a definition of season should be appropriate for the species. • Methods of increasing capture probabilities reduce bias in population estimates. See MacKenzie and Royle (2005) for more details on occupancy study design.
Range distribution	Maximum entropy modeling with presence-only data (Phillips et al. 2006); occupancy models with presence-absence data (MacKenzie et al. 2006)	Maxent (Phillips et al. 2005); PRESENCE (Hines 2006); R package **Unmarked** (Fiske and Chandler 2011)	• Yackulic et al. (2013) provide recommendations based on detection probability (equal to 1, less than 1, and constant; varies) and sampling probability (varies; constant). • Many assumptions (e.g., constant sampling and detection probability with respect to environmental covariates that determine occupancy) of presence-only data are difficult to meet in practice (Yackulic et al. 2013).
Genetic measures	Hybridization; geographic range; genetic variation; effective population size; movement; population substructure	STRUCTURE (Pritchard et al. 2000); GENEPOP (Raymond and Rousset 1995); ARLEQUIN (Schneider et al. 2000); GENELAND (Guillot et al. 2005)	• Methods rely on reduced genotyping errors (Bonin et al. 2004). The quality and quantity of DNA is important. • Genetic changes occurring with changes in the population are easier to detect after severe disturbances (Schwartz et al. 2007).
Diversity, species richness, species turnover, species evenness	Multispecies occupancy models (Dorazio et al. 2006; MacKenzie et al. 2006); hierarchical Bayesian analyses and MCMC (Markov chain Monte Carlo) (see Chapter 6)	JAGS (Plummer 2003); OpenBUGS (Lunn et al. 2000; Speigelhalter et al. 2007); R (R Development Core Team 2016)	• Requires repeat surveys (visits and sampling points) during a period of closure. • The sampling unit and a definition of season should be appropriate for all species in the community. • Optimal methods for species richness include focusing on the number of sampling occasions, while the sampled area percentage is more important for the occupancy probability by rare species in a community (Sanderlin et al. 2014). • Sampling design recommendations are needed with multiseason, multispecies occupancy models.

Source: Sanderlin et al. (2019)

Note: Sampling considerations were condensed from Williams et al. (2002), unless otherwise noted. Although there are several more, we list the most common analysis methods and programs used for monitoring state variables and vital rates. For most of these, the analytical approaches could be frequentist or Bayesian. If a Bayesian approach is deemed necessary, the state-space models and OpenBUGS (Lunn et al. 2000; Speigelhalter et al. 2007) and JAGS (Plummer 2003) programs are typically used with state variables and vital rates, with the exceptions of genetic measures.

[1] Program MARK (White and Burnham 1999)

[2] This analysis method results in a change metric, not a metric derived from repeated inventories that estimate state variables at discrete points in time and then extract trends

associated with counting animals, and we reiterate that poor appraisals of animal numbers will negate conclusions drawn on habitat relationships that are based on even the most carefully collected environmental variables.

The underlying premise of this volume is that it is insufficient to characterize habitat without understanding how it relates to the population being investigated (Boyce et al. 2015). The choice of which population parameters to study is critically important when making this linkage. In many situations, considering only the number of individuals may be a misleading measure of the population's status or health. For a species occupying an ecological sink, abundance could be considerable, yet reproduction and survival rates might be quite low. In this case, correlating habitat with population size might provide misleading results. The first step in linking habitats to populations should be to identify the appropriate parameters for analysis. We briefly summarize below many of the potential population measures to be considered in studies of animal ecology.

Abundance and Density

A tremendous amount of research has been devoted to developing abundance estimates for vertebrates. The validity of the results of these investigations, as well as their application in resource management and conservation, both hinge on reliable estimates of numerical abundance (Otis et al. 1978; Burnham et al. 1980; Smallwood and Schonewald 1996; Williams et al. 2002; Buckland et al. 2001; Buckland et al. 2015; Buckland et al. 2019). Additionally, the development of many wildlife habitat models is based, in part, on assessments of abundance (e.g., regression analysis; see below). It should be noted that abundance itself is an index of density, although it is a measurement without specific units, because the capture area is unknown.

Accurate estimates of density (or an index thereof) can vary widely, depending on the spatial scale of the study. Animals typically are not distributed uniformly in space. As a result, the plot size or sampling design must be adequate to capture this variation. For example, extrapolating bird abundance estimates from a 1.5 ha study site to the number of birds per 40 ha (often used as the standard area for reporting densities) runs the risk of a biased estimate, as the results from the 1.5 ha sampled area are unlikely to account for the variation to be found across 40 ha. Further, errors in estimating the size of the sample area (i.e., the area sampled will not be precisely 1.5 ha) will be multiplied, in this case, by 27. Guidance has been established, however, for plot size or the minimal number of detections for estimating abundance. For example, the International Bird Census Committee (1969) recommended that spot mapping occur within fixed plots, ranging in size from 10 to 20 hectares for woodland birds, and 50 to 100 hectares in more open habitats. Burnham et al. (1980) and Buckland et al. (2015) recommended a minimum of 40–60 detections for estimating density using distance sampling. In many cases, the number of detections is related to the size of the area being sampled.

Therefore, if density is to be related to ecological variables, then it should be estimated with the appropriate spatial scale(s). Otherwise, the pattern will be masked by a high variability in density among the scales. The choice of a spatial scale must be based on the species' relationship with limiting factors in the landscape, not on the spatial resolution of the technology that is used for observation, or on some artificial or convenient area (Morrison 2012). Variations in measured density are influenced by the research area's size, the year in which the study occurred, site selection, sampling method, trap type, and various other factors (Smallwood and Smith 2001). A determination of the effect of study area size on abundance or density values should be incorporated into the preliminary sampling phase of all investigations (Morrison et al. 2008). Ideally, this research would also include randomly placed spatial replicates to capture variations in estimates as they relate to changing conditions. Various sampling designs (e.g., stratification, cluster sampling, adaptive cluster sampling) can be applied, depending on the rarity and spatial distribution of the species.

Estimates of abundance or density are most common in the literature. These can range from simple counts of individuals that were seen, heard, or trapped; to counts adjusted for detection or capture probabilities; to counts adjusted for both detection and spatial location. Relying on simple counts to constitute a census hinges either on the assumption that all individuals are sampled, without missing any, or the assumption that the probability of detection is 1. These situations are rarely the case for free-ranging wild populations, because some portion is obscured from detection, or the observer simply fails to see, hear, or capture them. Counts are therefore mostly used for relative abundance estimates, where the assumption is that the proportion of undetected individuals is reasonably constant. Generally, count approaches stress the similarity in their sampling methodologies, in an attempt to standardize detection likelihoods. Modeling and analytical advances in capture-recapture methods (Otis et al. 1978; White and Burnham 1999; White 2019) and distance-based estimators (Burnham et al. 1980; Buckland et al. 2001) have provided valuable tools for incorporating detection probabilities into abundance and density estimates. A limitation of these approaches, however, is that they are data hungry. For example, testing the model selection algorithm from the original CAPTURE program (Otis et al. 1978) using population sizes of 50–100 with 5–10 trapping sessions, Menkens and Anderson (1988) found that CAPTURE selected the right model only 11 percent of the time, a rate slightly below random (12.5%) expectations. Choosing the wrong model leads to large (and largely unknown) errors in abundance estimation (Menkens and Anderson 1988; McKelvey and Pearson 2001). This often negates the ability to apply complex models to rare species, which typically are among those of primary conservation concern. Menkens and Anderson (1988) suggested using the Lincoln-Petersen estimator when data are sparse, but there are limits. This estimator is far from perfect (McCullough and Hirth 1988), and it tends to bias estimates on the low side when population sizes are small (Chapman 1951).

Spatial Density Estimators

An intrinsic problem with standard capture-recapture models is that they estimate abundance, rather than density. That is, given the caveats above, they provide unbiased estimates of the sampled population, but the space occupied by that population is undefined. Transforming abundance, which is actually an index of abundance, into a density assessment requires ad hoc approaches to approximate the "effective trapping area" (see Anderson et al. 1983). Because the biases associated with these ad hoc methods are unknown, even after correction, abundance remains an index of density. To address this problem, a group of maximum likelihood models (Efford 2004) and Bayesian alternatives (Royle and Young 2008; Royle et al. 2013) have been developed. These models take the spatial locations of recaptures and estimate a function that defines the pattern of individual use, thereby employing a combination of spatial data and capture-recapture data to directly estimate density. Because the majority of capture-recapture datasets contain the spatial coordinates of the trap locations, most of the modern datasets can, alternatively, use spatial capture-recapture models. A great deal of information is contained in the spatial arrangements of locations, but it is not utilized in traditional capture-recapture models. As a result, spatial capture-recapture models can be applied to partially marked populations, with modest reductions in accuracy (Chandler and Royle 2013). They can even be employed for unmarked populations (Ramsey et al. 2015), though these models require strong assumptions of space use by individuals and must be thought of as a bit speculative. Because these models produce density surfaces as a primary output, these surfaces can theoretically be used in lieu of telemetry data to construct resource selection models of habitat use (Royle et al. 2013). It should be noted that, although extremely popular, many of these methods are very new and have had little real world validation. Therefore, as of 2020, the use of these models is still somewhat controversial. For example, the models benefit from multiple recaptures, both because this allows fitting more-complex recapture models, and because it permits better data to define the areas of use. The optimal tradeoffs between length of sampling and population closure, however, have only recently begun to be explored (Dupont et al. 2019).

Occupancy

An option for rare species is to model occupancy instead of abundance (MacKenzie and Royle 2004; MacKenzie et al. 2017; Sutherland and Linden 2019). Occupancy is a variant of capture-recapture methodologies, where the site is treated as an individual and the pattern of detections at the site is used to estimate the detection probability. Whereas this approach does not require individual data, it nonetheless contains a number of strong assumptions. The basic model has 6 assumptions (MacKenzie et al. 2017:126):

1. The occupancy state of the units does not change during the survey period.
2. The probability of occupancy is equal across all units.
3. The probability of detection, given presence, is equal across all units.
4. Detections across survey units are independent.
5. Detection histories observed at each location are independent.
6. There are no species misidentifications.

While some of these elements can be relaxed, the strong assumption of closure within a unit (assumption 1) and independence between detections across units (assumption 4) remain problematic in open systems, where organisms move about (Otto et al. 2013). Some have suggested that in open systems, the term "use" be employed in place of "occupancy." While occupancy, in and of itself, can serve as a helpful metric, there is often a desire to use occupancy as a surrogate for abundance or density, an assumption that is rarely tested (Royle and Nichols 2003) and is only true under very specific circumstances (Efford and Dawson 2012; Noon et al. 2012). Whereas occupancy estimates may require reduced efforts in the field, they also provide less information about the population. For example, occupancy for a plot is a binary variable (0 or 1), and this does not account for the number of individuals that are present. Rossman et al. (2016), however, developed a promising procedure (which they called "dynamic *N* occupancy") to estimate local abundance, reproduction, immigration, and apparent survival probabilities, using only detection/non-detection data. We expect future developments of this approach in the future. See Chapter 5 for more information on occupancy modeling.

Survival, Reproduction, and the Finite Rate of Population Increases

Whereas abundance and density may be informative, they may be somewhat misleading without understanding survival and reproduction rates (Van Horne 1983). For example, abundance may be similar between 2 locations, but survival may differ. Thus a population of 1,000 at 1 location that has high survival and reproduction rates would have a very different trajectory than a population of 1,000 at another site with low rates in these elements. This is often the case with ecological traps or sink situations (Dunning et al. 1992; Battin 2004). Hence understanding rates of survival and reproduction are key pieces of information to more fully understand the status of a population.

Incorporating such estimates can provide insight into the stability of a population. Age-specific survival and reproduction rates can be used to populate a Leslie matrix model (Leslie 1945; Caswell 2001) and permit estimates of the finite rate of population increase (λ). Anthony et al. (2006) applied such models to threatened northern spotted owls (*Strix occidentalis caurina*) to determine if owl populations were stationary ($\lambda = 1$), increasing ($\lambda > 1$), or declining ($\lambda < 1$). They concluded that populations on 9 study areas were declining, and 4 were stationary. The authors were able to include covariates in their analysis, and the results suggested that fluctuations in fecundity and survival were related to weather and prey abundance. This information provided the basis for more-focused management and future monitoring efforts. Further meta-analyses on northern spotted owls were conducted by Dugger et al. (2016), which included competition effects from barred owls

(*Strix varia*), and on California spotted owls (*S. occidentalis occidentalis*) by Tempel et al. (2016), which identified the importance of forest canopy cover on that subspecies.

Multispecies Efforts

Often conservation efforts apply metrics that are related to some aspect of biodiversity or to a suite of species in an area during a specific time. These metrics range from the number of species found (i.e., species richness), the relative proportion that each species contributes to the overall population (i.e., diversity), how equally each species' population contributes to overall abundance (i.e., evenness), and colonization and extinction rates (see Heyer et al. 1994 for a review of these metrics). Multispecies monitoring is also becoming more frequent (see DeWan and Zipkin 2010; Noon et al. 2012), because of statistical advances with multispecies occupancy models and monitoring based on abundance estimates from capture-recapture models (Dorazio et al. 2006; MacKenzie et al. 2017; Baumgardt et al. 2019). These models have been extended to measures of species diversity and community dynamics (Iknayan et al. 2014).

Depending on the study objectives, monitoring multiple species may be preferable to a single-species approach, especially with limited resources. Indeed, understanding how populations of sympatric species co-vary—either positively or negatively—helps unravel how species' populations may interact and how this influences habitat use and spatial distributions. A single-species approach may mask biotic interactions that have a greater influence on habitat use than vegetative structure and composition.

Movement

Understanding how animals move across the landscape helps paint a more complete picture of the topography, vegetation types, and conditions that a species uses over time (van Moorter et al. 2015). This information can help identify areas requiring conservation protection, such as corridors, patches (e.g., size, shape, juxtaposition), migration routes, wintering grounds, and the like (see Morrison et al. 2020:Chapter 3 for a discussion of fragmentation and connectivity). Historically, researchers and managers have employed radio telemetry (White and Garrott 1990) and band recovery (Brownie et al. 1993) to track movements. The use of radio telemetry usually results in a more complete and accurate depiction of the activities and areas used by animals and provides a more reliable basis for developing habitat relationships than is available with visual observations and traps (e.g., Ribble et al. 2002; Briner et al. 2003). With the availability of increasingly tinier transmitters and batteries, researchers have been studying a greater number of species of smaller animals. Careful study designs, however, are needed to avoid introducing biases into this research, such as the methods used to select animals on which to attach the transmitter, the timing and frequency of obtaining relocations, the number of animals used, and other issues. See, for example, the "radio-handicapping" admonition leveled by Guthery and Lusk (2004) at northern bobwhite researchers who had reported really low survival rates. The old practice of tagging these birds and following them (i.e., "collar 'em and foller 'em") is giving way to much more sophisticated sampling strategies, built around a specific project goal. Many fine books and articles are available on radio tagging, including White and Garrott (1990), Kenward (2000), and Millspaugh and Marzluff (2001). See also Turchin (1998), van Toor et al. (2018), and Michelot et al. (2019) for analyses of animal movements.

Radio collars are now available that can transmit animal locations to satellites. Other collars can act as geographic positioning system (GPS) receivers and record animal movements at set intervals, or record lighting conditions from which latitudinal locations, at least, can be inferred. Thus the traditional process of observers locating (e.g., triangulating) radio-equipped animals using hand-held, vehicle-, or aircraft-mounted antenna/receivers is giving way to more-automated systems. Recent advances in global positioning systems and satellite telemetry have opened new possibilities for tracking movements over large distances. On the other hand, the high costs associated with both of these emerging technologies and the analysis of data collected through them often limit sample sizes, and the equipment constraints (battery requirements and a relatively large transmitter size) may restrict their application to bigger species (Hebblewhite and Haydon 2010).

Indirect methods (employing genetic monitoring and stable isotopes) and the use of genomics may be more cost effective and provide a greater amount of movement information through larger sample sizes or increased information from individuals that dispersed and reproduced. Non-invasive methods for collecting DNA samples can be deployed to passively collect information on the presence of target species or their place of origin. For example, Hawley et al. (2016) were able to use this technique for a cougar (*Puma concolor*) detected in Connecticut that came from South Dakota, a distance of more than 2,450 km. There are 2 kinds of genetic analyses commonly used to infer movement. The first is relatedness, which integrates movements associated with successful mating across some indefinite time period (see Morrison et al. 2020:Chapter 3 for additional discussion). The second, used in Hawley et al. (2016), is assignment testing, which, given a group of potential source populations, examines the DNA of an individual organism and asks which population it was most likely to be associated with.

Species Distribution Models

Species distribution models (SDM) relate field observations (e.g., presence-absence, abundance) to environmental predictor variables, based on statistically or theoretically derived responses (Guisan and Thuiller 2005; Elith and Leathwick 2009). Environmental predictors include limiting factors, disturbances affecting environmental conditions (natural or human-induced), and resources. SDM models can be constructed so that relationships between species and their overall environment are evident at different spatial scales.

The application of SDMs has emphasized plants and fish, with fewer employed for wildlife species. Combinations of focal-species distributions, measures of vegetation and topography, and climate were used in the development of many of these models (Lyet et al. 2013; Pikesly et al. 2013; Shirley et al. 2013; Bucklin et al. 2015; Young and Carr 2015). Williford et al. (2016) utilized species distribution modeling as an independent line of evidence to support a phylogeographic analysis of the bobwhite genus *Colinus*, based on molecular genetics. Missing from many of the SDMs, however, is an inclusion of sympatric species and how they might influence the distribution of the focal species. Given that other species can influence the realized niche, an awareness of these relationships underlies a more complete understanding of animal ecology. Recently, Miller et al. (2019) provided guidance on integrating data, including information on multiple species, into these models. We view this as a crucial advancement for improving the predictive capabilities of SDMs.

As developed above, an entire class of habitat models are predicated on correlating animal numbers to some features of that animal's biotic and abiotic environment. The purpose of these models is to develop adequate predictions of the presence, abundance, or density of the species, based on environmental features (discussed later in this chapter; also see Chapter 5). The methods vary, depending on the available information, since the techniques for handling presence-absence data are not necessarily the same as those suitable for density data. Researchers and managers can use these models to predict the shifts that will occur in animal abundance, given changes applied to the variables in the model.

Measurements of Environmental Features

Here we describe some common techniques used to measure environmental features, assuming that these features are related to a species' habitat. We recognize that *our* understanding of a particular species' habitat may or may not correspond with how *that species* perceives it. By this, we mean that the variables we measure are those we think best characterize what the species of interest actually selects. The array of variables we choose to measure are often based on previous work that has identified those factors that have strong relationships to the presence or abundance of a species. Through trial and error, we learn more about the relative strength of these variables, and we add or delete other variables as they are identified through new research and technological advances.

When we choose environmental variables to measure, we typically select those that might account for the presence and abundance of a species, rather than those that might relate to the species' absence. That is, we usually focus on selection for conditions, rather than the avoidance of conditions, which might have implications for different kinds of errors in predicting presence and absence. Yet definitively demonstrating avoidance itself can be logically problematic. If an organism makes use of 1 environmental condition but not another, is it selecting for the first, completely avoiding the second, or simply preferring the first over the second? In general, except in cases of sessile organisms in confined geographic areas, empirically determining absence—and, from that, inferring avoidance—can be very difficult. Some analyses and modeling approaches, therefore, compare presence to "lack of presence" or to random locations. Errors in conclusions or predictions of false presence or false absence are also sensitive to scales of space and time.

Habitat can be viewed along a continuum of temporal and spatial scale. It can range from site to landscape to region, and, in the case of some species, to global (Fig. 2.8). This continuum varies by species. In some cases, they are local endemics, tied to a specific

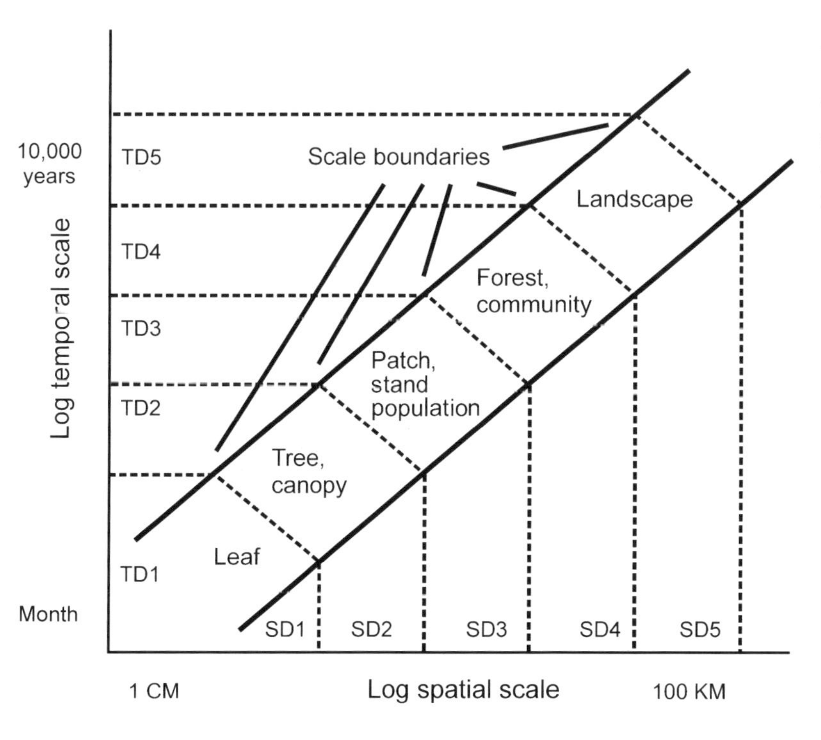

Figure 2.8. A graph of an approximate matching of spatial (SD) and temporal (TD) scales in ecological studies. (Reproduced from Bissonette (1997), with the permission of Springer Nature)

place or a particular environmental condition. In other instances, the distribution of a species—such as wolverines (*Gulo gulo*) or northern goshawks—can be circumpolar. Thus it can be instructive to consider habitat along a spatial gradient, ranging from a site to an entire geographic range.

Habitat can also vary by specific function. For example, the nesting habitat of Mexican spotted owls differs from their foraging habitat, which may not be the same as their roosting habitat, which can be other than their dispersal habitat (Timm et al. 2016). Thus it is incumbent upon researchers to be clear as to which aspect of habitat is being studied and described. Furthermore, one must also consider temporal variation in the habitat. Many species are mobile, and their needs change with time. Breeding areas for waterfowl and shorebirds, for example, may differ considerably from those encountered along flyways or found on their wintering areas. Therefore, investigators must clearly articulate exactly what they are measuring and how it relates to the requirements of the relevant species.

Many terrestrial animal habitat studies focus on vegetation measurements. Historically, ecologists emphasized vegetation structure (i.e., cover, height, density) and configuration (i.e., patch size, shape, juxtaposition, interspersion) (see Hildén 1965; Wiens 1969; James 1971; Anderson and Shugart 1974; Willson 1974; James and Wamer 1982; Rotenberry 1985; Rotenberry and Wiens 1998). Focusing on structure alone, however, may negate clear and established relationships that many animals have with plant species. Indeed, we know that sage-grouse (*Centrocercus urophasianus*) are associated with sagebrush (*Artemisia tridentata*), red tree voles (*Arborimus longicaudus*) with Douglas-fir (*Pseudotsuga menziesii*), and willow flycatchers (*Empidonax traillii*) with willows (*Salix* spp.). It was not until Rotenberry (1985) brought the debate between vegetation physiognomy and composition to the surface, however, that investigators began to consider both simultaneously.

Many earlier researchers failed to adequately place their studies into a specific spatial scale, thereby obscuring the relative (and proper) roles that structure and floristics can play in predicting habitat relationships at different scales (Levin 1992). As noted by Rotenberry (1985), the same species that appears to respond to the physical configuration of the environment at the continental scale may show little correlation with physiognomy at the regional or local scale. Thus many animals may be differentiating between vegetation types on the basis of physiognomy (i.e., they occupy a general area that is "proper" in its structural configuration), and then key in on particular species of plants. Note that Rotenberry's (1985) scenario relates closely to Hutto's (1985) hypothesized mode for the process of habitat "selection" in animals. Rotenberry (1985) quantified his ideas using data from the shrubsteppe studies of Wiens and coworkers (Wiens and Rotenberry 1981; Wiens et al. 1987), and his analysis of these data indicated that, when the correlation between physiognomy and flora was statistically separated, a significant relationship remained between bird abundance and flora, but not between birds and physiognomy (Fig. 2.9).

We return here to an important theme: the variables measured and the level of refinement required in a model should be based on the scale of interest. Simple presence-absence studies of animals at regional or broader scales generally do not require a floristic analysis of the vegetation. Broad categorization by physiognomy—most likely including a differentiation no lower than life form (e.g., deciduous and evergreen), or by general ecological classes or vegetative types—is probably adequate. Plant taxonomy becomes increasingly important, however, as investigations grow more and more site specific.

Historically, habitat studies have been restricted to a spatial scale, but the scale chosen may have been inappropriate for the species under investigation. A burgeoning area of inquiry is to consider multiple scales simultaneously to identify the scale that best predicts distribution, and also the variables that might influence survival, reproduction, and population persistence (Mayor et al. 2009; McGarigal et al. 2016). For example, Timm et al. (2016) conducted

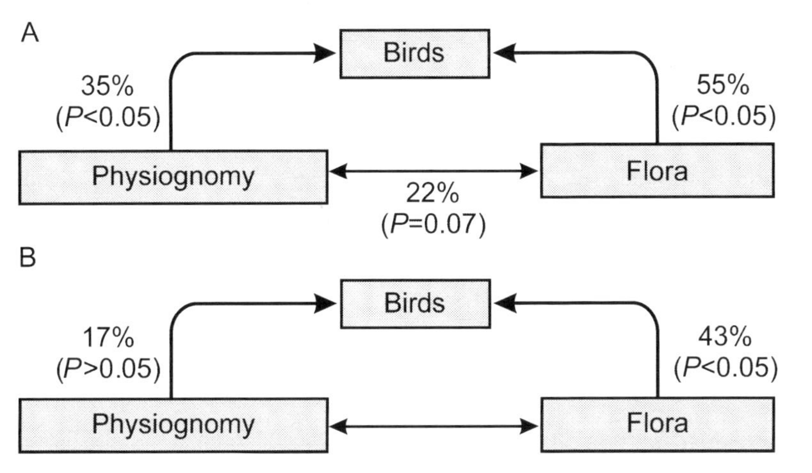

Figure 2.9. (*A*) A diagram of coefficients of determination ($r^2 \times 100$) between similarity and distance matrices, based on the avian, floristic, and physiognomic composition of 8 grassland study sites. Significance levels of association are given in parentheses. (*B*) A diagram of partial coefficients of determination, as above. The correlation between physiognomy and flora has been partialed out. (Reproduced from Rotenberry (1985:Fig. 1), with the permission of Springer Nature)

multiscaled habitat modeling to assess habitat selection by Mexican spotted owls in northern Arizona. The best predictor of nest success varied by spatial scale, with the percentage of canopy cover and slope emerging as the strongest predictors of habitat selection by this subspecies. Their multiscaled models consistently outperformed single-scaled model counterparts, with respect to the models' predictive performances.

Landscape

Over the past few decades, we have seen the emergence of landscape ecology. This expanding field of inquiry is, in many ways, an extension of island biogeography, reserve design, metapopulation theory, and patch connectivity (Soulé 1991). Landscape ecology includes methods for identifying large-scale patterns and understanding the processes that shape them (Urban et al. 1987). Disturbance drivers, such as weather, climate, fire, flood, insects, disease, and human development, often influence landscape patterns (see Morrison et al. 2020).

For example, high-frequency, low-severity fire regimes in the American Southwest were responsible for maintaining ponderosa pine (*Pinus ponderosa*) forest structure there (Moir et al. 1997). Pre-suppression wildfire acted to maintain a heterogeneous mosaic structure by limiting tree recruitment. With the application of a fire-suppression policy in the early twentieth century, ponderosa pine forest structure changed (Pyne 2015). With the cessation of fire and the presence of favorable climate conditions, this species experienced a large regeneration pulse in 1919, and millions of ponderosa pine trees became established. Without fire to reduce their densities, the tree seedlings flourished and grew, diminishing the heterogeneous mosaic structure of the forest, creating ladder fuels, and rendering the forest vulnerable to low-frequency, high-severity fire regimes. The reduced heterogeneity changed habitat conditions for many species of birds (Block and Finch 1997) and jeopardized the sustainability of the ponderosa pine forest, owing to stand-replacing wildfires.

With technological advances in remote sensing, such as satellite imagery and computing power, landscape ecology is rapidly advancing. This has enabled ecologists in this discipline to develop a plethora of variables to measure patterns in animal distribution (Bissonette 1997; Bissonette and Storch 2003; Millspuaugh and Thompson 2009; Cushman and Huettmann 2010; Keller and Smith 2014). Cushman et al. (2013) outlined the basic steps for landscape analyses: (1) establish objectives, (2) define the landscape, (3) identify the larger landscape context, (4) determine the scope of the analysis, (5) identify key habitat attributes, (6) select the metrics, and (7) interpret landscape patterns. Objectives could include describing landscape patterns and how they relate to a species' distribution over time, linking landscape patterns with community metrics, or projecting the effects of management activities (e.g., vegetation manipulation) or environmental fluctuations (e.g., climate change) on the distribution and abundance of the target species or an assemblage of species.

The definition of a landscape should be based on the extent or range of the target species, and it should be adequate in size to account for the population(s) in it (see Morrison et al. 2020:Chapter 1 for how to determine the population). An integral aspect of identifying and delineating a landscape is to select the appropriate grain size for the organism under investigation. "Grain" is defined as the smallest areal unit of study as it relates to the target species (Kotliar and Wiens 1990). As illustrated in Figure 2.10, selecting the appropriate grain can influence depictions of landscape patterns.

Regardless how a landscape is defined, the larger outlying areas must also be considered. "Landscapes" (as used here) are basically human constructs, designed to permit analysis. They typically do not include perimeter walls or barriers, and animals can move in and out of defined landscapes. Further, changes occurring outside of a study's landscape—as a result of human activity, weather, or disturbance—can influence processes within the landscape. Cushman et al. (2013) outlined 3 levels of analysis: (1) focal patch, (2) local landscape structure, and (3) global landscape structure. Focal-patch analysis involves characterizing the attributes of the patches, such as size and shape. Discussing local landscape structure entails describing the local pattern gradients that vary spatially and relating those to population distribution. This can entail analyzing a geographic area by using a moving window to describe landscape pattern in a way that resembles how an animal might move across the landscape. Global landscape structure refers to the entire landscape that is selected and used by a species or a set of species.

Analysis of this habitat element involves the application of many traditional landscape pattern metrics. The common metrics that are used focus on the size, configuration, and context of the environmental conditions that are present, such as vegetation patches (Keller and Smith 2014). A commonly used variable is the proportion of various patch types. Proportion

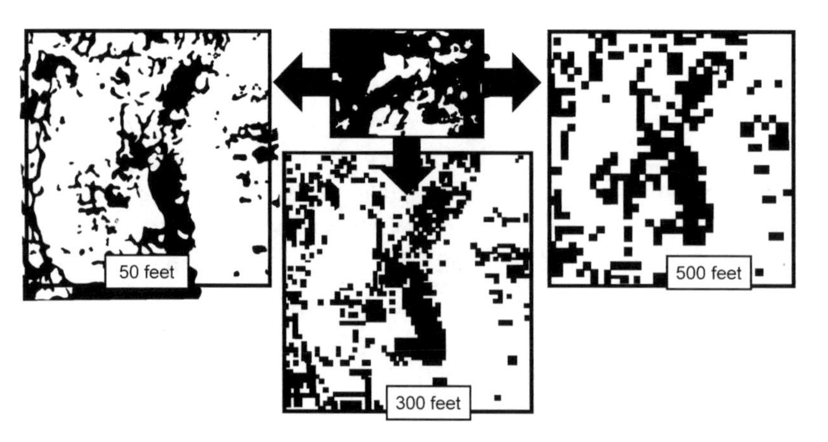

Figure 2.10. Images showing that spatial grain should be appropriate for the organism of interest. For American martens (*Martes americana*), a grain of 50 feet was finer than needed, whereas a grain of 500 feet might have ignored important landscape patterns. (Reproduced from Cushman et al. (2013:Fig. 6.1), courtesy of the US Department of Agriculture, Forest Service)

alone, however, provides little information on the size of the patches or how they arranged in space, since 2 different landscapes can have the same proportions yet differ in their patch sizes and arrangement. Regardless of the variables used, it is critical that patch size, shape, and arrangement are considered simultaneously, not separately.

Keller and Smith (2014) noted the growing use of landscape models in wildlife science but recognized that many models have limited predictive power, because of (1) a disconnect between image resolution, minimal mapping unit, and habitat selection by the organism; and (2) the types of metrics (e.g., proportion of patch types) used in a landscape analysis. Many theme coverages in geographic information systems (GIS) have resolutions ≥30 m. Whereas such resolution may be adequate for larger animals (e.g., bears or cougars), it is too big for littler species (e.g., passerine birds or small mammals). Many species may be closely tied to edges or ecotones, but the size of these edges may be scale variant, according to the species under investigation. A species that is present in the edge demarcating a grassland and a forest uses a more pronounced edge than another located in a canopy gap within that forest. Another consideration is "depth of edge," which describes how far into adjacent patches various kinds of edge influences penetrate. Thus, whereas a landscape analysis might indicate edge as being important for a grassland-forest species, it would be unable to identify this element at the canopy gap scale, and it might not account for differential depths of edges along several kinds of ecotones and adjacent types of patches. Alternatively, the entire concept of "edge" may simply be an artifact of vegetation patch size and the various different configurations of such patches (Guthery and Bingham 1992; Guthery 2008:66–70).

Macrohabitat

As established by Hildén (1965) in his seminal paper on habitat selection by birds, proximate and ultimate factors work in sequence to influence habitat use. Proximate factors are those that elicit a settling response, such as the use of a forest or grassland type, whereas ultimate factors are those that influence survival and reproduction. This distinction fits nicely into the hierarchical structure of habitat selection presented by Johnson (1980). Species first choose a general area in which to settle (first level); then establish a home range (second level); use sites for specific needs, such as denning or foraging (third level); and obtain resources from specific substrates, such as foraging on a leaf (fourth level). Here, we consider first- and second-level habitat selection, or macrohabitat, as being influenced by proximate factors (Block and Brennan 1993).

Many of the currently used habitat models operate at the macrohabitat scale, including most statewide wildlife-habitat relationship (WHR) constructs (Block et al. 1994), Gap Analysis models (Scott et al. 1993), and habitat suitability index (HSI) models (USDI Fish and Wildlife Service 1981). (See Rumble et al. 1999; Mitchell et al. 2002; and Chapter 5 for examples.) Most of these models use broad-scale categorizations of vegetation types (frequently mislabeled as "habitat types"), often broken into a seral stage as a predictor of animal presence or abundance.

Many developers of these models substantially mismatched scales in the variables used to create them, however. This is especially evident in HSI and WHR models. As a result, predictions of species occurrences contain numerous errors of commission (i.e., predicting species to be present when they are absent) and omission (i.e., not predicting species to be present when they are) (Block et al. 1994). A possible explanation is that considering macrohabitat alone obscures finer-scaled needs (e.g., microhabitat) of the species. In the case of errors of commission for species, those microhabitat features may be missing; for errors of omission, the models failed to recognize microhabitat conditions found within the macrohabitat. Additionally, it may be that the vegetation classification system has little to do with the habitat requirements of a particular species. Moreover, available maps for broad-scale analyses are almost always repurposed. For example, one needs to ask how closely "large sawtimber" relates to the habitat needs of forest animals prior to using it in WHR models.

How we categorize vegetation is not a trivial matter if we expect our macrohabitat models to perform adequately. Unfortunately, many workers classify vegetation types based on qualitative judgments that are difficult to replicate. The lack of any testing before most of these models are used to make management decisions further confounds our ability to evaluate their appropriateness (Block et al. 1994). Therefore, models developed at the macrohabitat scale are very useful in our understanding of broad habitat relationships, but their application should be limited to use at a wide scale. This should be regarded as the first step in a process encompassing analysis at multiple scales. For example, many investigations have been conducted on the habitats of all 3 subspecies of spotted owls. These studies have been done at various spatial scales and provide an example of the different levels at which habitat is studied. For example, early research based on extensive surveys showed that northern spotted owls were closely tied to old-growth Douglas-fir forests (Forsman et al. 1984; Marcot and Gardetto 1980). Similar investigations of Mexican spotted owls linked them to old-growth, mixed-conifer forests, as well as ponderosa pine–gambel oak (*Pinus ponderosa–Quercus gambelii*) forests and slickrock canyons of northern Arizona and Utah. These initial studies led to more-detailed research on home-range sizes and characteristics (Solis and Gutiérrez 1990; Ganey et al. 2003). Nest, roost, and foraging sites, which represented small patches with forested stands, were described within these home-range studies. Whereas investigators found extensive overlap between nest- and roost-site characteristics, foraging habitat was much more variable and provided a diversity of conditions needed by the suite of prey taken by spotted owls (Block et al. 2005).

That habitat is species-specific led researchers to evaluate the relationships between species diversity and environmental heterogeneity. The assumption was that a more heterogeneous environment provided habitats for more species (Tews et al. 2004). Perhaps the precursor to the development of the habitat heterogeneity hypothesis was seminal work by MacArthur and MacArthur (1961), evaluating the relationship between wood warblers (Parulidae) and foliage-height diversity. They found that the diversity of warbler species became greater with increasing stratification of the vegetation. Vegetation diversity is not always the result of vertical stratification, however. Roth (1976) developed a method by which the dispersion of clumps of vegetation—in this case, shrubs—formed the basis for a measure of habitat heterogeneity. He was able to relate bird species' diversity to the horizontal patchiness of the vegetation. The underlying premise of northern goshawk management is that horizontal patchiness in ponderosa pine forests provided a heterogeneity of habitats needed by a diversity of prey species (Reynolds et al. 1992). Others also built relationships between measures of horizontal vegetation development and animal communities (e.g., Wiens 1969; Morrison and Meslow 1983; Rotenberry and Wiens 1998).

Whereas the habitat heterogeneity hypothesis is a useful construct for describing animal diversity, vegetation diversity explains only a limited amount of the variation in species richness, as many other factors can influence the structure of an animal community. Island biogeography and metapopulation theories predict a positive relationship between species diversity and the size of an area (MacArthur and Wilson 1967; Harris 1984; Weins 1997). Habitat heterogeneity and areal size, however, may not work in a linear fashion to increase species diversity when considered together. Kadmon and Allouche (2007) and Allouche et al. (2012) noted a unimodal pattern when relating environmental heterogeneity to species diversity. They attributed this to a tradeoff between that form of heterogeneity and the amount of suitable area available for individual species. Their

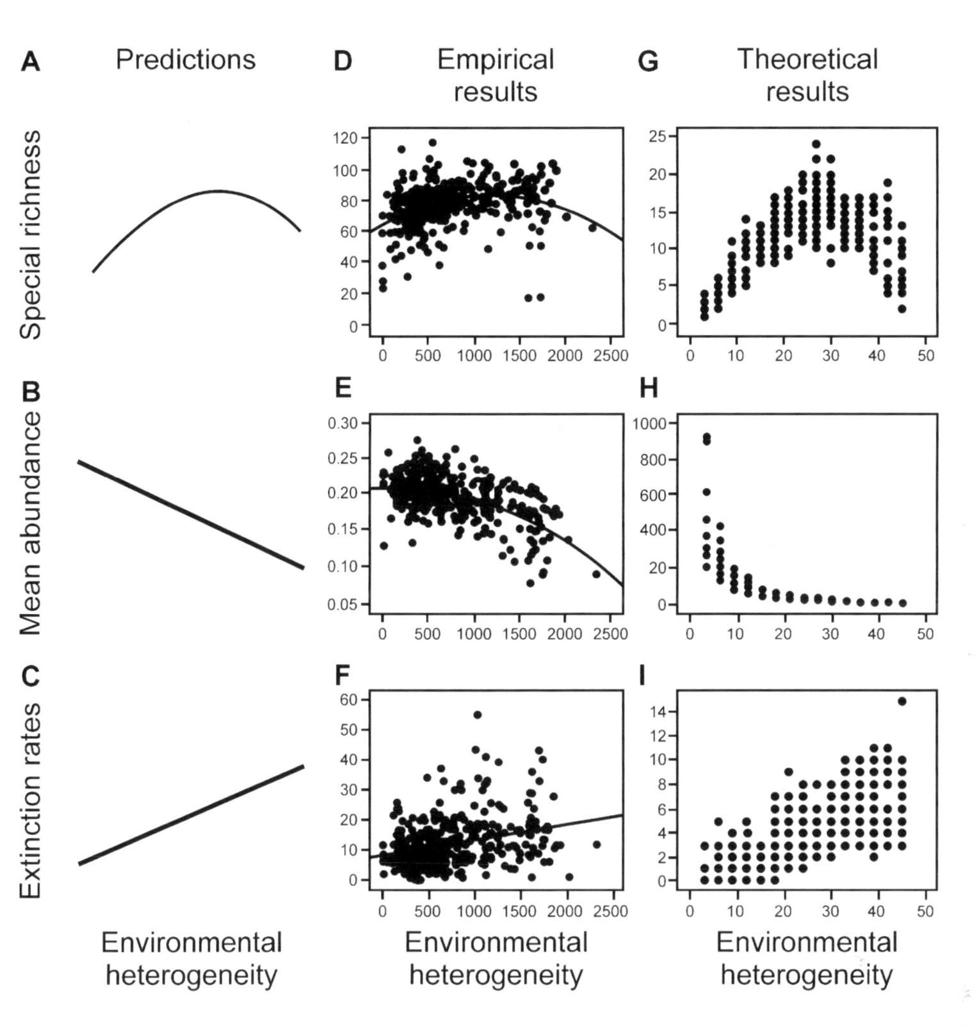

Figure 2.11. A sampling configuration of habitat variables used in a study of small mammal habitat use. (Reproduced from Dueser and Shugart (1978:Fig. 1), with the permission of John Wiley & Sons)

postulation was that as heterogeneity increases, the amount of effective area available for individual species decreases, thereby reducing population sizes and increasing the likelihood of extinctions Therefore any area, regardless of its size or environmental heterogeneity, has a threshold limit for the number of species it can support (see Fig. 2.11).

Microhabitat

Microhabitat includes third- and fourth-order selection from the Johnson (1980) model previously mentioned in this chapter. The selection of variables to measure needs to follow a thoughtful process. We see many early-career graduate students develop study plans to quantify about anything you can imagine. Their hope is that if they measure enough things, then at least 1 will turn out to be significant. The confusion here is that just because a variable may show up as statistically significant, it may have little relevance to the species being investigated. Ultimately, whether a variable is deemed to be important hinges on how that resource or condition contributes to survival and reproduction. As Krebs (1978:40) noted, "We must be careful here to define the perceptual world of the animal in question before we begin to postulate the mechanisms of habitat selection." Thus 2 basic factors must be kept separate when evaluating habitat selection: (1) evolutionary factors, which are those that contribute to reproduction and survival; and (2) behavioral factors, which provide the cues by which animals select habitat.

Researchers should keep 5 considerations in mind and use them to guide their selection of appropriate microhabitat variables, making sure that these variables (1) are biologically meaningful and relate to factors that influence the distribution and local abundance of the species, (2) are measured readily and precisely with mostly non-destructive procedures, (3) have little variation within the season where it relates to the species under study, (4) are measured at the appropriate spatial scale, and (5) should each describe the environment in the immediate vicinity of the animal (Dueser and Shugart 1978; Whitmore 1981). The first criterion emphasizes the importance of selecting variables based on prior knowledge of the species. Even so, we urge caution in blindly using the same variables as those employed by previous investigators. Variables that were deemed significant at 1 location may not be relevant at another, because of geographic differences in site conditions (e.g., vegetation, topography, soil, weather). If multiple studies, however, result in comparable relationships, confidence is enhanced in using that variable. In addition, we must be careful not to let our preconceived notions and biases eliminate potentially important variables. It is often worthwhile to let a biologist who is familiar with the area, but not the species, review your list of variables. A plant ecologist probably can also offer valuable advice to a wildlife biologist planning a study.

The second consideration hinges how quickly variables can be measured to provide precise point estimates. For example, counting every blade of grass might offer a good estimate of density, but this would be time prohibitive. A quicker approach might be to estimate cover using a line intercept, point intercept, or sampling frame. Further, Dueser and Shugart (1978) emphasized non-destructive sampling, because they planned on making repeated observations over time. Obviously, it would be difficult to repeat sampling within the same area if the site had been severely disturbed as a result of gathering data. The third concern, regarding sampling variation, is both a biological and a statistical issue. Measurements should also be sufficiently precise, so that variation within the season of study was not obscured by the probably much larger variations that occur between seasons. Regarding the fourth point, we note that multiscaled measurements may aid our understanding of a species' habitat, but it may be a mistake to include them in the same analysis, unless a clear biological and statistical justification is provided, or unless the study is designed with some manner of hierarchical analysis or modeling. As for the fifth consideration, both Dueser and Shugart (1978) and Whitmore (1981) emphasized that microhabitat should be measured in close proximity to the animal. If not, the results may not reflect microhabitat, as defined by the behavior of the organism.

Indirect Measurements of Habitat

In trying to piece together how animals use their environment, ecologists have explored various indirect measures. Leopold (1933) established that animals require 3 primary resources for survival: food, cover, and water. As mentioned above, cover and water are correlates that are often measured at all spatial scales to describe habitat. A more indirect measure of habitat use is to assess the amount of available food. Indeed, a major area of inquiry for animal ecologists is in understanding food habits. This entails describing 3 aspects: (1) what animals eat, and how and where they obtain it (see Chapter 4); (2) how food is distributed spatially and temporally; and (3) the overlap between the focal species' habitat, including where food occurs. Various types of information are used to piece this together, ranging from direct to indirect measures. Direct measures include food-habits studies, which allow ecologists to describe and quantify the types of foods consumed. This information can be obtained by direct observation of foraging animals or by analyzing feces (van Dijk et al. 2007), the crops of birds (Dalke 1937), excreted pellets (Errington 1930), stomach contents (Hartley 1948), or stable isotopes present in animal hairs or feathers (Hobson et al. 2000) to determine their diet. Assuming a representative, unbiased sample of that diet—that is, knowing where food occurs—can help an investigator hone in on the focal species' habitat.

Reynolds et al. (1992) used information on the primary prey of northern goshawks to develop management recommendations for this species of concern (Fig. 2.2). From the literature, they pieced together descriptions of the habitats of the prey species to develop a broader delineation of goshawk habitat. Although somewhat appealing, their approach was indirect and fraught with various shortcomings. First, their prey habitat descriptions were based on literature summaries, not on actual information from where prey co-occurred with goshawks. Second, the resulting depiction of goshawk habitat was based on compiling information about and summarizing prey habitat. This synthesis was not definitive, but it represented a useful conceptual framework for future empirical work that could test and validate their model.

Block et al. (2005) conducted an investigation of the prey ecology of Mexican spotted owls in northern Arizona ponderosa pine–gambel oak forests. Their study involved ascertaining the diet of these owls through pellet analyses and then describing their habitats by characterizing the conditions where they were caught (by live trapping). This information was then used to fine tune an understanding of owl foraging habitat and provide information for future management considerations.

We should note here that most of the approaches taken by animal ecologists are indirect, correlative reflections of what an animal may be responding to in making decisions about habitat selection. For example, few would argue against the premise that insectivorous birds must have an adequate supply of the proper prey. Yet many studies of such birds do not quantify their prey, because of the extreme difficulty in doing so (Morrison et al. 1990). Thus we often make indirect measurements of prey by assuming that insect prey biomass varies positively with increased foliage. Moreover, indirect measures do not end here. Measuring foliage volume is also extremely difficult, so we normally use some indirect measure of it. What we have, then, is an indirect measure of foliage volume being used to indirectly index insect abundance. Hence errors are grossly compounded, and it is difficult to know the biological meaning of our results in terms of specific causes.

Another problem is that all foliage is not created equal. For example, Litt et al. (2014) showed decreased arthropod abundance and richness as exotic plants replaced native vegetation. Flanders et al. (2006) observed that in their bird-count plots containing up to 40 percent buffelgrass (*Pennisetum ciliare*), northern bobwhite abundance was reduced by 50 percent, with similar reductions in arthropod biomass. Therefore, gauging foliage volume or biomass may be a biased calculation of insect biomass, unless researchers account for the plant species being sampled (Sands et al. 2009). Other examples are big-game biologists making indirect estimates of food requirements, including micronutrient needs, through various indirect assessments of forage abundance (e.g., clipping and weighing, ocular estimates or line intercepts of shrub cover); and herpetologists using numerous measures of ground, shrub, and canopy cover to indirectly describe microsite conditions (e.g., temperature, soil moisture) available to salamanders. The prevalence of such imprecise measurements—and indirect means of doing so with other variables—ranks among the primary weaknesses in the area of habitat relationship research and modeling. Moreover, it will be difficult to increase the accuracy of our models and our management decisions until we begin to make more-direct measurements of the factors that influence the distribution, abundance, and behaviors of wildlife. We have done a lot of the relatively easy work, but any substantial advances await implementation of this much-more-difficult task. We are not calling for the abandonment of indirect assessments, since science could not exist without them. For example, we do not actually measure temperature, but rather the expansion of liquid or metal that correlates with it. When using indirect measures, the key factor is the quality of the relationship between indirect and direct measurements. It is pretty accurate for thermometers, but in wildlife habitat definitions, the relationships between, say, insect biomass and a broad vegetative cover type are tenuous, at best. Such measurements, therefore, have limited predictive power, and we should openly acknowledge the limits they place on the quality of our results.

As discussed above, indirect assessments of foliage structure (e.g., foliage height diversity) are gross approximations and provide only limited predictive power. Yet determinations of canopy closure are being used as the centerpieces of management guidelines for many species of wildlife, including federally threatened Mexican spotted owls, northern goshawks, and general references to "forest-interior species" (e.g., Fedrowitz et al. 2014). A disconnect here is that although research has clearly shown strong relationships between species' occurrence and canopy cover, many management agencies do not measure canopy cover when they sample habitat. Using other options, such as tree basal area, to index canopy cover have been unsuccessful. If management guidelines are to be interpreted and applied adequately, we must know that our methodologies are accurate and truly reflect some type of ecological causation.

A more useful measure of foliage structure for many species may be foliage volume by height—that is, the actual surface area of foliage available for consumption or as substrates for insects and other prey at various levels in the understory, subcanopy, and canopy. It is extremely tedious, however, to quantify actual foliage volume. Plant ecologists have done so by cutting down trees and then measuring and counting their leaves, needles, and other parts, as such data are necessary for accurate predictions of photosynthetic rates and calculations of tree growth (see Carbon et al. 1979). After ecologists have completed such work, they can develop statistical models that relate some more easily measured aspect of the plant to foliage volume (VanDeusen and Biging 1984). Plant ecologists also use the dry volume of plant material as a way to gauge total volume.

Field methods for measuring the total surface area of foliage have been used for decades, as they provide an estimate for a leaf-area index (Marshall and Waring 1986). Such methods are rapidly being replaced, however, by the use of low-altitude aerial light detection and ranging, or LiDAR (Almeida et al. 2019). Similarly, ground-based LiDAR provides detailed, 3-dimensional structural data on understory vegetation, which may obviate the need for destructive sampling (Yao et al. 2011). We present more about LiDAR later on in this chapter.

Researchers have also developed measures of foliage volume that are even more indirect. In a comparison of the habitats of chestnut-backed chickadees (*Parus* [*Poecile*] *rufescens*) and black-capped chickadees (*P. atricapillus*), Sturman (1968) used equations for shapes that approximated the structure of conifers and hardwoods. Similar methods (e.g., Mawson et al. 1976) can be viewed as compromises between labor-intensive techniques that involve sampling whole trees and the qualitative estimates of foliage-height diversity (FHD).

Many other methods of estimating foliage development have been devised (see a review by Ganey and Block 1994; also see Campbell and Norman 1989). A spherical densiometer—most easily described as a round, concave, or convex mirror scribed with a grid—has been used to estimate canopy closure directly over this device. Ganey and Block (1994) tested the relative results between a spherical densiometer and another device, a sighting tube. The latter is a simply a piece of PVC pipe (or similar material) with a level fastened onto it, and an observer can look at canopy closure through it. These authors found that the densiometer resulted in higher estimates of canopy closure than the sighting tube—57 percent of all estimates showed >50 percent cover for the densiometer, whereas only 38 percent did so for the sighting tube—but they could not determine which method was the most accurate. Part of the reason why is that a densiometer measures both vertical crown cover and crown length. Their results do show, however, that the method being used can have a significant impact on conclusions regarding canopy cover, and on subsequent management efforts, particularly if different methods were used in the application of such actions than were employed in the original research. This, in turn, could have substantial ecological ramifications.

These issues come to the fore when hard-and-fast habitat rules are used to evaluate what are approximate measures. For example, for spotted owls, a ≥40 percent crown cover rule for foraging habitat is often invoked. Whether that number was derived by spherical densiometer, a sighting tube, the proportion of light penetration, or canopy structure determined from ground- or air-based LiDAR—all of which will give different estimates—can make a huge difference in the proportion of the landscape considered as habitat for these owls.

The Focal-Animal Approach

Most studies of microhabitat selection are variations of the focal-animal approach (see Chapter 4 for focal-animal sampling in behavioral research). These methods use the location of individual animals to demarcate an area being used by that species, from which a set of environmental variables are then measured. Typically, 1 animal's specific location serves as the center of a sampling plot. Or a series of observations for an individual animal might be used to delineate an area from which samples are then made (e.g., Wenny et al. 1993). The major assumption in this approach is that the measurements sample the habitat of that species.

Animals engage in a variety of activities: feeding, drinking, calling, resting, grooming, nesting, denning, and others. These actions consume varying amounts of time and energy, and they often take place in different locations within the home range (e.g., calling from exposed locations, or resting in the shade). The amount of time spent in any 1 activity may not indicate the importance of that activity to an animal. For example, drinking may take only a few minutes each day, but without water, that animal is unlikely to survive. A researcher must be aware of these behaviors when designing a study based on the focal-animal approach. (Chapter 4 provides a more detailed description of behavioral sampling.) This adds strength to the argument made above regarding the importance of considering time-activity budgets in an analysis of habitat relationships.

Many avian studies have used the location of a singing or displaying male, a nest site, a roost, or a foraging bird as the center of plots describing the habitat of the species (e.g., James 1971; Holmes 1981; Morrison 1984; Brennan et al. 1987; VanderWerf 1993). Do such activity-centered plots extend to the broader habitat of the species, or do they describe only a habitat subset? And, especially for larger-size plots, might they include areas of non-habitat—that is, places not used by the species of interest—thus biasing the picture of what actually constitutes that species' habitat? Collins (1981) examined habitat data collected at perch and nest sites for several species of warblers (Parulidae). He discovered that 29 percent of the nest sites had vegetation structures that significantly differed from the corresponding perch sites within their territories, so habitat calculations based on perch sites overestimated the tree component. In a similar study, Petit et al. (1988) noted differences between nest and perch sites for both hooded warblers (*Setophaga citrina*) and wood thrushes (*Hylocichla mustelina*) and found that that vegetation was denser at nest sites than at perch sites. They attributed the disparity to these birds needing better concealment for their nests. This does not imply that a study based only on perches or on nests is flawed. Rather, it means that such investigations might only describe part of what might be called the "breeding habitat" of the species. Reynolds et al. (1992) distinguished among nest sites, post-fledging areas, and foraging habitat for northern goshawks, because each had unique structural attributes for its vegetation (e.g., tree density, cover). Similar examples can be found in the literature for all major groups of wildlife, such as the bedding sites of deer (Ockenfels and Brooks 1994); the locations of small mammals learned through trapping (Morrison and Anthony 1989; Kelt et al. 1994; Morzillo et al. 2003) and telemetry (Hall and Morrison 1997; Ucitel et al. 2003); and movements of amphibians (Griffin and Case 2001).

Use Versus Availability: Basic Designs

Many studies of resource selection (e.g., vegetation, food) are based on comparisons between what a species uses versus what generally occurs in that locale. Some authors assume all of a given resource is available, but that argument is on shaky ground. Resource occurrence can be constrained by various forces, such as intra- or interspecific competition, predation, resource distribution, weather, access, and numerous other influences. Resources that might be broadly suitable within the fundamental niche of a species may be reduced or not obtainable within the realized niche. Hence the availability of a given resource should be carefully defined, to ensure that comparisons with its use by the target species have some validity (Jenkins et al. 2019).

Even so, the comparison of use of a habitat element (e.g., food, a habitat characteristic) relative to its abundance or accessibility is a cornerstone of wildlife-habitat analyses. Thomas and Taylor (1990) reviewed 54 papers published in the *Journal of Wildlife Management* that analyzed these factors and concluded that use-availability studies could be categorized into 3 basic designs: (1) the availability and use of all items are estimated for all animals (i.e., the population), but individual animals are not distinguished, and a determination is only made for the item that is actually used (for about 28% of the studies reviewed); (2) individual animals are identified, and the use of each item is estimated for each animal (for 46% of the studies); and (3) the same as the second design, except that availability as well as use of the items are estimated for each individual animal (for 26% of the studies). The third design probably provides the greatest detail and thus may be the most useful way to construct a picture of the overall habitat of the population, although it is likely to also be the most time-intensive and expensive type of investigation to conduct.

Manley et al. (2002) outlined 7 assumptions that underlie resource-selection research: (1) the variables that are measured remain constant during the sampling period, (2) the probability function of resource selection remains constant, (3) available resources are identified correctly, (4) used and unused resources are identified correctly, (5) the variables influencing selection are determined, (6) animals are not impeded from accessing the resources, and (7) units are sampled randomly and independently. Yet many of these assumptions are probably violated in reality. Much of what we know about foraging theory (see Chapter 4) suggests that animals will exploit a resource patch until it reaches a lower threshold, causing them to move on to a more lucrative patch. Hence it is not just the presence of the resource that is important, but also its abundance. Moreover, determining and verifying availability is problematic. A resource may be present, but factors such as territoriality, inter- and intraspecific competition, predation risk, and the like may impede its availability. This also ties into the sixth assumption, as animals may not have free access to a resource. Regardless of these limitations, an analysis of resource selection offers a framework from which one can evaluate whether a species uses certain resources disproportionately to its occurrence. This allows the investigator to relax some of the above 7 assumptions and helps skirt the issue of availability.

Thomas and Taylor (1990) and Manly et al. (2002) provided thorough reviews of studies that fit each of these categories. They also supplied guidelines for the sample sizes necessary to conduct such analyses. Virtually all classes of statistical techniques have been used to analyze use-availability (or use-nonuse) data, depending on the objectives of the researcher, the structure of the data, and adherence to statistical assumptions (i.e., univariate parametric or non-parametric comparisons, multivariate analyses, Bayesian statistics, and various indices). These techniques have been thoroughly reviewed (e.g., Johnson 1980; Alldredge and Ratti 1986, 1992; Thomas and Taylor 1990; Aebischer et al. 1993; Manly et al. 2002). Aebischer et al. (1993) presented a compositional analysis widely used in wildlife home-range studies, although it can also be employed in diet or time-budget analyses. The proportional use of a resource (e.g., food item, vegetation type, etc.) by individual animals is the basis for an analysis. Given that these data are multivariate, the analysis is done using multivariate linear models. Applications can examine relationships with food abundance or compare use versus occurrence of vegetation classes to test for preferences. Compositional analyses, however, can give strongly biased results, especially Type I errors, when habitat matrix cells with zeroes use small decimal values to make the division work (Bingham and Brennan 2004). Reorganizing the habitat matrix so that no zero-use cells are present is the only way to get around this shortcoming.

Caution is warranted in interpreting results from use-availability studies. Garshelis (2000) found that 90 (58%) of the 156 habitat papers published in the *Journal of Wildlife Management* from 1985 through 1995 employed use-availability comparisons. He noted some common issues that may make it difficult to interpret their results. As the number of categories being considered increases, the number of observations within each category decreases and may mask habitat selection. Also, there may be a mismatch between the scale at with categories are defined, and the scale at which individuals utilize the habitat. For example, members of a species may make use of a certain cover type but are actually selecting smaller patches within that type. Another problematic issue is observing habitat use and selection during the wrong time of day. For example, a study of foraging owls would provide little insight into their choice of nesting habitat. Other sources of variation that can cloud selection studies are differences among sexes or ages in habitat use or social status, or whether the individuals of interest are residents or transients. The point here is that researchers must be cognizant of these sorts of disparities and frame their investigations accordingly.

In many of these designs, however, the composition of an available habitat is measured only once. This is then compared with observations of habitat used by the animals that are collected over a longer period of time (e.g., a season) to achieve an adequate sample size. Habitat used by animals is, in essence, a composite, put together over a defined period. Resource availability changes over time (e.g., depletion of a food resource), which could substantially influence habitat use. The ramifications depend largely on the goals of the investigation. To rectify this potential problem in use-availability studies, Arthur et al. (1996) developed a method of estimating habitat use by compiling multiple observations of habitat availability. Their method allows quantitative comparisons among habitat categories and is not affected by arbitrary decisions about which categories to include in a study (see also Manly et al. 2002).

What to Measure

Habitat Features

As we established earlier, habitat is defined by the behavior of animals at multiple spatial and temporal scales. Relationships for many species will fluctuate with the scale of the observation, and the strength of those relationships may vary concordantly. Wiens et al. (1987) concluded that the study of shrubsteppe birds was facilitated by considering relationships within a nested hierarchy of scales for a "reasonably long period of time." Such a general approach might allow a researcher to determine the appropriate scale for a study, both in space and time.

Many investigations of animal habitat relationships have focused on vegetation, because animals generally are linked in some fashion to plants for food (either directly or indirectly), cover, or both. The key question here is, to what specific aspect of vegetation are animals responding? Is it structure, composition, or some combination of both? Also, at what scale does this happen: macrohabitat (i.e., broad vegetation patterns) or microhabitat (i.e., specific arrangements and sizes of plants)? Carnivores generally respond directly and indirectly to vegetation, based on the habitat requirements of their prey. For example, cougars use forest edge cover to ambush some species from trees, and they also stalk proximal to grass or shrublands, due to the feeding behavior of ungulates. Therefore, for carnivores, their relationships to vegetation are generally complex. Or perhaps they are not responding to vegetation at all, as is the case with many nesting seabirds or denning wolverines (Magoun and Copeland 1998; Jones and Kress 2012).

To answer these questions, we now review variables collected by researchers seeking to describe the habitat-use patterns of animals. We offer these examples as good starting points for students planning similar evaluations of habitat use by a particular species. The first step in selecting variables is to evaluate studies done on the species of interest or on those with comparable ecologies in analogous systems. You should not be constrained by previous investigations on similar species, however. Consider variables collected by other researchers studying different taxa, but in related systems. For example, a person planning a study to examine habitats of ground-foraging birds could gain valuable information on the selection of types of variables and sampling design by reviewing papers on small mammals and reptiles. Below we discuss some general categories of vegetation variables to collect at different spatial scales. Regardless of the choices that are made, they should closely follow the study's objectives.

Macrohabitat

The 2 primary variables to describe macrohabitat are vegetation type and seral stage. "Vegetation type" refers to broad hierarchies of plant associations used to characterize the extant vegetation. A number of classification systems have been established by vegetation ecologists (e.g., Küchler 1964; Bailey 2009), which include descriptions of existing vegetation or potential natural vegetation. Vegetation types can be broad categorizations (e.g., desert, grassland, shrubland, woodland, forest, and tundra), or they can be more specific and identify associations of plant species. Often they are based on the dominant plant species (using attributes such as basal area or cover). They can also be predicated on the dominant and subdominant species, or on overstory and understory species. For example, forests can be differentiated as ponderosa pine, mixed conifer, spruce-fir, and the like. Some examples of vegetation classification schemes include ecological systems in the United States (Comer et al. 2003), rangelands (Shiflet 1994), forests (Eyre 1980), ecoregions (Bailey et al. 1978; Bailey 2009), and potential natural vegetation (Küchler 1964). Other broad vegetation descriptions are couched as "ecoregions," using the framework of James Omernik (e.g., by Sarr et al. 2015), and the unfortunately named "habitat types" of Rexford Daubenmire (e.g., by Pfister et al. 1977).

Describing the vegetation type alone may not provide the level of resolution needed to depict a species' habitat. We know that some animals are regarded as early successional species, using vegetation that is in an early seral stage (e.g., grasses/forbs or shrubs). Many brush-inhabiting birds, for example, occupy forests that were disturbed as the result of fires or timber harvests and use shrub fields that are the precursor to forest establishment. In contrast, other species may require old-forest conditions (i.e., later seral stages).

Abiotic factors, such as topography and climate, may also be strong predictors of where species are found. For example, Bowden et al. (2003) modeled occupancy by Mexican spotted owls on 25 quadrats, ranging from 40 to 75 km^2 in size, and found that incorporating topography improved the model's fit. They estimated the true surface area by developing triangulated irregular networks (TIN) for each quadrat, using the US Geological Survey's digital elevation models. Surface area was used as an index for topographic "ruggedness"—that is, the more surface area there was within a quadrat, the greater the ruggedness. They also calculated the ratio of surface area to planar area, which they termed the "TIN ratio." Ganey et al. (2004) found that the TIN ratio was highly correlated with owl abundance on sampled quadrats, and it improved the precision of the estimate considerably. Copeland et al. (2010) evaluated the influence of August temperatures and spring snow cover as predictors for denning wolverines in Fennoscandia and North America. All 562 reproductive wolverine dens occurred at sites with persistent spring snow cover, and 95 percent of summer and 86 percent of winter telemetry locations were concordant with spring snow coverage.

Microhabitat

The number of microhabitat variables used by researchers is probably only constrained by their imagination. Variables run the gamut of measurements and indices of vegetation structure and composition, volumes and decay classes of standing (i.e., snags) and fallen (i.e., logs) dead trees, edaphic characteristics, substrate environment, weather, and the like. Our intent here is not to provide an exhaustive list of all variables, but to hone in on those that have been shown to be strong covariates of species presence.

Studies of microhabitat have transitioned from qualitative descriptions (e.g., Grinnell and Miller 1944) to quantitative, multivariate descriptions (James 1971; and many other since). Birds initially received the most attention with regard to analyses of habitat-use patterns. This is probably a reflection of the conspicuousness of birds. Most are active during the day, give at least some vocalizations during all parts of the year, and are inexpensive to observe (requiring only binoculars and a notebook). James (1971) pioneered the application of multivariate statistics to differentiate the microhabitats of multiple bird species. Her approach was largely based on the conceptual framework of the niche and a bird's niche gestalt. Using 15 variables of vegetation structure that she had developed with a fellow researcher (James and Shugart 1970), she was able to describe the multidimensional "habitat space" of a bird community in Arkansas.

Since the late 1980s and continuing to the present, other vertebrate groups have received increasing attention in studies of microhabitat. Dueser and Shugart (1978) were among the first to investigate multivariate microhabitat differences among small mammal species, and their study is a good example of the pattern-seeking nature of most habitat descriptions. Its specific objectives were to characterize and compare the microhabitats of species within an upland forest in eastern Tennessee, and then to examine the relationships species abundances and distributions had with the relative availability of the selected microhabitats. They gathered information on vertical strata at each capture site of a small mammal—overstory, understory, shrub level, forest floor, and litter-soil level—but they did not include floristic variables (Table 2.2). This was an unfortunate omission in their microhabitat analysis—a missed opportunity to improve their descriptions of microhabitat associations. The authors did, however, record the number of woody and herbaceous species. They also paid special attention to features of the forest floor, such as litter-soil compaction, fallen log density, and short herbaceous stem density. They found that certain soil variables played a significant role in describing the differences in microhabitats of the species of interest. Except for the authors' lack of detailed information on plant taxonomy, we consider this study to be a good example of a very detailed set of variables used to differentiate among species of co-occurring animals.

Floristics are important in characterizing the habitat of small mammals, Block et al. (2005) recorded vegetation cover and basal area of trees at trap stations where they caught small mammals in an Arizona ponderosa pine–gambel oak forest. They found close associations of both Mexican woodrats (*Neo-*

Table 2.2. Designation, descriptions, and sampling methods for variables measuring forest habitat structure

Variable	Methods
1. Percentage of canopy closure	Percentage of points with overstory vegetation, from 21 vertical ocular tube sightings along the center lines of 2 perpendicular 20 m^2 transects centered on the trap
2. Thickness of woody vegetation	Average number of shoulder height contacts (trees and shrubs) from 2 perpendicular 20 m^2 transects centered on the trap
3. Shrub cover	Same as (1), for presence of shrub-level vegetation
4. Overstory tree size	Average diameter (in cm) of nearest overstory tree
5. Overstory tree dispersion	Average distance (m) from trap to nearest overstory tree, in quarters
6. Understory tree size	Average diameter (cm) of nearest understory tree, in quarters around the trap
7. Understory tree dispersion	Average distance (m) from trap to nearest understory tree, in quarters
8. Woody stem density	Live woody stem count at ground level within a 1.00 m^2 ring centered on the trap
9. Short woody stem density	Live woody stem count within a 1.00 m^2 ring centered on the trap (stems (0.40 m in height)
10. Woody foliage profile density	Average numbers of live woody stem contacts with a 0.80 cm diameter metal rod rotated 360°, describing a 1.00 m^2 ring centered on the trap and parallel to the ground at heights of 0.05, 0.10, 0.20, 0.40, 0.60 . . . 2.00 m above ground level
11. Number of woody species	Woody species count within a 1.00 m^2 ring centered on the trap
12. Herbaceous stem density	Live herbaceous stem count at ground level within a 1.00 m^2 ring centered on the trap
13. Short herbaceous stem density	Live herbaceous stem count within a 1.00 m^2 ring centered on the trap (stems <0.40 m in height)
14. Herbaceous foliage profile density	Same as (10), for density of live herbaceous stem contacts
15. Number of herbaceous species	Herbaceous species count within a 1.00 m^2 ring centered on the trap
16. Evergreenness of overstory	Same as (1), for presence of evergreen canopy vegetation
17. Evergreenness of shrubs	Same as (1), for presence of evergreen shrub-level vegetation
18. Evergreenness of herb stratum	Percentage of points with evergreen herbaceous vegetation, from 21 step-point samples along the center lines of 2 perpendicular 20 m^2 transects centered on the trap
19. Tree stump density	Average number of tree stumps $\geq$7.50 cm in diameter, per quarter
20. Tree stump size	Average diameter (cm) of nearest tree stump $\geq$7.50 cm in diameter, in quarters around the trap
21. Tree stump dispersion	Average distance (m) to nearest tree stump $\geq$7.50 cm in diameter, in quarters around the trap
22. Fallen log density	Average number of fallen logs $\geq$7.50 cm in diameter, per quarter
23. Fallen log size	Average diameter (cm) of nearest fallen log $\geq$7.50 cm in diameter, in quarters around the trap
24. Fallen log dispersion	Average distance (m) from trap to nearest fallen log $\geq$7.50 cm in diameter, in quarters around the trap
25. Fallen log abundance	Average total length (+0.50 m) of fallen logs $\geq$7.50 cm in diameter, per quarter
26. Litter-soil depth	Depth of penetration ($<$10.00 cm) into litter-soil material of a hand-held core sampler with a 2.00 cm diameter barrel
27. Litter-soil compactability	Percentage of compaction of litter-soil core sample (26)
28. Litter-soil density	Dry weight density (g/cm^2) of litter-soil core sample (26), after oven drying at 45°C for 48 hours
29. Soil surface exposure	Same as (18), for percentage of points with bare soil or rock

Source: Dueser and Shugart (1978:Appendix)

toma mexicana) and brush mice (*Peromyscus boylii*) with gambel oaks, a prominent midstory tree species. Another noteworthy result of this study was to document that these woodrats and mice were found close to rock outcrops. Moreover, non-vegetation features may have figured prominently into the habitat of these ground-dwelling mammals, as was shown by Deuser and Shugart (1978).

Reinert (1984) sought to differentiate the microhabitats of timber rattlesnakes (*Crotalus horridus*) and northern copperheads (*Agkistrodon contortrix*), which occur sympatrically in temperate deciduous

Table 2.3. Structural and climatic microhabitat variables

Mnemonic	Variable	Sampling method
ROCK	Rock cover	Coverage (%) within 1 m^2 quadrant centered on the snake location
LEAF	Leaf litter cover	Same as ROCK
VEG	Vegetation cover	Same as ROCK
LOG	Fallen log cover	Same as ROCK
WSD	Woody stem density	Total number of woody stems within a 1 m^2 quadrant
WSH	Woody stem height	Height (cm) of tallest woody stem within a 1 m^2 quadrant
MDR	Distance to rocks	Mean distance (m) to nearest rocks (>10 cm maximum length) in each quarter
MLR	Length of rocks	Mean maximum length (cm) of rocks used to calculate MDR
DNL	Distance to log	Distance (m) to nearest log (≥7.5 cm maximum diameter)
DINL	Diameter of log	Maximum diameter (cm) of nearest log
DNOV	Distance to overstory	Distance (m) to nearest tree (≥7.5 cm dbh [diameter at breast height])
DBHOV	Dbh of overstory tree	Mean dbh (cm) of nearest overstory tree within each quarter
DNUN	Distance to understory	Same as DNOV (trees <7.5 cm dbh, >2.0 m height)
CAN	Canopy closure	Canopy closure (%) within a 45° cone with an ocular tube
SOILT	Soil temperature	Temperature (°C) at 5 cm depth within 10 cm of the snake
SURFT	Surface temperature	Temperature (°C) of substrate within 10 cm of the snake
IMT	Ambient temperature	Temperature (°C) of air at 1 m above the snake
SURFRH	Surface relative humidity	Relative humidity (%) at humidity substrate within 10 cm of the snake
IMRH	Ambient relative humidity	Relative humidity (%) 1 m humidity above the snake

Source: Reinert (1984:Table 1)

forests of eastern North America. His approach was a multivariate habitat description consistent with the Hutchinsonian definition of the niche (Green 1971) and based on the concept of the niche gestalt (James 1971). Reinert (1984) was interested not only in features pertaining to the ground, but also in the weather conditions immediately surrounding the snakes. He measured temperature and humidity at several locations near an animal, as well as the structure of the surrounding vegetation. Here again, however, no information on plant taxa was included (Table 2.3), but that might be less of a problem with reptiles than with other taxonomic groups.

Morrison et al. (1995) used time-constrained surveys to describe the microhabitats of amphibians and reptiles in mountains in southeastern Arizona. A stopwatch ran while observers walked slowly, searching the ground and tree trunks and turning over moveable rocks, logs, and litter to examine protected locations. After a predetermined time, a survey was stopped and a 5 m diameter plot was then centered on the animal's location which served as the site where microhabitat conditions were measured. The authors quantified substrate temperature, various descriptors of the vegetation, and other habitat characteristics. Block and Morrison (1998) used time-constrained searches (as described above) and pitfall traps to sample the herpetofaunal community in oak woodlands in California. They compared habitat attributes at pitfall traps where a species was captured with attributes where it was not. The authors were able to describe both structural and floristic relationships for many of the species encountered: slender salamanders (*Batrachoseps nigriventris*) and yellow-blotched ensatinas (*Ensatina eschscholtzii croceater*) were both closely tied to canyon live oaks (*Quercus chrysolepis*), and western fence lizards (*Sceloporus occidentalis*) and Gilbert's skinks (*Plestiodon gilberti*) were associated with California black oaks (*Quercus kelloggii*). Again, this is an example of floristic relationships for species from different taxa.

Welsh and Lind (1995) analyzed the habitat affinities of Del Norte salamanders (*Plethodon elongatus*) in relation to landscape, macrohabitat, and micro-

Table 2.4. Hierarchical arrangement of ecological components in forest environments

Hierarchical scale[1]	Variable category	Variables[2]
Landscape scale	Geographic relationships	Latitude (degrees)
		Longitude (degrees)
		Longitude (degrees)
		Slope (%)
		Aspect (degrees)
Macrohabitat or stand scale	Trees: density by size	Small conifers_C
		Small hardwoods_C
		Large conifers_C
		Large hardwoods_C
		Forest age (in years)
	Dead and down wood: surface area and counts	Stumps_B
		All logs, decayed_C
		Small logs, sound_C
		Sound log area_L
		Conifer log–decay area_L
		Hardwood log–decay area_L
	Shrub and understory composition (>0.5 m)	Understory conifers_L
		Understory hardwoods_L
		Large shrub_L
		Small shrub_L
		Bole_L
		Height II–ground vegetation_B (0.5–2 m)
	Ground-level vegetation (<0.5 m)	Ferns_L
		Herbs_L
		Grasses_B
		Height I–ground vegetation_B (0–0.5 m)
	Ground cover	Mosses_L
		Lichens_B
		Leaves_B
		Exposed soil_B
		Litter depth (cm)
		Dominant rock_B
		Co-dominant rock_B
	Forest climate	Air temperature (°C)
		Soil temperature (°C)
		Solar index
		Percentage of canopy closed
		Soil pH
		Soil relative humidity (%)
		Relative humidity (%)
Microhabitat scale	Substrate composition	Pebble_P (% of 32–64 mm diameter rock)
		Cobble_P (% of 64–256 mm diameter rock)
		Cemented_P (% of rock cover embedded in soil/litter matrix)

Source: Welsh and Lind (1995:Table 1)

Note: The study involved 43 measurements of the forest environment, taken in conjunction with sampling for Del Norte salamanders (*Plethodon elongatus*)

[1] Spatial scales are arranged in descending order, from coarse to fine resolution (see Wiens 1989). Level I relationships (biogeographic scale) were not analyzed, because all sampling occurred within the range.

[2] C = count variables, which are numbers per hectare; B = Braun-Blanquet variables, which are the percentage of cover in a 1/10 ha circle; L = line transect variables, which are the percentage of 50 m line transects; P = percentage of the 49 m^2 salamander search area. Small trees = 12–53 cm dbh (diameter at breast height), large trees = >53 cm dbh.

habitat scales. They presented a detailed rationale for their selection of methods, including the choice of analytical techniques, data screening, and interpretation of the output. The variables they measured, separated by spatial scale, are shown in Table 2.4. A similar example for multiple species of amphibians was given by Welsh and Lind (2002).

How to Measure

Sampling Principles

In this section we review some of the common methods used to measure wildlife habitat. Many methods used to assess vegetation—long established by plant ecologists (e.g., Daubenmire 1959; Mueller-Dombois and Ellenberg 1974; Cook and Stubbendieck 1986; Bonham 1989; Elzinga et al. 2001)—have provided techniques for animal ecologists to use or modify.

Even a cursory review of the "Methods" sections in animal ecology publications shows a reliance on standard, or classical, methods of quantifying the structure and floristics of vegetation: point quarter, circular plots and nested circular plots, sampling squares, line intercepts, and so on. But we should also be innovative and consider other variables not used in the past. Standard methods do, however, provide an established starting point from which wildlife biologists can adapt specific methods as needed, and they allow easy comparability between studies.

Previous investigations are a fine starting point for the development of a new study, but we should also learn from their mistakes. In such instances, a rather brief pilot study, designed to evaluate methods and variables applied in earlier work, is well worth your time and effort. Unfortunately, the reverse is more often the case. As Green noted (1979:31), "Carry out some preliminary sampling to provide a basis for evaluation of sampling design and statistical analysis options. Those who skip this step because they do not have enough time usually end up losing time."

Preliminary Sampling

The first step in any investigation is to document the objectives, or rationale, for the research. A study whose purpose is to characterize the habitat of a selected species may differ from another that seeks to compare the habitats of multiple species. We must remember that the variables we measure—and how we measure them—will play a substantial role in addressing the overall objectives. For example, if we do not record vegetation data by taxonomic classification, then floristics can play no role in our analyses or in subsequent management applications. As cautioned above, when designing a study, researchers must walk a fine line between repeating what others have done and inventing new methods, and between efficiently measuring only the most pertinent variables and more finely dividing the variables to test for new associations.

Unfortunately, the "Methods" sections in most papers provide little, if any, information on *why* the author(s) chose the methods used. Often researchers duplicate habitat protocols published in previous papers without critically evaluating whether the methodologies are applicable to their own study. For example, a common procedure is to record vegetation within 11.3 m radius plots. This is a holdover from the non-metric English system: an 11.3 m radius results in a 0.04 ha plot, which corresponds to the 0.1-acre plot that has been the standard for vegetation measurements. Modern researchers use this radius because it is what has traditionally been used in the literature, and it probably does afford some degree of sampling efficiency and comparability with similar studies. Petit et al. (1988), however, concluded that 0.04 ha plots were too large to characterize the nest sites of hooded warblers and wood thrushes, because the birds selected locales at a smaller scale. Larger plots included too much variation and masked the characteristics these warblers and thrushes selected for nesting. In this case, smaller plots would be more representative of their nesting habitat.

There is clearly a need for preliminary sampling

to establish the most predictive measurements of an animal's habitat. Most likely, every ecology student is exposed to the notion that one should conduct a pilot study before collecting data in earnest. Although we certainly understand the limits that time and money place on the intensity and duration of a research project, it makes little sense to base one's efforts in the field primarily on established dogma and untested techniques. As a measure of the need for such investigations, we can safely state that our most popular publications (as determined by the number of times papers are cited) are consistently those that evaluate field methods and analytical techniques, rather than those that center on descriptors of an animal's habitat use in a particular area.

Preliminary sampling allows one to both test the predictability of the proposed field methods and ensure that adequate sample sizes are being accumulated (see Chapter 3 for a discussion of sample size analyses). The literature is replete with examples of researchers measuring dozens of variables, with only a few emerging as significant in describing or modeling the habitat of the species under investigation. Yet all of these variables were collected for the duration of the study. Recording such data is time consuming. Wouldn't a more efficient procedure involve collecting initial samples during the same phenological period, and then conducting preliminary analyses to determine which variables were duplicative (i.e., highly intercorrelated) and which had little or no predictive power? Box 2.1 presents the details of such an analysis. Granted, collecting enough samples in a pilot study to permit this pre-analysis runs the risk of becoming a full-fledged study in itself. An alternative is to review previous research to help inform the variables included in a planned investigation.

Sampling Methods

The most popular methods for measuring microhabitat originated with a protocol created by James and Shugart (1970). They developed a quantitative method of obtaining vegetation data "in a simple and regular manner." Their original intent was to provide a method that could simply augment the data on bird populations being gathered in the National Audubon Society's breeding-bird censuses and the winter bird population studies conducted throughout the United States. Nonetheless, as noted earlier, their strategy has had an extremely wide applicability throughout the entire ecological community. As we detail later on in this chapter, these authors started by evaluating the relative efficiencies of various sampling methods. They gathered data on the density, basal area, and frequency of tree species, and on canopy height, shrub density, percentage of ground cover, and percentage of canopy cover. They established 0.04 ha (0.1-acre) plots to estimate tree density and frequency. To estimate shrub density, they made 2 transects at right angles to each another across the 0.1-acre plots, counting the number of woody stems that were intercepted by their outstretched arms. An ocular tube was used to estimate vegetation cover. In addition, they provided details on how the sampling equipment could be constructed and gave examples of data sheets.

The authors also compared 4 of the standard methodologies recommended by plant ecologists for making quantitative estimates of vegetation. These were the plotless methods, such as the quarter method (Cottam and Curtis 1956; Phillips 1959) and wandering quarter method (Catana 1963); and areal approaches, such as arm-length transects (Rice and Penfound 1955; Penfound and Rice 1957) and circular plots (Lindsey et al. 1958). They then compared the averages from work accomplished using these 4 sampling methods in 30 minutes of field effort by 1 observer, assuming he or she was familiar with the technique and the species of plants in the study area (see Table 2.5). (Obviously, this will vary with the terrain, the density of the vegetation, and the actual number of observers involved.) The authors found that results from the 2 plotless methods (quarter and wandering quarter) tended to overestimate the total tree density and underestimate the tree density by species, but those from the 2 areal methods they used—1/100th-acre rectangles and 1/10-acre

Box 2.1

Ecologists have a tendency to want to collect everything possible when in the field. After all, since we are out there, should we not spend our time fruitfully? The environment is complex, with many interacting factors. As such, does it not follow that we are setting ourselves up for failure by restricting the types of data collected when in the field? Once that sampling period is over—be it in the summer of 1996 or the winter of 1996–1997—it will never occur again.

Although these sentiments are understandable on the surface, they fail when put to the test by both logic and reality. First, it is not logical to try and collect everything. Many variables are intercorrelated, essentially measuring the same phenomenon. Second, thousands of wildlife-habitat studies have been conducted. Might we not expect that some furthering of knowledge has already been gained? Let us not try to reinvent the wheel with every study. From a realistic standpoint, time is simply not available to collect an adequate number of samples on every variable you think might be important.

Although the preceding comments seem trivial on the surface, both new graduate students and seasoned professionals fall into the "collect everything" trap. You need only note that most multivariate papers spend some time in their "Methods" sections explaining how the dataset was reduced. Principal components analysis and methods using the stepwise inclusion of variables are 2 examples. Most multivariate models include only a few independent variables in the final model, but the majority of multivariate studies of wildlife-habitat relationships collect 20 or 30—or even over 100—variables.

Another issue concerns the intensity of sampling, conducted both in terms of the numbers of samples and the intensity with which each sampling method is applied. That is, if you are using a point intercept, should sampling points be 1 m or 2 m apart along the line? If you are using a circular plot to record tree density, should the plot be 5 m or 10 m in radius? There are few studies available to help guide such decisions. Most "standards" became such through repeated use of a convenient metric (e.g., 0.1-acre plots).

As we discuss in this chapter, preliminary data collection and analysis can prevent wasted effort. Some specific steps in such a design are provided below, and they can easily be generalized to most situations.

Step 1. Conduct a thorough literature review of similar studies.

- —Do not restrict yourself to your own taxonomic family. For example, studies on mammals in oak (*Quercus*) woodlands provide insight into environmental conditions that could affect herpetofauna or birds in oaks.
- —Make a list of both the variables collected and those that were found to be important predictors.
- —Make note of the methods used.

Step 2. Conduct preliminary sampling.

- —Sample as intensively as could be necessary (according to the literature), and sample as many variables as the literature indicates might be important (you should have a manageable list of variables, say 10–20). If you are not sure if the plot should have a 10 m or 15 m radius, record your data so they can later be analyzed by both radii.
- —Run a correlation analysis to determine the redundancy that exists among your independent variables.
- —Analyze the preliminary data to determine what independent variables are the best predictors of your dependent variable.
- —Make sure to include sample size analyzes.

(Preliminary sampling can be done during the first part of your first field season. If you are conducting a study during different seasons, then this procedure should be repeated at the start of each different season.)

Step 3. Conduct the study.

By following these or similar procedures, your field time can be used to collect a larger sample size, or include another sex or age group or even species, rather than gather data that are likely to be excluded from a later analysis.

Table 2.5. Average recording results accomplished in 30 minutes of field effort

Sampling method	Number of units	Number of trees identified and measured
Quarter method	12 quarters	48
Wandering quarter method	40 trees	40
1/10-acre circles	2 circles	57
1/100-acre rectangles	6 rectangles	19

Source: James and Shugart (1970:Table 1)

Note: The work involved recording the species and diameters of trees in an Ozark Mountain upland forest in Arkansas

circles—gave fairly accurate estimates of total density and density by species.

Circular plots are easy to establish, mark, measure, and relocate, and estimates of animal numbers within such plots can be statistically related to vegetation data in a straightforward manner. Such sites allow vegetation and animals to be sampled at specific locations in space and time. Thus plots are also easy to pinpoint using GPS, and data from them can subsequently be input into GIS programs. If plots can be considered independent data points (a function of the sampling design and the behavior of the animals), then the sample size is equal to the number of sites sampled. Or, if the plots are used to sample from a single study area, then the sampled sites can be averaged and the associated measures of variance calculated. Noon (1981) presented a useful description and examples of transect and areal plot-sampling systems. The problem with transects is that they cover relatively large areas, thus making it difficult to relate particular animal observations (or abundances) to specific sections of the transect. Transects are, however, widely used to provide an overall description of the vegetation over an entire study areas.

In summary to this point, fixed-area plots and transects can provide site-specific, detailed analyses of wildlife-habitat relationships. The majority of sampling methods employed since the 1970s to develop such relationships—thus supplying data for subsequent multivariate analyses—have used fixed-area (usually circular) plots as the basis for creating a sampling scheme that could then incorporate subplots, sampling squares, and transects.

Next, we describe examples of some of the more widely used methods. Dueser and Shugart (1978) developed a detailed sampling scheme that combined plots of various sizes and shapes, as well as short transects (see Fig. 2.11). Although designed for analyses of habitats for small mammals, these techniques can easily be adapted for most terrestrial vertebrates. The authors established 3 independent sampling units, centered on each trap: a 1.0 m^2 ring, 2 perpendicular 20 m^2 arm-length transects, and a 10 m radius circular plot. The 1.0 m^2 circular plot provided a measure of the vertical foliage profile (from the ground through 2 m in height) for both herbaceous and woody vegetation. Also, 4 replicate core-sample estimates of litter-soil depth, compactability, and dry-weight density were made on the perimeter of this central ring. The 2 arm-length transects supplied measurements of the cover type, surface characteristics, density, and evergreenness of the 4 vegetation strata. Data recorded for each quarter of the 10 m radius plot included the species, diameter at breast height (dbh), distance from the trap to the nearest understory and overstory trees, numbers of stumps and fallen logs, the basal diameter of and distance to the nearest stump and closest fallen log, and the total length of the fallen logs.

We consider the study by Dueser and Shugart (1978) to be an excellent starting point for any investigation being established in forested areas, and their basic sampling strategy could also be adapted for other vegetation types. Although the variables they measured (see Table 2.2) all relate to microhabitat, some involve the specific trap location (e.g., litter samples, soil compactability), whereas others apply more to the conditions surrounding the trap (e.g., data collected along the arm-length transects, such as the distance to trees). Thus, to a minor degree, the authors were mixing scales of measurement within the general construct of measuring microhabitat. As discussed previously, microhabitat itself is a general term that should be clearly defined for each research application. In their publications, investigators should explain how each variable fits within the overall concept of spatial scale, and how it meets the study's objectives and pertains to the specific target species in under investigation.

We have previously discussed the importance of James's (1971) paper to our conceptualization of how animals perceive their environment—the niche gestalt. She felt that the size of the plot used in her investigations would give an adequate description of the vegetation within an individual bird's territory. We will see below that she actually tested this assumption by comparing several methods (i.e., preliminary sampling). She also acknowledged a potential bias in concentrating on song perches and assumed that song-perch habitat reflected the fuller array of habitat elements for each bird species. As we have mentioned earlier in this chapter when discussing studies by Collins (1981) and Petit et al. (1988), this assumption may not hold. Despite the problems inherent in using only 1 behavior as the basis for habitat evaluation, the methods used by James (1971) have had a positive and pronounced influence on most of the analyses of wildlife habitat that followed.

In his research on snake populations, Reinert (1984) adopted techniques similar to those used in the bird study by James (1971) and research on small mammals by Dueser and Shugart (1978). He applied the basic conceptual framework employed by these earlier authors—the niche gestalt and multivariate representation of the niche—in developing the rationale for his procedures. Here again, we see the commonality in methods running across studies of wildlife-habitat relationships. Reinert (1984), however, made several modifications in the sampling methods used by James (1971) and Dueser and Shugart (1978). Notably, he used a 35 mm camera, equipped with a 28 mm wide-angle lens, to photograph 1 m^2 plots from directly above the location of a snake. He then determined the various surface cover percentages by superimposing each slide onto a 10 × 10 square grid. This author, then, more rigorously quantified his measure of cover values than most researchers, who usually use ocular estimates. As discussed in the following chapter, considerable error can be entered into a dataset when ocular estimates are used to measure plant cover (Block et al. 1987). The specific sampling scheme used by Reinert (1984) is summarized in Figure 2.12, and his vari-

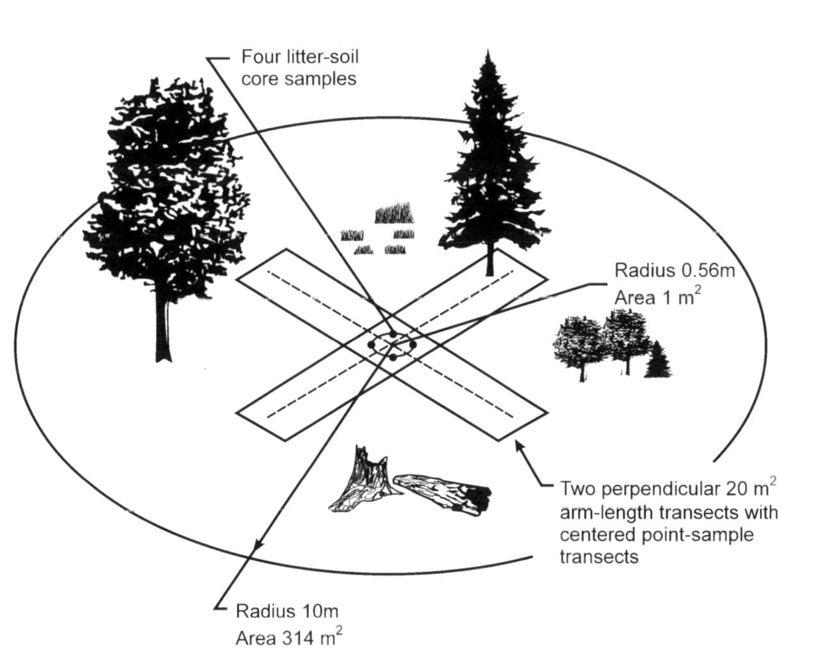

Figure 2.12. A drawing of a sampling arrangement for snake locations. (Reproduced from Reinert (1984:Fig. 1), with the permission of the Ecological Society of America)

able list is shown in Table 2.3. Note the similarity between his design and that of Dueser and Shugart (1978), including the minor mixing of spatial scales. Reinert (1984) added several environmental variables that addressed air, surface, and soil temperature and humidity. The values these variables take are obviously dependent on the time of day and the general weather conditions when the measurements were made. But such constraints do not influence (i.e., are not correlated with) the other variables that are measured. In addition, temperature data have a much different statistical distribution than most vegetation variables. Such mixing of variable types in the same analysis must be undertaken with caution and should probably be avoided (unless partitioned out in a step-wise analysis).

While variable distance sampling is not commonly used in wildlife studies, in forested areas it provides an accurate and rapid way to measure trees. These techniques, which preferentially sample larger trees, have excellent sample size attributes in a distribution in which large trees are widely spaced and small trees are more closely spaced, which is common. Importantly, the methodologies use optical devices to determine which trees are sampled and, therefore, do not require a formal plot layout. There are 2 formats: point based, which, from a sampling standpoint, is equivalent to an infinite number of nested circular plots; and transect based, which is equivalent to an infinite number of nested, long, rectangular plots. Here we will discuss variable transect sampling, because it has greater utility for sampling large, rare trees, and these can be very important for wildlife. In variable transect sampling, the transect width is set by the diameter of the tree and the "factor," which is a triangular wedge extending from the transect center outward. If the tree is larger or equal to the wedge, it is included in the sample. Thus very big trees can be a long distance from the transect center and still be part of the sample. Because a formal transect does not need to be laid out, it is easy to make long transects and, for large trees, create huge, precisely delineated plots. For sampling elements like big snags, variable transect methods allow a fast and accurate way to collect a meaningful sample. (See Husch et al. 2002 for a complete description of variable plot sampling.)

LiDAR

Over the past couple of decades, the development of LiDAR has provided new opportunities to quantify animal habitat without needing to measure plots on the ground. Basically, LiDAR is a remote-sensing technology where lasers emit beams of coherent light that are shot from an aircraft or spacecraft to the earth's surface. These rays intercept trees, branches, leaves, shrubs, rocks, and the ground, and they are reflected back to their point of origin (Dubayah and Drake 2000). The time elapsed between when a laser beam is initiated and when it comes back to its point of origin is a measure of the distance traveled. This technology enables a characterization of topography and of forest cover and layering—overstory trees, understory trees, and shrubs—that is far more detailed than is possible with traditional remote-sensing tools.

The use of LiDAR in studies of animal ecology is gaining traction quickly. When you consider that LiDAR provides data in 3 dimensions, and that all species occupy 3-dimensional space, the connection to animal ecology is readily apparent (Davies and Asner 2014). Moreover, LiDAR can supply data at a hierarchy of spatial scales, from a tree branch to entire landscapes. Much of the early research regarding this methodology focused on validating the accuracy of LiDAR, which resulted in a consensus that LiDAR is both accurate and is a valuable tool for advancing animal ecology (Davies and Asner 2014).

Many animal ecology studies that used LiDAR were reviewed by Davies and Asner (2014). They broke these into 3 broad taxonomic groups: flying vertebrates (birds and bats), non-flying mammals, and invertebrates. They then related the use of LiDAR to various environmental attributes (Table 2.6, although invertebrates are not included), which

Table 2.6. Published LiDAR-based vegetation and topographic structure studies in relation to animal ecology

Taxonomic group	Structural attribute	Response
Birds and bats	Vegetation canopy heterogeneity	22 of 44 species had increased occupancy with increasing heterogeneity 2 of 44 species had decreased occupancy Species richness increased Bat activity and occurrence increased
	Canopy vertical distribution	2 of 2 species had increased abundance and occupancy with increasing vertical distribution Species diversity increased
	Canopy height	Chick mass increased in blue tits, decreased in great tits, was climate dependent for great tit chick mass Native to exotic species ratio increased with increasing height Forest species richness increased Shrub species richness decreased 21 of 49 species had increased abundance or occupancy with increasing height 9 of 49 species had decreased abundance Species diversity increased Bat activity and occurrence increased
	Canopy cover	Ratio of native to exotic species increased with increasing cover Species diversity increased 11 of 23 species had increased abundance and occurrence with increased cover 6 of 23 species had decreased abundance
	Understory density	Species diversity increased with increasing density 12 of 34 species had increased abundance or occupancy with increasing understory density 7 of 37 species had decreased abundance with increasing understory density Foraging bat abundance decreased with increasing density
	Horizontal structure	2 of 2 species preferred intermediate or mixed levels of horizontal structure Species richness increased with increasing patch diversity
	Contiguous forest	Ratio of native to exotic species increased with larger forest patches 1 of 1 species preferred larger forest patches
	Elevation	Species richness increased with increasing elevation
	Slope	Species richness decreased with increasing steepness
Nonflying mammals	Canopy vertical distribution	2 of 3 species preferred increased vertical distribution
	Canopy height	3 of 5 species preferred increased height Moose used increased height during higher temperatures
	Canopy cover	3 of 4 species preferred increased cover Roe deer preferred increased cover in cold weather, but not when foraging
	Understory density	2 of 2 species preferred increased density when hunting 3 of 4 species species preferred increased density when foraging Roe deer avoided the understory when resting
	Contiguous forest	1 of 1 species preferred large forest patches
	Elevation	Mule deer preferred higher elevations in the winter
	Aspect	Mule deer preferred warmer slopes in the winter
	Ruggedness	2 of 2 species preferred increased slopes for hunting

Source: Modified from Davies and Ashner (2014). See their publication for a list of studies referenced.

crossed spatial scales, ranging from site to landscape, and included both vegetative and topographic measures. The variety of attributes that were considered attests to the flexibility of LiDAR for applications in wildlife research. Most bird and bat studies found a positive relationship between bird species' richness and the structural diversity of the associated vegetation. Canopy cover, understory density, and horizontal patchiness all emerged as variables correlated with bird occurrence. Fewer investigations of non-flying mammals considered 3-dimensional vegetation structure, but 2 of them—a study by Zhao et al. (2012) on Pacific fishers (*Martes pennanti*) and another by Coops et al. (2010) on mule deer (*Odocoileus hemionus*)—showed a positive relationship with vertical structure. Topography appeared frequently in non-flying mammal studies, with many of them evaluating predator-prey interactions. Much of this research concluded that topographic ruggedness afforded refugia for prey species, enabling coexistence with their predators.

LiDAR results in a voluminous amount of data. This, coupled with a limited amount of available software, may result in prolonged processing time to analyze the data (Tattoni et al. 2012). The cost of LiDAR, however, was comparable to that of conducting traditional forest stand inventories (Hummell et al. 2011). The advantages are that LiDAR is able to sample a far greater area in less time than other methods, and the resulting data can be used for multiple purposes. LiDAR represents a useful and flexible source of data for addressing objectives related to animal ecology, both now and into the future.

Summary

Clearly, a modern-day wildlife researcher has a vast and rich array of methods and tools from which to draw when crafting field studies of wildlife habitat and animal ecology. The most important first step is to clearly and fully articulate the study's objectives, which, in turn, should define the scale and scope of the space and time over which the research needs to be conducted, and from which the results will be interpreted and used. The next step is the choice of study methods, including the possibility of initially conducting a pilot study to best determine the sample size and decide which variables to measure. Let the occurrence, abundance, vital rates, and behavior of the animals under investigation define their habitat. Successful research on wildlife-habitat relationships, and in animal ecology in general, depends on not forgetting the organism-centric definitions of habitat, resource, and landscape.

LITERATURE CITED

Abadi, F., O. Gimenez, R. Arlettaz, and M. Schaub. 2010. An assessment of integrated population models: Bias, accuracy, and violation of the assumption of independence. Ecology 91:7–14.

Adriaensen, F., J. P. Chardon, G. de Blust, E. Seinnen, S. Villalba, H. Gulinck, and E. Matthysen. 2003. The application of "least-cost" modelling as a functional landscape model. Landscape and Urban Planning 64:233–247.

Aebischer, N. J., P. A. Robertson, and R. E. Kenward. 1993. Compositional analysis of habitat use from animal radio-tracking data. Ecology 74:1313–1325.

Alldredge, R. J., and J. T. Ratti. 1986. Comparison of some statistical techniques for analysis of resource selection. Journal of Wildlife Management 50:157–165.

Alldredge, R. J., and J. T. Ratti. 1992. Further comparison of some statistical techniques for analysis of resource selection. Journal of Wildlife Management 56:1–9.

Allouche, O., M. Kalyuzhny, G. Moreno-Rudueda, M. Pizarro, and R. Kadmon. 2012. Area-heterogeneity tradeoff and the diversity of ecological communities. Proceedings of the National Academy of Sciences 109:17495–17500.

Almeida, D. R. A., S. C. Stark, R. Chazdon, B. W. Nelson, R. G. Cesar, P. Meli, E. B. Gorgens, M. M. Duarte, R. Valbuena, V. S. Moreno, A. F. Mendes, N. Amazonas, N. B. Gonçalves, C. A. Silva, J. Schietti, and P. H. S. Brancalion. 2019. The effectiveness of LiDAR remote sensing for monitoring forest cover attributes and landscape restoration. Forest Ecology and Management 438:34–43.

Amstrup, S. C., T. L. McDonald, and B. F. J. Manly, eds. 2005. Handbook of capture-recapture analysis. Princeton University Press, Princeton, NJ.

Anderson, D. R. 2001. The need to get the basics right in wildlife field studies. Wildlife Society Bulletin 29:1294–1297.

Anderson, D. R., K. P. Burnham, G. C. White, and D. L.

Otis. 1983. Density estimation of small-mammal populations using a trapping web and distance sampling methods. Ecology 64:674–680.

Anderson, S. H., and H. H. Shugart Jr. 1974. Habitat selection of breeding birds in an east Tennessee deciduous forest. Ecology 55:828–837.

Anthony, R. G., E. D. Forsman, A. B. Franklin, D. R. Anderson, K. P. Burnham, G. C. White, C. J. Schwarz, J. D. Nichols, J. E. Hines, G. S. Olson, S. H. Ackers, L. S. Andrews, B. L. Biswell, P. C. Carlson, L. V. Diller, K. M. Dugger, K. E. Fehring, T. L. Fleming, R. P. Gerhardt, S. A. Gremel, R. J. Gutiérrez, P. J. Happe, D. R. Herter, J. M. Higley, R. B. Horn, L. L. Irwin, P. J. Loschl, J. A. Reid, and S. G. Sovern. 2006. Status and trends in demography of northern spotted owls, 1985–2003. Wildlife Monographs 163:1–48.

Arnason, A. N. 1972. Parameter estimates from mark-recapture experiments on two populations subject to migration and death. Researches on Population Ecology 13:97–113.

Arnason, A. N. 1973. The estimation of population size, migration rates, and survival in a stratified population. Researches on Population Ecology 15:1–8.

Arnqvist, G., and D. Wooster. 1995. Meta-analysis: Synthesizing research findings in ecology and evolution. Trends in Ecology & Evolution 10(6):236–240.

Arthur, S. M., B. F. J. Manly, L. L. McDonald, and G. W. Garner. 1996. Assessing habitat selection when availability changes. Ecology 77:215–227.

Bailey, R. G. 2009. Ecosystem geography: From ecoregions to sites, 2nd edition. Springer, New York.

Bailey, R. G., R. D. Pfister, and J. A. Henderson. 1978. Nature of land and resource classification—a review. Journal of Forestry, 76:650–655.

Barker, R. J. 1997. Joint modeling of live-recapture, tag-resight, and tag-recovery data. Biometrics 53:666–677.

Battin, J. 2004. When good animals love bad habitats: Ecological traps and the conservation of animal populations. Conservation Biology 18:1482–1491.

Baumgardt, J. A., M. L. Morrison, L. A. Brennan, B. L. Pierce, and T. A. Campbell. 2019. Development of multispecies, long-term monitoring programs for resource management. Rangeland Ecology and Management 72:168–181.

Besbeas, P., S. N. Freeman, B. J. T. Morgan, and E. A. Catchpole. 2002. Integrating mark-recapture-recovery and census data to estimate animal abundance and demographic parameters. Biometrics 58:540–547.

Bingham, R. L., and L. A. Brennan. 2004. Comparison of Type I error rates for statistical analyses of resource selection. Journal of Wildlife Management 68:206–212.

Bissonette, J. A. 1997. Scale-sensitive ecological properties: Historical context, current meaning. Pp. 3–31 in J. A. Bissonette, ed. Wildlife and landscape ecology: Effects of pattern and scale. Springer, New York.

Bissonette, J. A., and I. Storch, eds. 2003. Landscape ecology and resource management: Linking theory with practice. Island Press, Washington, DC.

Block, W. M., and L. A. Brennan. 1993. The habitat concept in ornithology: Theory and application. Current Ornithology 11:35–91.

Block, W. M. and D. M. Finch. 1997. Songbird ecology in ponderosa pine forests: A literature review. General Technical Report RM-292. US Department of Agriculture, Forest Service, Rocky Mountain Forest and Range Experiment Station, Fort Collins, CO.

Block, W. M., J. L. Ganey, P. E. Scott, and R. King. 2005. Prey ecology of the Mexican spotted owl in pine-oak forests of northern Arizona. Journal of Wildlife Management. 69:618–629.

Block, W. M., and M. L. Morrison. 1998. Habitat relationships of amphibians and reptiles in California oak woodlands. Journal of Herpetology 32:51–60.

Block, W. M., M. L. Morrison, J. Verner, and P. N. Manley. 1994. Assessing wildlife-habitat relationships models: A case study with California oak woodlands. Wildlife Society Bulletin 22:549–561.

Block, W. M., K. A. With, and M. L. Morrison. 1987. On measuring bird habitat: Influence of observer variability and sample size. Condor 72:182–189.

Bonham, C. D. 1989. Measurements for terrestrial vegetation. John Wiley & Sons, New York.

Bookhout, T. A., ed. 1994. Research and management techniques for wildlife and habitats, 5th edition. The Wildlife Society, Bethesda, MD.

Bowden, D. C., G. C. White, A. B. Franklin, and J. L. Ganey. 2003. Estimating population size with correlated sampling unit estimates. Journal of Wildlife Management 67:1–10.

Boyce, M. S., C. J. Johnson. E. H. Merrill, S. E. Nielson, E. J. Solberg, and B. van Moorter. 2015. Review: Can habitat selection predict abundance? Journal of Animal Ecology 85:11–20.

Brennan, L. A., and W. M. Block. 2018. Population ecology. Pp. 685–710 *in* M. L. Morrison, A. D. Rodewald, G. Voelker, M. R. Colon, and J. F. Prather, eds. Ornithology: Foundation, analysis, and application. Johns Hopkins University Press, Baltimore.

Brennan, L. A., W. M. Block, and R. J. Gutiérrez. 1987. Habitat use by mountain quail in northern California. Condor 89:66–74.

Brennan, L. A., A. N. Tri, and B. G. Marcot. 2019. Quantitative analyses in wildlife science. Johns Hopkins University Press, Baltimore.

Briner, T., J.-P. Airoldi, F. Dellsperger, S. Eggimann, and W. Nentwig. 2003. A new system for automatic radio-tracking of small mammals. Journal of Mammalogy 84:571–578.

Brownie, C., D. R. Anderson, K. P. Burnham, and D. R. Robson. 1978. Statistical inference from band recovery data: A handbook. Resource Publication No. 156. US Department of the Interior, Fish and Wildlife Service, Washington, DC.

Brownie, C., J. E. Hines, J. D. Nichols, K. H. Pollock, and J. B. Hestbeck. 1993. Capture-recapture studies for multiple strata including non-Markovian transition probabilities. Biometrics 49:1173–1187.

Buckland, S. T., D. R. Anderson, K. P. Burnham, J. Laake, D. L. Borchers, and L. Thomas. 2001. Introduction to distance sampling: Estimating abundance of biological populations. Oxford University Press, Oxford.

Buckland, S. T., D. L. Miller, and E. A. Rexstad. 2019. Distance sampling. Pp. 79–98 *in* L. A. Brennan, A. N. Tri, and B. G. Marcot, eds. Quantitative analyses in wildlife science. Johns Hopkins University Press, Baltimore.

Buckland, S. T., E. A. Rexstad, R. A. Marques, and C. S. Oedekoven. 2015. Distance sampling: Methods and applications. Springer International, Cham, Switzerland.

Bucklin, D. N., M. Basille, A. M. Benscoter, L. A. Brandt, F. J. Mazzotti, S. S. Romanach, C. Speroterra, and J. I. Watling. 2015. Comparing species distribution models constructed with different subsets of environmental predictors. Diversity and Distributions 21:23–35.

Burnham, K. P., D. R. Anderson, and J. L. Laake. 1980. Estimation of density from line transect sampling of biological populations. Wildlife Monographs 72:3–202.

Campbell, G. S., and J. M. Norman. 1989. The description and measurement of plant canopy structure. Pp. 1–19 *in* G. Russell, B. Marshall, and P. G. Jarvis, eds. Plant canopies: Their growth, form and function. Cambridge University Press, Cambridge.

Capen, D. E., ed. 1981. The use of multivariate statistics in studies of wildlife habitat. General Technical Report RM-87. US Department of Agriculture, Forest Service, Rocky Mountain Forest and Range Experiment Station, Fort Collins, CO.

Carbon, B. A., G. A. Bartle, and A. M. Murray. 1979. A method for visual estimation of leaf area. Forest Science 25:53–58.

Caswell, H. 2001. Matrix population models: Construction, analysis, and interpretation, 2nd edition. Sinauer Associates, Sunderland, MA.

Catana, A. J., Jr. 1963. The wandering quarter method of estimating population density. Ecology 44:349–360.

Chandler, R. B., and J. A. Royle. 2013. Spatially explicit models for inference about density in unmarked or partially marked populations. Annals of Applied Statistics 7:936–954.

Chapman, D. G. 1951. Some properties of the hypergeometric distribution with application to zoological censuses. University of California Publications in Statistics 1(7):131–160.

Collins, S. L. 1981. A comparison of nest-site and perch-site vegetation structure for seven species of warblers. Wilson Bulletin 93:542–547.

Comer, P., D. Faber-Langendoen, R. Evans, S. Gawler, C. Josse, G. Kittel, S. Menard, M. Pyne, M. Reid, K. Schulz, K. Snow, and J. Teague. 2003. Ecological systems of the United States: A working classification of US terrestrial systems. NatureServe, Arlington, VA.

Cook, C. W., and J. Stubbendieck, eds. 1986. Range research: Basic problems and techniques. Society for Range Management, Denver.

Cooke, B. J., and A. L. Carroll. 2017. Predicting the risk of mountain pine beetle spread to eastern pine forests: Considering uncertainty in uncertain times. Forest Ecology and Management 396:11–25.

Cooperrider, A. Y., R. J. Boyd, and H. R. Stuart, eds. 1986. Inventory and monitoring of wildlife habitat. US Department of the Interior, Bureau of Land Management Service Center, Denver.

Coops, N. C., J. Duffe, and C. Koot. 2010. Assessing the utility of lidar remote sensing technology to identify mule deer winter habitat. Canadian Journal of Remote Sensing, 36:81–88.

Copeland J. P., K. S. McKelvey, K. B. Aubry, A. Landa, J. Persson, R. M. Inman, J. Krebs, E. Lofroth, H. Golden, J. R. Squires, A. Magoun, M. K. Schwartz, J. Wilmot, C. L. Copeland, R. E. Yates, I. Kojola, and R. May. 2010. The bioclimatic envelope of the wolverine (*Gulo gulo*): Do climatic constraints limit its geographic distribution? Canadian Journal of Zoology. 88:233–246.

Cormack, R. M. 1964. Estimates of survival from the sighting of marked animals. Biometrika 51:429–438.

Cottam, G., and J. T. Curtis. 1956. The use of distance measures in phytosociological sampling. Ecology 37:451–460.

Cushman, S. A., and F. Huettmann, eds. 2010. Spatial complexity, informatics, and wildlife conservation. Springer, New York.

Cushman, S. A., K. McGarigal, K. S. McKelvey, C. D. Vojta, and C. M. Regan. 2013. Landscape analysis for habitat monitoring. Chapter 5 *in* M. M. Rowland and C. D. Vojta, technical eds. A technical guide for habitat monitoring. General Technical Report WO-89. US Department of Agriculture, Forest Service, Washington Office, Washington, DC.

Dalke, P. M. 1937. Food habits of adult pheasants in Michigan based on crop analysis method. Ecology 18:199–213.

Daubenmire, R. 1959. Canopy coverage method of vegetation analysis. Northwest Science 33:39–64.

Davies, A. B., and G. P. Asner. 2014. Advances in animal ecology from 3D-LiDAR in ecosystem mapping. Trends in Ecology & Evolution 29:681–691.

DeWan, A. A., and E. F. Zipkin. 2010. An integrated sampling and analysis approach for improved biodiversity monitoring. Environmental Management 45:1223–1230.

Dorazio, R. M., J. A. Royle, B. Söderström, and A. Glimskär. 2006. Estimating species richness and accumulation by modeling species occurrence and detectability. Ecology 87:842–854.

Dubayah, R. O., and J. B. Drake. 2000. Lidar remote sensing for forestry. Journal of Forestry 46(6):44–46.

Dueser, R. D., and H. H. Shugart Jr. 1978. Microhabitats in a forest-floor small mammal fauna. Ecology 59:89–98.

Dugger, K. M., E. D. Forsman, A. B. Franklin, R. J. Davis, G. C. White, C. J. Schwarz, K. P. Burnham, J. D. Nichols, J. E. Hines, C. B. Yackulic, P. F. Doherty Jr., L. Bailey, D. A. Clark, S. H. Ackers, L. S. Andrews, B. Augustine, B. L. Biswell, J. Blakesley, P. C. Carlson, M. J. Clement, L. V. Diller, E. M. Glenn, A. Green, S. A. Gremel, D. R. Herter, J. M. Higley, J. Hobson, R. B. Horn, K. P. Huyvaert, C. McCafferty, T. McDonald, K. McDonnell, G. S. Olson, J. A. Reid, J. Rockweit, V. Ruiz, J. Saenz, and S. G. Sovern. 2016. The effects of habitat, climate, and barred owls on long-term demography of northern spotted owls. Condor 118:57–116.

Dunning, J. B., B. J. Danielson, and H. R. Pulliam. 1992. Ecological processes that affect populations in complex landscapes. Oikos 65:169–75.

Dupont, P., C. Milleret, O. Gimenez, and R. Bischof. 2019. Population closure and the bias-precision trade-off in spatial capture-recapture. Methods in Ecology and Evolution 10:661–672.

Efford, M. 2004. Density estimation in live-trapping studies. Oikos 106:598–610.

Efford, M. G., and D. K. Dawson. 2012. Occupancy in continuous habitat. Ecosphere 3(4):1–15.

Elith, J., and J. R. Leathwick. 2009. Species distribution models: Ecological explanation and prediction across space and time. Annual Review of Ecology, Evolution, and Systematics 40:677–697.

Elzinga, C. L., D. W. Sazler, J. W. Willoughby, and J. P. Gibbs. 2001. Monitoring plant and animal populations. Blackwell, Malden, MA.

Errington, P. L. 1930. The pellet analysis method of raptor food habits study. Condor 32:292–296.

Eyre, F. H. 1980. Forest cover types of the United States and Canada. Society of American Foresters, Washington, DC.

Fedrowitz, K., J. Koricheva, S. C. Baker, D. B. Lindenmayer, B. Palik, R. Rosenvald, W. Beese, J. F. Franklin, J. Kouki, E. Macdonald, C. Messier, A. Sverdrup-Thygeson, and L. Gustafsson. 2014. Can retention forestry help conserve biodiversity? A meta-analysis. Journal of Applied Ecology 51:1669–1679.

Fiske, I., and R. Chandler. 2011. **Unmarked**: An R package for fitting hierarchical models of wildlife occurrence and abundance. Journal of Statistical Software 43:1–23.

Flanders, A., W. P. Kuvlesky Jr., D. C. Ruthven III, R. E. Zaiglin, R. L. Bingham, T. Fulbright, F. Hernández, and L. A. Brennan. 2006. Impacts of invasive exotic grasses on South Texas breeding bird communities. Auk 123:171–182.

Forsman, E. D., E. C. Meslow, and H. M. Wight. 1984. Distribution and biology of the spotted owl in Oregon. Wildlife Monographs 87:3–64.

Franklin, A. F., R. J. Gutiérrez, J. D. Nichols, M. E. Seamans, G. C. White, G. S. Zimmerman, J. E. Hines, T. E. Munton, W. D. LaHaye, J. A. Blakesley, G. N. Steger, B. R. Noon, D. W. H. Shaw, J. J. Keane, T. L. McDonald, and S. Bitting. 2004. Population dynamics of the California spotted owl (*Strix occidentalis occidentalis*): A meta-analysis. Ornithological Monographs 54:1–54.

Ganey, J. L., and W. M. Block. 1994. A comparison of two techniques for measuring canopy closure. Western Journal of Applied Forestry 9:21–23.

Ganey, J. L., W. M. Block, and S. H. Ackers. 2003. Structural characteristics of forest stands within home ranges of Mexican spotted owls in Arizona and New Mexico. Western Journal of Applied Forestry 18(3):189–198.

Ganey, J. L., G. C. White, D. C. Bowden, and A. B. Franklin. 2004. Evaluating methods for monitoring populations of Mexican spotted owls: A case study. Pp. 337–385 *in* W. L. Thompson, ed. Sampling rare or elusive species: Concepts, designs, and techniques for estimating population parameters. Island Press, Washington, DC.

Garshelis, D. L. 2000. Delusions in habitat evaluation: Measuring use, selection, and importance. Pp. 111–164 *in* L. Boitana and T. K. Fuller, eds. Research techniques in animal ecology: Controversies and consequences. Columbia University Press, New York.

Golley, F. B. 1993. A history of the ecosystem concept in ecology: More than a sum of the parts. Yale University Press, New Haven, CT.

Gopalaswamy A.M., P. Singh, D. Jathanna, N. S. Kumar, K. U. Karanth, J. A. Royle, and J. E. Hines. 2012. Program SPACECAP: Software for estimating animal density using spatially explicit capture-recapture models. Methods in Ecology and Evolution 3:1067–1072.

Green, R. H. 1971. A multivariate statistical approach to

the Hutchinsonian niche: Bivalve molluscs of central Canada. Ecology 52:543–556.

Green, R. H. 1979. Sampling design and statistical methods for environmental biologists. John Wiley & Sons, New York.

Griffin, P. C., and T. J. Case. 2001. Terrestrial habitat preferences of adult arroyo southwestern toads. Journal of Wildlife Management 65:633–644.

Grinnell, J., and A. H. Miller. 1944. The distribution of the birds of California. Pacific Coast Avifauna 27. Cooper Ornithological Club, Berkeley, CA.

Guillot, G., F. Mortier, and A. Estoup. 2005. Geneland: A computer package for landscape genetics. Molecular Ecology Notes 5:708–711.

Guisan, A., and W. Thuiller. 2005. Predicting species distribution: Offering more than simple habitat models. Ecology Letters 8:993–1009.

Gundersen, G., E. Johannesen, H. P. Andreassen, and R. A. Ims. 2008. Source-sink dynamics: How sinks affect demography of sources. Ecology Letters 4:14–21.

Gurevitch, J., P. S. Curtis, and M. H. Jones. 2001. Meta-analysis in ecology. Advances in Ecological Research. 32:199–247.

Guthery, F. S. 1999. Slack in the configuration of habitat patches for northern bobwhites. Journal of Wildlife Management 63:245–250.

Guthery, F. S. 2008. A primer on natural resource science. Texas A&M University Press, College Station.

Guthery, F. S., and R. L. Bingham. 1992. On Leopold's principle of edge. Wildlife Society Bulletin 20:340–344.

Guthery, F. S., and J. J. Lusk. 2004. Radiotelemetry studies: Are we radio-handicapping northern bobwhites? Wildlife Society Bulletin 32:194–201.

Hale, R., M. A. Colton, P. Peng, and S. E. Swearer. 2019. Do spatial scale and life history affect fish-habitat relationships? Journal of Animal Ecology 88:439–449.

Hall, L. S., and M. L. Morrison. 1997. Den and relocation site characteristics and home ranges of *Peromyscus truei* in the White Mountains of California. Great Basin Naturalist 57:124–130.

Harris, L. D. 1984. The fragmented forest. University of Chicago Press, Chicago.

Hartley, P. H. T. 1948. The assessment of the food of birds. Ibis 90:361–381.

Hawley, J. E., P. W. Rego, A. P. Wydeven, M. K. Schwartz, T. C. Viner, R. Kays, K. L. Pilgrim, and J. A. Jenks. 2016. Long-distance dispersal of a subadult male cougar from South Dakota to Connecticut documented with DNA evidence. Journal of Mammalogy 97:1435–1440.

Hebblewhite, M., and D. T. Haydon. 2010. Distinguishing technology from biology: A critical review of the use of GPS telemetry data in ecology. Philosophical Transactions of the Royal Society B, Biological Sciences 365:2303–2312.

Hedrick, P. W. 2005. A standardized genetic differentiation measure. Evolution 59:1633–1638.

Herman, S. G. 2002. Wildlife biology and natural history: Time for a reunion? Journal of Wildlife Management 66:933–946.

Heyer, W. R., M. A. Donnelly, R. W. McDiarmid, L. C. Hayek, and M. S. Foster. 1994. Measuring and monitoring biological diversity: Standard methods for amphibians. Smithsonian Institution Press, Washington, DC.

Hildén, O. 1965. Habitat selection in birds. Annales Zoologici Fennici 2:53–75.

Hines, J. E. 2006. PRESENCE: Software to estimate patch occupancy and related parameters, version 13.3. US Geological Survey, Patuxent Wildlife Research Center, Laurel, MD. http://www.mbr-pwrc.usgs.gov/software/presence.html.

Hobson K. A., B. N. McLellan, and J. G. Woods. 2000. Using stable carbon ($\delta^{13}C$) and nitrogen ($\delta^{15}N$) isotopes to infer trophic relationships among black and grizzly bears in the Upper Columbia River Basin, British Columbia. Canadian Journal of Zoology 78:1332–1339.

Holmes, R. T. 1981. Theoretical aspects of habitat use by birds. Pp. 33–37 *in* D. E. Capen, ed. The use of multivariate statistics in studies of wildlife habitat. General Technical Report RM-87. US Department of Agriculture, Forest Service, Rocky Mountain Forest and Range Experiment Station, Fort Collins, CO.

Hummel, S., A. T. Hudak, E. H. Uebler, M. J. Falkowski, and K. A. Megown. 2011. A comparison of accuracy and cost of LiDAR versus stand exam data for landscape management on the Malheur National Forest. Journal of Forestry 109:267–273.

Husch, B., T. W. Beers, and J. A. Kershaw Jr. 2002. Forest mensuration. John Wiley & Sons, New York.

Huston, M. A. 2002. Introductory essay: Critical issues for improving predictions. Pp. 7–21 *in* J. M. Scott, P. J. Heglund, M. L. Morrison, J. B. Haufler, M. G. Raphael, W. A. Wall, and F. B. Samson, eds. Predicting species occurrences: Issues of accuracy and scale. Island Press, Washington, DC.

Hutto, R. L. 1985. Habitat selection by nonbreeding, migratory land birds. Pp. 455–476 *in* M. L. Cody, ed. Habitat selection in birds. Academic Press, New York.

International Bird Census Committee. 1969. Recommendations for an international standard for a mapping method in bird census work. Bird Study 16:249–255.

Iknayan, K. J., M. W. TIngley, B. J. Furnas, and S. R. Beissinger. 2014. Detecting diversity: Emerging methods to estimate species diversity. Trends in Ecology & Evolution 29:97–106.

Irwin, L. L., and J. G. Cook. 1985. Determining appropriate variables for a habitat suitability model for pronghorns. Wildlife Society Bulletin 13:434–440.

James, F. C. 1971. Ordinations of habitat relationships among breeding birds. Wilson Bulletin 83:215–236.

James, F. C., and H. H. Shugart Jr. 1970. A quantitative method of habitat description. Audubon Field Notes 24:727–736.

James, F. C., and N. D. Wamer. 1982. Relationships between temperate forest bird communities and vegetation structure. Ecology 63:159–171.

Jenkins, J. M. A., D. B. Lesmeister, and R. J. Davis. 2019. Resource selection analysis. Pp. 199–215 *in* L. A. Brennan, A. N. Tri, and B. G. Marcot, eds. Quantitative analyses in wildlife science. Johns Hopkins University Press, Baltimore.

Johnson, D. H. 1980. The comparison of usage and availability measurements for evaluating resource preference. Ecology 61:65–71.

Johnson, D. H. 1981. The use and misuse of statistics in wildlife habitat studies. Pp. 11–19 *in* D. E. Capen, ed. The use of multivariate statistics in studies of wildlife habitat. General Technical Report RM-87. US Department of Agriculture, Forest Service, Rocky Mountain Forest and Range Experiment Station, Fort Collins, CO.

Jolly, G. M. 1965. Explicit estimates from capture-recapture data with both death and immigration-stochastic model. Biometrika 52:225–247.

Jones, H. P., and S. W. Kress. 2012, A review of the world's active seabird restoration projects. Journal of Wildlife Management 76:2–9.

Kadmon, R., and O. Allouche. 2007. Integrating the effects of area, isolation, and habitat heterogeneity on species diversity: A unification of island biogeography and niche theory. American Naturalist 170:443–454.

Kaplan, E. L., and P. Meier. 1958. Nonparametric estimation from incomplete observations. Journal of the American Statistical Association 53:457–481.

Keller, J. K., and C. R. Smith. 2014. Improving GIS-based wildlife-habitat analysis. Springer, New York.

Kelt, D. A., P. L. Meserve, and B. K. Lang. 1994. Quantitative habitat associations of small mammals in a temperate rainforest in southern Chile: Empirical patterns and the importance of ecological scale. Journal of Mammalogy 75:890–904.

Kenward, R. E. 2000. A manual of wildlife radio tagging, 2nd edition. Academic Press, San Diego.

Kotilar, N. B., and J. A. Wiens. 1990. Multiple scales of patchiness and patch structure: A hierarchical framework for the study of heterogeneity. Oikos 59:253–260.

Krebs, C. J. 1978. Ecology: The experimental analysis of distribution and abundance. Harper & Row, New York.

Küchler, A. W. 1964. Potential natural vegetation of the coterminous United States. Special Publication No. 36. American Geographical Society, New York.

Leopold, A. 1933. Game management. C. Scribner's Sons, New York.

Leslie, P. H. 1945. On the use of matrices in certain population mathematics. Biometrika 33:183–212.

Levin, S. A. 1992. The problem of pattern and scale in ecology. Ecology 73:1943–1967.

Lindsey, A. A., J. D. Barton, and S. R. Miles. 1958. Field efficiencies of forest sampling methods. Ecology 39:428–444.

Litt, A. R., E. E. Cord, T. E. Fulbright, and G. L. Schuster. 2014. Effects of invasive plants on arthropods. Conservation Biology 28:1532–1549.

Lunn, D. J., A. Thomas, N. Best, and D. Spiegelhalter. 2000. WinBUGS—a Bayesian modeling framework: Concepts, structure, and extensibility. Statistics and Computing 10:325–337.

Lyet, A., W. Thuiller, M. Cheylan, and A. Besnard. 2013. Fine-scale regional distribution modelling of rare and threatened species: Bridging GIS tools and conservation in practice. Diversity and Distributions 19:651–663.

MacArthur, R. H., and J. W. MacArthur. 1961. On bird species diversity. Ecology 42:594–598.

MacArthur, R. H., and E. O. Wilson. 1967. The theory of island biogeography. Monographs in Population Biology 1. Princeton University Press, Princeton, NJ.

MacKenzie, D. I., J. D. Nichols, J. A. Royle, K. H. Pollock, L. L. Bailey, and J. E. Hines. 2017. Occupancy estimation and modeling. Academic Press, New York.

MacKenzie, D. R., and J. A. Royle. 2004. Occupancy as a surrogate for abundance estimation. Animal Biodiversity and Conservation 27:461–467.

MacKenzie, D. R., and J. A. Royle. 2005. Designing occupancy studies: General advice and allocating survey effort. Journal of Applied Ecology 42:1105–1114.

Magoun, A. J., and J. P. Copeland. 1998. Characteristics of wolverine reproductive dens. Journal of Wildlife Management 62:1313–1320.

Manly, B. F. J., L. L. McDonald, D. L. Thomas, T. L. McDonald, and W. P. Erickson. 2002. Resource selection by animals: Statistical design and analysis for field studies, 2nd edition. Kluwer Academic, Dordrecht, Netherlands.

Marcot, B. G., and J. Gardetto. 1980. Status of the spotted owl in Six Rivers National Forest, California. Western Birds 11:79–87.

Marshall, J. D., and R. H. Waring. 1986. Comparison of methods of estimating leaf-area index in old-growth Douglas-fir. Ecology 67:975–979.

Mawson, J. C., J. W. Thomas, and R. M. DeGraaf. 1976. Program HTVOL: The determination of tree crown volume by layers. Research Paper NE-354. US Department of Agriculture, Forest Service, Northeastern Forest Experiment Station, Upper Darby, PA.

Mayor, S. J., D. C. Schneider, J. A. Schaefer, and S. P. Mahoney. 2009. Habitat selection at multiple scales. Ecoscience 16:238–247.

McCullough, D. R., and D. H. Hirth. 1988. Evaluation of the Petersen-Lincoln estimator for a white-tailed deer population. Journal of Wildlife Management 52:534–544.

McGarigal, K., H. Y. Wan, K. A. Zeller, B. C. Timm, and S. A. Cushman. 2016. Multi-scale habitat selection modeling: A review and outlook. Landscape Ecology 31:1161–1175.

McKelvey, K. S., and D. E. Pearson. 2001. Population estimation with sparse data: The role of estimators versus indices revisited. Canadian Journal of Zoology 79:1754–1765.

Menkens, G. E., Jr., and S. H. Anderson. 1988. Estimation of small mammal population size. Ecology 69:1952–1959.

Michelot, T., P. G. Blackwell, and J. Matthiopoulos. 2019. Linking resource selection and step selection models for habitat preferences in animals. Ecology 100(1):e02452.

Miller, D. A. W., K. Pacifici, J. S. Sanderlin, and B. J. Reich. 2019. The recent past and promising future for data integration methods to estimate species' distributions. Methods in Ecology and Evolution 2019:22–37.

Millspaugh, J. J., and J. M. Marzluff, eds. 2001. Radio tracking and animal populations. Academic Press, San Diego.

Millspaugh, J. J., and F. R. Thompson III, eds. 2009. Models for planning wildlife conservation in large landscapes. Academic Press, San Diego.

Mitchell, M. S., J. W. Zimmerman, and R. A. Powell. 2002. Test of a habitat suitability index for black bears in the southern Appalachians. Wildlife Society Bulletin 30:794–808.

Moir, W. H., B. Geils, M. A. Benoit, and D. Scurlock. 1997. Ecology of southwestern ponderosa pine forests. Pp. 3–27 *in* W. M. Block and D. M. Finch, eds. Songbird ecology in southwestern ponderosa pine forests. General Technical Report RM-292. US Department of Agriculture, Forest Service, Rocky Mountain Forest and Range Experiment Station, Fort Collins, CO.

Morrison, M. L. 1984. Influence of sample size on discriminant function analysis of habitat use by birds. Journal of Field Ornithology 55:330–335.

Morrison, M. L. 2012. The habitat sampling and analysis paradigm has limited value in animal conservation: A prequel. Journal of Wildlife Management 76:1–13.

Morrison, M. L., and R. G. Anthony. 1989. Habitat use by small mammals on early-growth clear-cuttings in western Oregon. Canadian Journal of Zoology 67:805–811.

Morrison, M. L., W. M. Block, L. S. Hall, and H. S. Stone. 1995. Habitat characteristics and monitoring of amphibians and reptiles in the Huachuca Mountains, Arizona. Southwestern Naturalist 40:185–192.

Morrison, M. L., W. M. Block, M. D. Strickland, B. Collier, and M. J. Peterson. 2008. Wildlife study design, 2nd edition. Springer-Verlag, New York.

Morrison, M. L., L. A. Brennan, B. G. Marcot, W. M. Block, and K. S. McKelvey. 2020. Foundations for Advancing Animal Ecology. Johns Hopkins University Press, Baltimore.

Morrison, M. L., and E. C. Meslow. 1983. Bird community structure on early-growth clearcuts in western Oregon. American Midland Naturalist 110:129–137.

Morrison, M. L., C. J. Ralph, J. Verner, and J. R. Jehl Jr., eds. 1990. Avian foraging: Theory, methodology, and applications. Studies in Avian Biology 13. Cooper Ornithological Society, Los Angeles.

Morzillo, A. T., G. A. Feldhamer, and M. C. Nicholson. 2003. Home range and nest use of the golden mouse (*Ochrotomys nuttalli*) in southern Illinois. Journal of Mammalogy 84:553–560.

Mueller-Dombois, D., and H. Ellenberg. 1974. Aims and methods of vegetation ecology. John Wiley & Sons, New York.

Nichols, J. D. 2016. And the first one now will later be last: Time-reversal in Cormack-Jolly-Seber models. Statistical Science 31:175–190.

Nichols, J. D., J. E. Hines, J.-D. Lebreton, and R. Pradel. 2000. The relative contributions of demographic components to population growth: A direct estimation approach based on reverse-time capture-recapture. Ecology 81:3362–3376.

Noon, B. R. 1981. Techniques for sampling avian habitats. Pp. 42–53 *in* D. E. Capen, ed. The use of multivariate statistics in studies of wildlife habitat. General Technical Report RM-87. US Department of Agriculture, Forest Service, Rocky Mountain Forest and Range Experiment Station, Fort Collins, CO.

Noon, B. R., L. L. Bailey, T. D. Sisk, and K. S. McKelvey. 2012. Efficient species-level monitoring at the landscape scale. Conservation Biology 26:432–441.

Ockenfels, R. A., and D. E. Brooks. 1994. Summer diurnal bed sites of Coues white-tailed deer. Journal of Wildlife Management 58:70–75.

O'Connor, R. J. 2002. The conceptual basis of species distribution modeling: Time for a paradigm shift? Pp. 25–33 *in* J. M. Scott, P. J. Heglund, M. L. Morrison, J. B. Haufler, M. G. Raphael, W. A. Wall, and F. B. Samson, eds. Predicting species occurrences: Issues of accuracy and scale. Island Press, Washington, DC.

Otis, D. L., K. P. Burnham, G. C. White, and D. R. Anderson. 1978. Statistical inference from capture data on closed animal populations. Wildlife Monographs 62:3–135.

Otto, C. R., L. L. Bailey, and G. J. Roloff. 2013. Improving species occupancy estimation when sampling violates the closure assumption. Ecography 36:1299–1309.

Penfound, W. T., and E. L. Rice. 1957. An evaluation of the arms-length rectangle method in forest sampling. Ecology 38:660–661.

Petit, K. E., D. R. Petit, and L. J. Petit. 1988. On measuring vegetation characteristics in bird territories: Nest sites vs. perch sites and the effect of plot size. American Midland Naturalist 119:209–215.

Pfister, R. D., B. L. Kovalchik, S. F. Arno, and R. C. Presby. 1977. Forest habitat types of Montana. General Technical Report INT-34. US Department of Agriculture, Forest Service, Intermountain Forest and Range Experiment Station, Ogden, UT.

Phillips, E. A. 1959. Methods of vegetation study. Holt, Rinehart, & Winston, New York.

Phillips, S. J., M. Dudik, and R. E. Schapire. 2005. Maxent software for species distribution modeling. https://biodiversityinformatics.amnh.org/open_source/maxent/.

Pickett, S. T. A., J. Kolasa, and C. G. Jones. 1994. Ecological understanding. Academic Press, San Diego.

Pikesley, S. K., S. M. Maxwell, K. Pendoley, D. P. Costa, M. S. Coynel, A. Formia, B. J. Godley, W. Klein, J. Makanga-Bahouna, S. Maruca, S. Ngouessono, R. J. Parnell, E. Pemo-Makaya, and M. J. Witt. 2013. On the front line: Integrated habitat mapping for olive ridley sea turtles in the southeast Atlantic. Diversity and Distributions 19:1518–1530.

Plummer, M. 2003. JAGS: A program for analysis of Bayesian graphical models using Gibbs sampling. P. 125 *in* Proceedings of the 3rd International Workshop on Distributed Statistical Computing, vol. 124. Technische Universität Wien, Vienna, Austria.

Plummer, M., N. Best, K. Cowles, and K. Vines. 2006. CODA: Convergence diagnosis and output analysis for MCMC. R News 6:7–11.

Pollock, K. H. 1982. A capture-recapture design robust to unequal probability of capture. Journal of Wildlife Management 46:757–760.

Pollock, K. H., D. L. Solomon, and D. S. Robson. 1974. Tests for mortality and recruitment in a *K*-sample tag-recapture experiment. Biometrics 40:329–340.

Pollock, K. H., S. R. Winterstein, C. M. Bunck, and P. D. Curtis. 1989a. Survival analysis in telemetry studies: The staggered entry design. Journal of Wildlife Management 53:7–15.

Pollock, K. H., S. R. Winterstein, and M. J. Conroy. 1989b. Estimation and analysis of survival distributions for radio-tagged animals. Biometrics 45:99–109.

Pritchard, J. K., M. Stephens, and P. Donnelly. 2000. Inference of population structure using multilocus genotype data. Genetics 155:945–959.

Pyne, S. J. 2015. Between two fires. University of Arizona Press, Tucson.

Ralph, C. J., G. R. Geupel, P. Pyle, T. E. Martin, and D. F. DeSante. 1993. Handbook for field methods for monitoring landbirds. General Technical Report PSW-144. US Department of Agriculture, Forest Service, Pacific Southwest Research Station, Albany, CA.

Ramsey, D. S., P. A. Caley, and A. Robley. 2015. Estimating population density from presence-absence data using a spatially explicit model. Journal of Wildlife Management 79:491–499.

Raymond, M., and F. Rousset. 1995. GENEPOP, version 1.2: Population genetics software for exact tests and ecumenicism. Journal of Heredity 86:248–249.

R Development Core Team. 2016. R: A language and environment for statistical computing. R Foundation for Statistical Computing, Vienna, Austria.

Reinert, H. K. 1984. Habitat separation between sympatric snake populations. Ecology 65:478–486.

Rettie, W. J., and F. Messier. 2000. Range use and movement rates of woodland caribou in Saskatchewan. Canadian Journal of Zoology 79:1933–1940.

Rexstad, E. A., and K. P. Burnham. 1991. User's guide for interactive program CAPTURE: Abundance estimation of closed animal populations. Colorado State University, Fort Collins.

Reynolds, R. T., R. T. Graham, M. H. Reiser, R. L. Bassett, P. L. Kennedy, D. A. Boyce, G. Goodwin, R. Smith, and E. L. Fisher. 1992. Management recommendations for the northern goshawk in the southwestern United States. General Technical Report RM-217. US Department of Agriculture, Rocky Mountain Forest and Range Experiment Station, Fort Collins, CO.

Reynolds, R. T., J. D. Wiens, and S. R. Salasky. 2006. A review and evaluation of factors limiting northern goshawk populations. Studies in Avian Biology 31:260–273.

Ribble, D. O., A. E. Wurtz, E. K. McConnell, J. J. Bueggie, and K. C. Welch Jr. 2002. A comparison of home ranges of two species of *Peromyscus* using trapping and radio-telemetry data. Journal of Mammalogy 83:260–266.

Rice, E. L., and W. T. Penfound. 1955. An evaluation of the variable-radius and paired-tree methods in the blackjack–post oak forest. Ecology 36:315–320.

Rossman, S., C. B. Yackulic, S. P. Saunders, J. Reid, R. Davis, and E. F. Zipkin. 2016. Dynamic *N*-occupancy models: Estimating demographic rates and local abundance from detection-nondetection data. Ecology 97:3300–3307.

Rotenberry, J. T. 1985. The role of habitat in avian community composition: Physiognomy or floristics? Oecologia 67:213–217.

Rotenberry, J. T., and J. A. Wiens. 1998. Foraging patch selection by shrubsteppe sparrows. Ecology 79:1160–1173.

Roth, R. R. 1976. Spatial heterogeneity and bird species diversity. Ecology 57:773–782.

Royle, J. A., R. B. Chandler, R. Sollmann, and B. Gardner. 2013. Spatial capture-recapture. Academic Press, New York.

Royle, J. A., R. B. Chandler, C. C. Sun, and A. K. Fuller. 2013. Integrating resource selection information with spatial capture-recapture. Methods in Ecology and Evolution 4:520–530.

Royle, J. A., and J. D. Nichols. 2003. Estimating abundance from repeated presence-absence data or point counts. Ecology 84:777–790.

Royle, J. A., and K. V. Young. 2008. A hierarchical model for spatial capture-recapture data. Ecology 89:2281–2289.

Rumble, M. A., T. R. Mills, and L. D. Flake. 1999. Habitat capability model for birds wintering in the Black Hills, South Dakota. Research Paper RMRS-RP-19. US Department of Agriculture, Forest Service, Rocky Mountain Research Station, Fort Collins, CO.

Sanderlin, J. S., W. M. Block, and J. L. Ganey. 2014. Optimizing study design for multi-species avian monitoring programmes. Journal of Applied Ecology 51:860–870.

Sanderlin, J. S., M. L. Morrison, and W. M. Block. 2019. Analysis of population monitoring data. Pp. 131–148 *in* L. A. Brennan, A. N. Tri, and B. G. Marcot, eds. Quantitative analyses in wildlife science. John Hopkins University Press, Baltimore.

Sands, J. P., L. A. Brennan, F. Hernández, and W. P. Kuvlesky. 2009. Impacts of buffelgrass (*Pennisetum ciliare*) on a forb community in south Texas. Invasive Plant Science and Management 2:130–140.

Sarr, D. A., A. Duff, E. C. Dinger, S. L. Shafer, M. Wing, N. E. Seavy, and J. D. Alexander. 2015. Comparing ecoregional classifications for natural areas management in the Klamath Region, USA. Natural Areas Journal 35(3):360–377.

Schneider, S., D. Roessli, and L. Excoffier. 2000. ARLEQUIN, user manual ver. 2.000: A software for population genetics data analysis. Genetics and Biometry Lab, Department of Anthropology, University of Geneva, Switzerland.

Schwartz, M. K., G. Luikart, and R. S. Waples. 2007. Genetic monitoring as a promising tool for conservation and management. Trends in Ecology & Evolution. 22:25–33.

Schwarz, C. J., and A. N. Arnason. 1996. A general methodology for the analysis of capture-recapture experiments in open populations. Biometrics 52:860–873.

Scott, J. M., F. Davis, B. Csuti, R. Noss, B. Butterfield, C. Groves, H. Anderson, S. Caicco, F. D'Erchia, T. C. Edwards Jr., J. Ulliman, and R. G. Wright. 1993. Gap analysis: A geographic approach to protection of biological diversity. Wildlife Monographs 123:3–41.

Seber, G. A. F. 1965. A note on the multiple-recapture census. Biometrika 52:249–259.

Shiflet, T. N., ed. 1994. Rangeland cover types of the United States. Society for Range Management, Denver.

Shirley, S. M., Z. Yang, R. A. Hutchinson, J. D. Alexander, K. McGarigal, and M. G. Betts. 2013. Species distribution modelling for the people: Unclassified landsat TM imagery predicts bird occurrence at fine resolutions. Diversity and Distributions 19:855–866.

Smallwood, K. S., and C. Schonewald. 1996. Scaling population density and spatial pattern for terrestrial, mammalian carnivores. Oecologia 105:329–335.

Smallwood, K. S., and T. R. Smith. 2001. Study design and interpretation of shrew (*Sorex*) density estimates. Annals Zoologica Fennici 38:149–161.

Solis, D. M., and R. J. Gutiérrez. 1990. Summer habitat ecology of northern spotted owls in northwestern California. Condor 92:739–748.

Soulé, M. E. 1991. Land use planning and wildlife maintenance: Guidelines for conserving wildlife in an urban landscape, Journal of the American Planning Association 57:313–323.

Spiegelhalter, D., A. Thomas, N. Best, and D. Lunn. 2007. OpenBUGS user manual, version 3.0.2. MRC Biostatistics Unit, Institute of Public Health, Cambridge.

Sturman, W. A. 1968. Description and analysis of breeding habitats of the chickadees, *Parus atricapillus* and *P. rufescens*. Ecology 49:418–431.

Sutherland, C., and D. W. Linden. 2019. Occupancy modeling applications. Pp. 131–148 *in* L. A. Brennan, A. N. Tri, and B. G. Marcot, eds. Quantitative analyses in wildlife science. Johns Hopkins University Press, Baltimore.

Taberlet, P., L. P. Waits, and G. Luikart. 1999. Noninvasive genetic sampling: Look before you leap. Trends in Ecology & Evolution 14:323–327.

Tattoni, C., F. Rizzolli, and P. Pedrini. 2012. Can LiDAR data improve bird habitat suitability models? Ecological Modelling 245:103–110.

Tempel, D. J., J. J. Keane, R. J. Gutiérrez, J. D. Wolfe, G. M. Jones, A. Koltunov, C. M. Ramirez, W. J. Berigan, C. V. Gallagher, T. E. Munton, P. A. Shaklee, S. A. Whitmore, and M. Z. Peery. 2016. Meta-analysis of California spotted owl (*Strix occidentalis occidentalis*) territory occupancy in the Sierra Nevada: Habitat associations and their implications for forest management. Condor 118(4):747–765.

Tews, J., U. Brose, V. Grimm, K. Tielbörger, M. C. Wichmann, M. Shwager, and F. Jeltsch. 2004. Animal species diversity driven by habitat heterogeneity/diversity: The importance of keystone structures. Journal of Biogeography 31:79–92.

Thomas, D. L., and E. Y. Taylor. 1990. Study designs and tests for comparing resource use and availability. Journal of Wildlife Management 54:322–330.

Thornton, T. F., ed. 2012. Haa Léelk'w Hás Aaní Saax'ú = Our grandparents' names on the land. University of Washington Press, Seattle.

Timm, B. C., K. McGarigal, S. A. Cushman, and J. L. Ganey. 2016. Multi-scale Mexican spotted owl (*Strix occidentalis lucida*) nest/roost habitat selection in Arizona and a comparison with single-scale modeling results. Landscape Ecology 31:1209–1225.

Turchin, P. 1998. Quantitative analysis of movement: Measuring and modeling population redistribution in animals and plants. Sinauer Associates, Sunderland, MA.

Tyre, A. J., H. P. Possingham, and D. B. Lindenmayer. 2001. Inferring process from pattern: Can territory occupancy provide information about life history parameters? Ecological Applications 11:1722–1737.

Ucitel, D., D. P. Christian, and J. M. Graham. 2003. Vole use of coarse woody debris and implications for habitat and fuel management. Journal of Wildlife Management 67:65–72.

Urban, D. L., R. V. O'Neill, and H. H. Shugart Jr. 1987. A hierarchical perspective can help scientists understand spatial patterns. Bioscience 37:119–127.

USDI Fish and Wildlife Service. 1981. Standards for the development of suitability index models. Ecological Services Manual 103. US Department of the Interior, Fish and Wildlife Service, Washington, DC.

VanderWerf, E. A. 1993. Scales of habitat selection by foraging 'elepaio in undisturbed and human-altered forests in Hawaii. Condor 95:980–989.

VanDeusen, P. C., and G. S. Biging. 1984. Crown volume and dimension models for mixed conifers of the Sierra Nevada. Northern California Forest Yield Cooperative Research Note No. 9. Department of Forestry and Resource Management, University of California, Berkeley.

van Dijk, J., K. Hauge, A. Landa, R. Andersen, and R. May. 2007. Evaluating scat analysis methods to assess wolverine *Gulo gulo* diet. Wildlife Biology 13:62–67.

Van Horne, B. 1983. Density as a misleading indicator of habitat quality. Journal of Wildlife Management 47:893–901.

Van Horne, B. 2002. Approaches to habitat modeling: The tensions between pattern and process and between specificity and generality. Pp. 63–72 *in* J. M. Scott, P. J. Heglund, M. L. Morrison, J. B. Haufler, M. G. Raphael, W. A. Wall, and F. B. Samson, eds. Predicting species occurrences: Issues of accuracy and scale. Island Press, Washington, DC.

Van Moorter B., C. M. Rolandsen, M. Basille, and J.-M. Gaillard. 2015. Movement is the glue connecting home ranges and habitat selection. Journal of Animal Ecology 85:21–31.

van Toor, M. L., B. Kranstauber, S. H. Newman, D. J. Prosser, J. Y. Takekawa, G. Technitis, R. Weibel, M. Wikelski, and K. Safi. 2018. Integrating animal movement with habitat suitability for estimating dynamic migratory connectivity. Landscape Ecology 33(6):879–893.

Waits, L. P. 2004. Using noninvasive genetic sampling to detect and estimate abundance of rare wildlife species. Pp. 211–228 *in* W. L. Thompson, ed. Sampling rare or elusive species. Island Press, Washington, DC.

Waser, P. M., and C. Strobeck. 1998. Genetic signatures of interpopulation dispersal. Trends in Ecology & Evolution 13:43–44.

Weir, R. D., and A. S. Harestad. 2003. Scale-dependent habitat selectivity by fishers in south-central British Columbia. Journal of Wildlife Management 67:73–82.

Welsh, H. H., and A. J. Lind. 1995. Habitat correlates of Del Norte salamander, *Plethodon elongatus* (Caudata: Plethodontidae), in northwestern California. Journal of Herpetology 29:198–210.

Welsh, H. H., and A. J. Lind. 2002. Multiscale habitat relationships of stream amphibians in the Klamath–Siskiyou region of California and Oregon. Journal of Wildlife Management 66:581–602.

Wenny, D. G., R. L. Clawson, J. Faaborg, and S. L. Sheriff. 1993. Population density, habitat selection and minimum area requirements of three forest-interior warblers in central Missouri. Condor 95:968–979.

Wester, D. B. 2019. Regression: Linear and nonlinear, parametric and nonparametric. Pp. 9–31 *in* L. A. Brennan, A. N. Tri, and B. G. Marcot, eds. Quantitative analyses in wildlife science. Johns Hopkins University Press, Baltimore.

White, G. C. 2001. Statistical models: Keys to understanding the natural world. Pp. 35–56 *in* T. M. Shenk and A. B. Franklin, eds. Modeling in natural resources management. Island Press, Washington, DC.

White, G. C. 2019. Estimation of population parameters using marked animals. Pp. 79–96 *in* L. A. Brennan, A. N. Tri, and B. G. Marcot, eds. Quantitative analyses in wildlife science. Johns Hopkins University Press, Baltimore.

White, G. C., and K. P. Burnham. 1999. Program MARK: Survival estimation from populations of marked animals. Bird Study 46:S120–S139.

White, G. C., and R. A. Garrott. 1990. Analysis of wildlife radio-tracking data. Academic Press, San Diego.

Whitmore, R. C. 1981. Applied aspects of choosing variables in studies of bird habitats. Pp. 38–41 *in* D. E. Capen, ed. The use of multivariate statistics in studies of wildlife habitat. General Technical Report RM-87. US Department of Agriculture, Forest Service, Rocky Mountain Forest and Range Experiment Station, Fort Collins, CO.

Wiens, J. A. 1969. An approach to the study of ecological relationships among grassland birds. Ornithological Monographs No. 8. American Ornithologists' Union, n.p.

Wiens, J. A. 1989. The ecology of bird communities. Vol. 1, Foundations and patterns. Cambridge University Press, Cambridge.

Wiens, J. A. 1997. Metapopulation dynamics and landscape ecology. Pp. 43–62 *in* I. A. Hanski and M. E. Gilpin, eds. Metapopulation biology. Academic Press, San Diego.

Wiens, J. A. 2016. Ecological challenges and conservation conundrums. John Wiley & Sons, Hoboken, NJ.

Wiens, J. A., and J. T. Rotenberry. 1981. Habitat associations and community structure of birds in shrubsteppe environments. Ecological Monographs 51:21–41.

Wiens, J. A., J. T. Rotenberry, and B. Van Horne. 1987. Habitat occupancy patterns of North American shrubsteppe birds: The effects of spatial scale. Oikos 48:132–47.

Williams, B. K., J. D. Nichols, and M. J. Conroy. 2002. Analysis and management of animal populations. Academic Press, San Diego.

Williford, D., R. W. DeYoung, R. L. Honeycutt, L. A. Brennan, and F. Hernández. 2016. Phylogeography of the bobwhite (*Colinus*) quails. Wildlife Monographs 193:1–49.

Willson, M. F. 1974. Avian community organization and habitat structure. Ecology 55:1017–1029.

Wright, J. A., R. J. Barker, M. R. Schofield, A. C. Frantz, A. E. Byrom, and D. M. Gleeson. 2009. Incorporating genotype uncertainty into mark-recapture–type models for estimating abundance using DNA samples. Biometrics 65:833–840.

Yackulic, C. B., R. Chandler, E. F. Zipkin, J. A. Royle, J. D. Nichols, E. H. Campbell Grant, and S. Veran. 2013. Presence-only modelling using MAXENT: When can we trust the inferences? Methods in Ecology and Evolution 4:236–243.

Yao, T., X. Yang, F. Zhao, Z. Wang, Q. Zhang, D. Jupp, J. Lovell, D. Culvenor, G. Newnham, W. Ni-Meister, and C. Schaaf. 2011. Measuring forest structure and biomass in New England forest stands using Echidna ground-based lidar. Remote Sensing of Environment 115:2965–2974.

Young, M., and M. H. Carr. 2015. Application of species distribution models to explain and predict the distribution, abundance and assemblage structure of nearshore temperate reef fishes. Diversity and Distributions 21:1428–1440.

Zhao, F., R. A. Sweitzer, Q. Guo, and M. Kelly. 2012. Characterizing habitats associated with fisher den structures in the southern Sierra Nevada, California using discrete return lidar. Forest Ecology and Management 280:112–119.

3 Animal Habitat

When and Where to Measure and How to Analyze It

Introduction

We reviewed various habitat variables to measure and how to measure them in Chapter 2. Here we consider when to measure and how to analyze the resultant data. The timing for measurements involves among- and within-season analyses, as well as a decision on how long (i.e., over how many years) data should be collected (Wiens 1984). The resource requirements of animals vary considerably among seasons. Migratory birds, for example, do not necessarily use the same resources on sites within their breeding territories, during migration, and on their wintering grounds. Many big-game ungulate species consume different types of forage on their summer range than on their winter range. Even species regarded as permanent residents in an area frequently switch to various foraging substrates and food sources as the seasons change. For instance, many bird species in the family Fringillidae have specialized beak morphologies to consume plant seeds, yet their breeding-season diet is dominated by insects, to meet protein requirements for egg formation and chick rearing. During much of the fall and winter seasons, their diet reverts largely to seeds. In addition, within a given time period, the resource-use and activity patterns of some organisms often vary with the time of day, the stage of the breeding cycle, and other divisions in an animal's life cycle. Thus it is critical that one understands the temporal aspects of resource use prior to designing a habitat study.

The distribution and abundance of animals changes with variations in their food supply, weather conditions, predator activity, and many other biotic and abiotic factors. As we show throughout this book, these factors must drive how we design and then interpret our research on wildlife habitat. Few studies, however, have explicitly recognized these shifts in species distributions and abundances, and instead have based their conclusions on investigations that are too limited in their duration and spatial extent.

In the final sections of this chapter we first review commonly used methods to analyze animal habitat data. The field of biometrics is advancing rapidly, providing researchers with numerous options for examining these data. Yet, regardless of the tools used, the results are only as good as the underlying raw data and the statistical sampling framework on which the analysis is based. Hence a carefully designed study that provides a representative sample is paramount (Morrison et al. 2008). Ideally, its design should identify the types of data collection, the structure for data analyses, and required sample sizes. Second, we address the all-important assump-

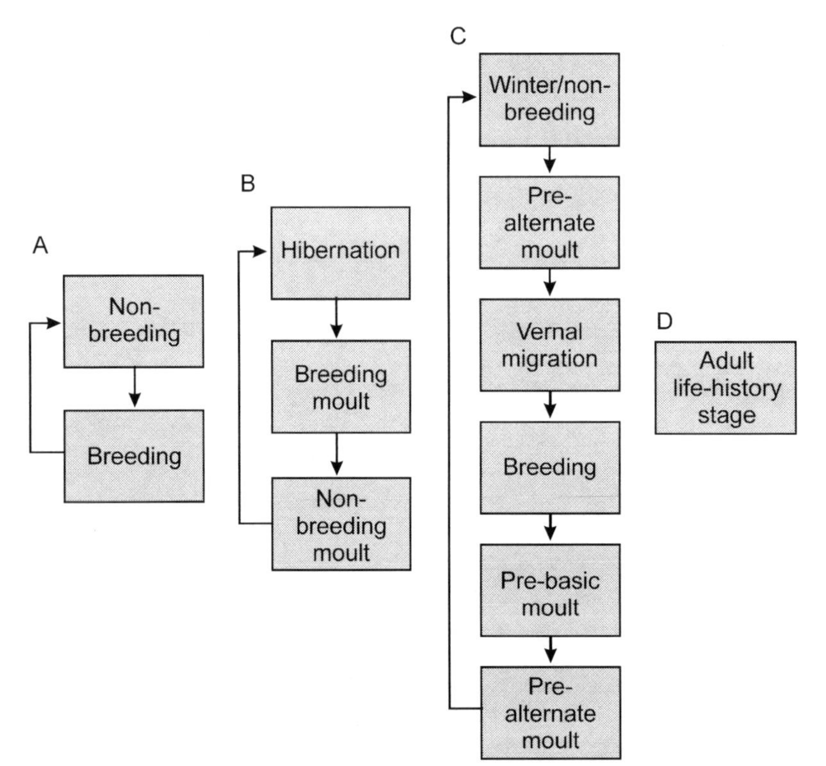

Figure 3.1. An example of adult life-history stages from selected vertebrates: (*A*) reef fish, (*B*) reptiles, (*C*) migratory birds, and (*D*) humans. (Reproduced from Wingfied (2007), with the permission of the Royal Society)

tions associated with these statistical techniques. Third, we follow this with a classification and discussion of several of the more commonly used methods and their applications to wildlife studies, including sample-size requirements and a review of statistical computer packages. A formal course in statistics, however, is not required for understanding the major points of this chapter. Many good texts are available and should be consulted for a general overview of this class of analyses (e.g., Morrison 1967; Cooley and Lohnes 1971; Pimentel 1979; Afifi and Clark 1984; Dillon and Goldstein 1984; McGarigal et al. 2000; Manley and Navarro Alberto 2016; Brennan et al. 2019).

When and Where to Measure

As is commonly known, the behavior, location, and needs of animals change—often substantially—throughout the year. Species frequently exhibit different life-history strategies over the course of their annual cycle (Wingfield 2007;). For example, some species of reef fish occupy a restricted space, and their annual cycle may correspond to breeding and non-breeding seasons (Fig. 3.1*A*). Many reptiles exhibit a 3-stage cycle, including breeding and non-breeding intervals, as well as a period of estivation or brumation (Fig. 3.1*B*). Migratory birds have a more complicated cycle, which includes spring and fall migration and periods of moult (Fig. 3.1*C*; also see Berkhead et al. 2016:Chapter 4). In contrast, humans can modify their environment using clothing, heat, and structures, which results in 1 basic annual life cycle (Fig. 3.1*D*).

Intra- and Interseasonal Variations in Habitat Use

Thus when to measure is a crucial question, which is entirely driven by a study's objectives. For example, the question might be, "What sorts of conditions are preferred by the species to den or nest?" Here, research confined to the period of the year when these activities occur is perfectly adequate. Another possibility could be to ask, "What conditions limit population size?" In this case, an investigation concentrating on reproduction would assume that reproductive potential was the limiting factor, rather than, say, adult survival. For any species with limited reproductive potential and long generation times, however, adult survival will dominate life-history sensitivity analyses (e.g., Noon and Biles 1990).

Even if the study's goals indicate that an investigation over just a single season is appropriate, conditions will vary annually, and habitat-use patterns can be expected to alter, as well. Research that extends through several seasons invariably finds that behaviors and habitat-use patterns change. For example, Jorde et al. (1984) observed that winter roost site selection for mallards (*Anas platyrhynchos*) was not the same when the weather was especially cold. Indeed, habitat use may be entirely different in extreme years. Many studies of winter forage for Rocky Mountain elk (*Cervus canadensis nelsoni* [*C. elaphus nelsonii*]) focus on the quantity and quality of perennial grasses on their winter range. In severe winters, however, when grasses are unavailable, due to heavy snow cover, elk shift to the consumption of arboreal lichens, which are present and available to them in densely forested areas (Ward and Marcum 2005). Long-term studies are particularly helpful in teasing apart factors that underlie population regulation, as Martínez-Padilla et al. (2013) did for red grouse (*Lagopus lagopus scoticus*). Clearly, seasonal use by organisms cannot be fully gleaned from data gathered in a single season. Ideally, the temporal frame of such a study would encompass the full range of seasons.

Marra et al. (2015) evaluated seasonality and the length of investigations in articles published in 5 high-impact ecology journals and 4 taxon-specific journals for the years 1994, 2000, 2006, and 2012. Their results confirmed a seasonal bias toward conducting research during the breeding season for all of the taxa considered (amphibians, reptiles, birds, and mammals; Fig. 3.2). Over 73 percent of all investigations occurred in only 1 cycle of 1 year, and only 5.5 percent of the studies examined seasonal interactions. Further, these trends remained somewhat consistent over the 18-year timespan when the papers were published (Fig. 3.3). These results confirm that researchers often sample from such a narrow period—usually only during spring and summer—that their results are very time and location specific, thus only minimally applying either to other situations or to different parts of an organism's life cycle and annual resource-use patterns. Without knowledge of an animal's total requirements, management recommendations from such investigations may have circumscribed and perhaps faulty implications.

As demonstrated by Marra et al. (2015), because of limited funding, constraints on the availability of

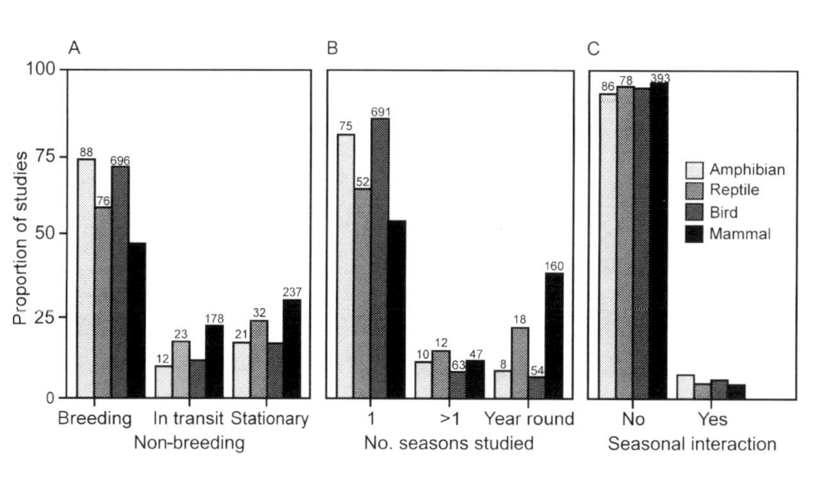

Figure 3.2. Bar graphs of proportions of studies on 4 vertebrate orders conducted (*A*) during each period of the annual cycle, and (*B*) during 1 annual cycle, more than 1, or year-round, as well as (*C*) those examining a seasonal interaction. (Reproduced from Marra et al. (2015), with the permission of the Royal Society)

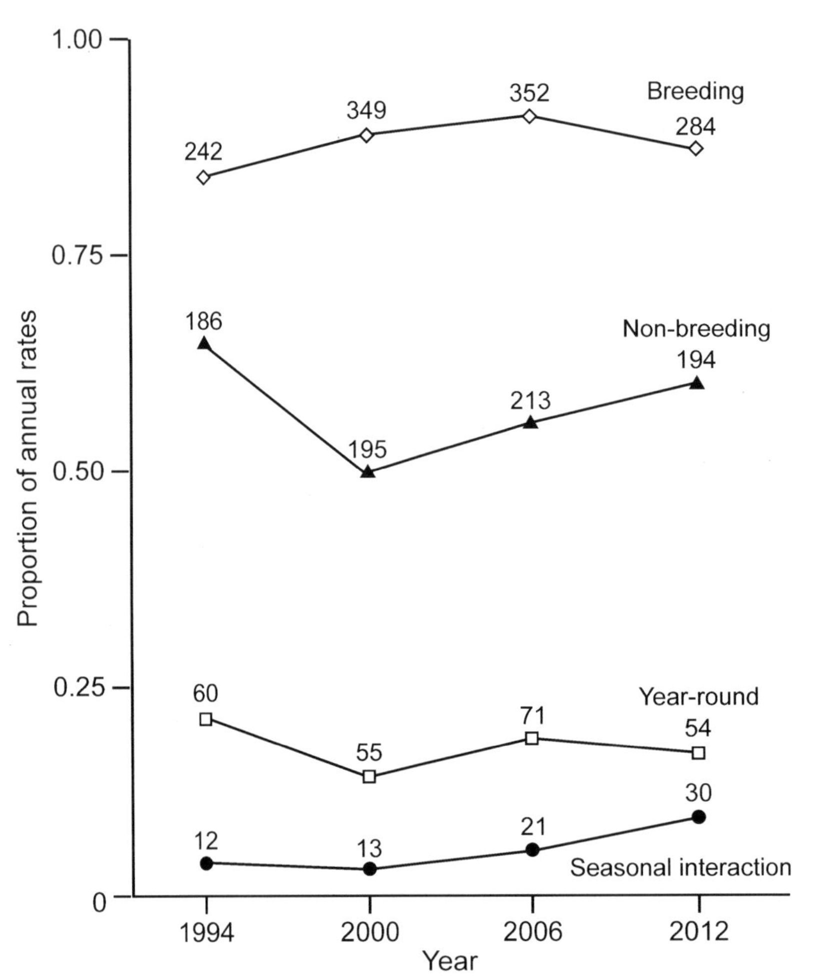

Figure 3.3. A graph of changes in the annual proportion of studies published from 1994 to 2012 that included research during the breeding season, during a stationary non-breeding period, or year-round, as well as those that included seasonal interactions. (Reproduced from Marra et al. (2015), with the permission of the Royal Society)

personnel, cold weather, and other problems related to the execution of studies, most researchers have concentrated their investigations from late spring through the summer (but see Lozano-Cavasos et al. 2016 for a notable exception to this pattern). It is now well known that the survival of an animal may depend in large part on the non-breeding season, particularly the wintering period. Intuitively, we would expect that fall and winter—when populations are at their biggest numbers (because of offspring), resources are declining (trees and arthropods are dying or going dormant), animals are physiologically stressed by dispersal or migration movements, and the weather is becoming harsher—are times when organisms would encounter the most difficult circumstances (Bock and Jones 2004). Further, because of these fluctuations in environmental conditions, aspects of the habitat change, even if an individual animal did not shift its location. Large-scale migrations by numerous species provide good indications of the lengths to which animals must go to find favorable living conditions. Species that migrate may also benefit by using different locations from year to year. Lafontaine et al. (2017), for example, found that over-winter survival of female woodland caribou (*Rangifer tarandus caribou*) using different wintering areas between years was greater than that of caribou using the same area over multiple years. The importance of non-breeding periods was popularized by Fretwell (1972), although he was certainly not the

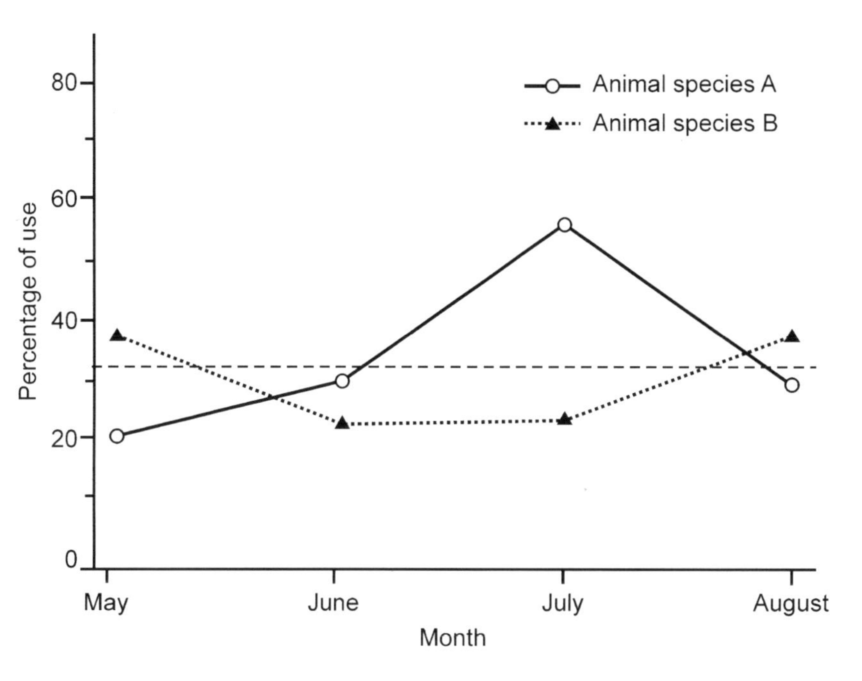

Figure 3.4. A graph showing the use of a species of tree by 2 hypothetical animal species during summer. The *dashed horizontal line* represents an approximate average.

first researcher to recognize this. If an animal varies its pattern of habitat use, or a change occurs in the environment itself, then a summer-based study cannot be used with any confidence to predict the subsequent responses by that animal.

The desired temporal granularity is a second vital factor to consider in decisions about when to measure. The crucial variability in an animal's use patterns disclosed through adequate coverage over a sufficient span of time can still be easily hidden, due to an excessive averaging of observations across seasons or life-history stages. It is easy to conceptualize what inappropriate aggregation may mean in obscuring habitat relationships. Figure 3.4 shows that the average use of 1 kind of tree over time by 2 hypothetical species is not a close approximation of their actual behavior, as it indicates that the animals use this tree species basically identically. Here, we see that the finer the temporal granularity, the more it reveals about habitat-use behaviors. This phenomenon is closely linked to the frequency of observations. Collecting data more often affords us greater flexibility in evaluating trends that happen over time. Changes in habitat-use patterns can occur in extremely short timeframes. Vilizzi et al. (2004) showed that habitat use by barbels (*Barbus barbus*), a species of freshwater fish, could be attributed to rapid environmental variables (e.g., discharge, water depth, time of day) and when sampling occurred. Working with red deer (*Cervus elaphus*) in Norway, Godvik et al. (2009) demonstrated that patterns of habitat use may change during an animal's daily foraging and resting rhythms.

Past research provides poor guidance in this regard. Schooley (1994) reviewed 43 papers published between 1988 and 1991 in the *Journal of Wildlife Management* that examined the habitat use of terrestrial vertebrates. Most studies pooled their data on habitat use among years, evidently without testing for annual variations. Using black bears (*Ursus americanus*) as an example, he illustrated the misleading inferences that can result from this practice, as patterns of yearly fluctuations in habitat use were lost (Fig. 3.5).

Difficulties in choosing the appropriate timeframe and the sampling density needed to uncover targeted patterns of habitat use are particularly evident in a multispecies context. Even within relatively small areas, seasonal patterns can differ widely for

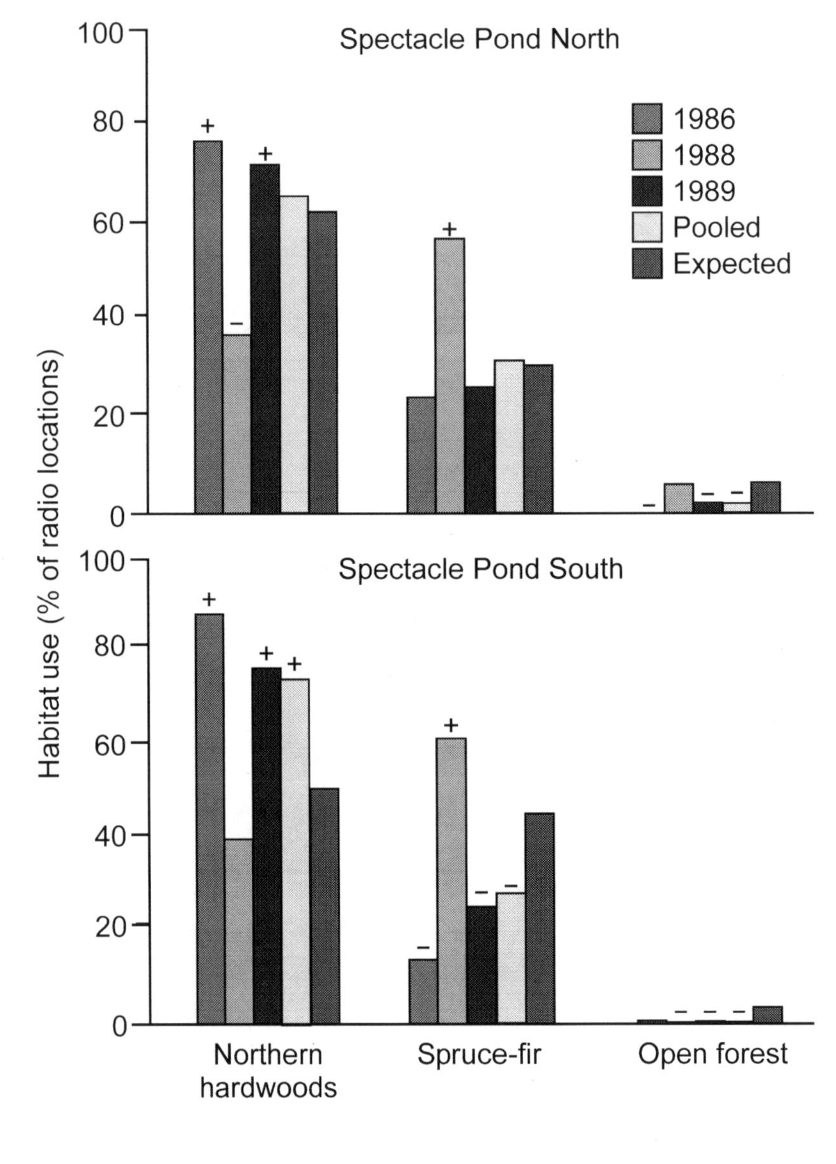

Figure 3.5. Graphs of habitat selection by female black bears (*Ursus americanus*) at 2 separate study areas (*top* and *bottom*) in northern Maine during fall (den entry on 1 September) 1986–1988. Habitat use in each area was presented for individual years and for all years pooled. Expected use was estimated from the availability of habitats. (While habitat availability was estimated separately for each year, only 1987 data are presented here, because availability differed little among years.) A + indicates use greater than availability, a – indicates use less than availability, and no symbol indicates use equal to availability. Selection was based on Bonferroni-adjusted 90 percent CIs. (Reproduced from Schooley (1994:Fig. 1), with the permission of The Wildlife Society)

species, both in and (especially) between vertebrate groups. For example, in most regions, the breeding season for many rodents can start as early as February, months before many bird species even arrive to breed in the area. Further, many resident bird species begin breeding several months before non-residents. Few studies acknowledge these sampling problems, or, if they do, the investigators gather samples from the middle of the season (which, for logistical reasons, usually conforms to the summer recess of most universities). In addition to temporal patterns, such as seasonality, habitat use can change because of shifting population densities within a community. Density-dependent factors may influence where conspecifics establish their breeding territories; where they forage, rest, and display; and how they move across the landscape.

At a minimum, researchers should acknowledge these sampling limitations, as well as the ramifications of such weaknesses on management recommendations. Using information gathered from 1 location, season, or year as the basis for prescribing management activities should be done cautiously, if at all. Ideally, investigators will avoid many of these problems by designing studies that actually determine if such potential temporal variations are influencing their results in major ways.

Long-Term Temporal Changes

The magnitude of temporal variation is a question of scale, which can run across as well as within years. Relationships developed inside of a year or over a span of a few years may not hold for a different timeframe. Moreover, environmental conditions can change drastically across years. "Normal" weather conditions (from a daily or extremely localized perspective) seldom occur, because they are, in reality, a crude average of extremes in rainfall, temperature, and wind. An animal's reaction to more-severe conditions over time could provide the most important data in evaluating habitat relationships.

Reviews by others (e.g., Likens 1983; Wiens 1984; Strayer et al. 1986; Leigh and Johnston 1994; Lindenmayer et al. 2012; Marra et al. 2015), revealed a tradition that has developed over the past several decades, especially among North American scientists, toward the pursuit of short-term studies. This situation arose from constraints imposed by limited-duration funding; the need to finish graduate programs within short periods of time (actually a noble goal); the pressure placed on researchers to publish; a bias against the publication of place-based research; trends toward modeling in lieu of field studies; a shift away from data-based management; and a focus on funding that is targeted toward equipment, rather than people (Lindemayer 2018). Restricting the timespan of investigations results in a snapshot approach to studying nature. Wildlife studies usually run from 1 to 3 years and, at best, give only a partial view of most ecological situations. At worst, they provide a false interpretation. Although we normally pose this issue as a dichotomy between short- versus long-term research, we usually fail to discuss the duration of an investigation relative to how long it *should* be, given the question(s) being asked and the probable variability in both the environmental setting and responses by organisms. Thus we should really be discussing the necessary study length: a few years, in the 3–10 year range, or much longer. Here, we are placing this discussion within human concepts of time, but for other aspects of animal ecology research, we need to relate our work to the timeframe of the organism of interest.

Investigations of longer duration—more than a few years—are especially suited to exploring 4 major classes of ecological phenomena: slow processes, rare or extreme events, subtle processes, and complex phenomena (Strayer et al. 1986; also discussed in Morrison et al. 2020:Chapter 3, on disturbance dynamics). Forest succession, the invasion of exotic species, and vertebrate population cycles are prominent examples of slow processes that have obvious importance in formulating management decisions. The substantial impact of El Niño oscillations and infrequent natural events (e.g., floods, disease outbreaks, hurricanes, fires), as well as vegetation succession and population eruptions, are all examples of happenings that are certainly missed by short-term studies. Slow processes and rare or extreme events are often associated with each other. Consider the relationship between precipitation and fire frequency in the American Southwest. Grissino-Mayer and Swetnam (2000) showed that the years from 1850 to 2000 had been wetter than average, resulting in fewer fires (Fig. 3.6). We are not suggesting that studies of fires and similar events need to extend for 150+ years, but we include this example to illustrate the length of time over which some processes endure.

Schradin and Hayes (2017) summarized information presented in a special issue of the *Journal of Mammalogy* on the value of long-term research. They argued that such investigations of free-ranging mammals were particularly useful for understanding the factors that regulate the natural populations and social systems of these species. Their conclusion probably also applies to other taxa, a prime example being the long-term demographic studies conducted on northern spotted owls (*Strix occidentalis caurina*) (see Morrison et al. 2002:Chapter 4). Processes that

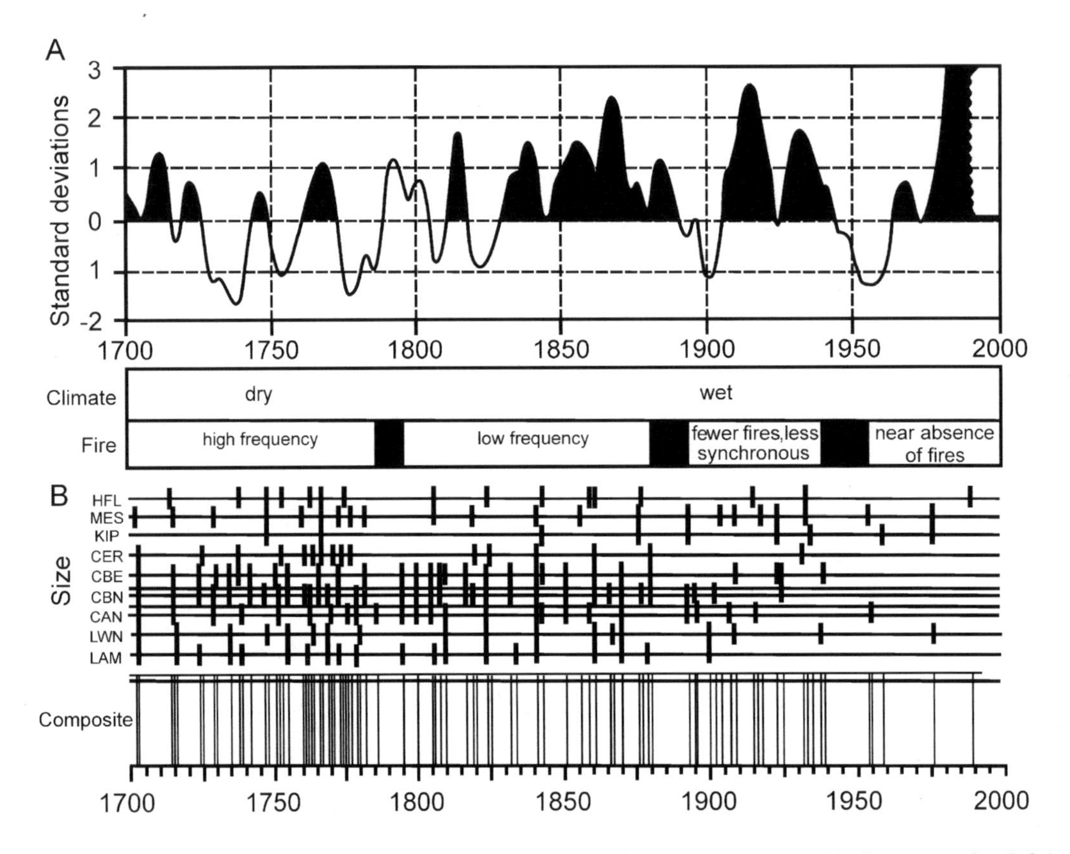

Figure 3.6. A reconstruction of precipitation in the American Southwest since AD 1700, illustrating the shift between below-normal to above-normal rainfall, both in terms of standard deviations (*A*) and size (*B*). (Reproduced from Grissino-Meyer and Swetnam (2000), with the permission of SAGE Publications)

change over time in a regular fashion, but where the year-to-year variance is large relative to the magnitude of the longer-term trend, are examples of the subtle series of events that short-term studies cannot evaluate. Although scientists certainly realize that nature is complex, seldom do we provide the necessary time span for a phenomenon to reveal enough of its characteristics to allow a meaningful interpretation.

Yet how long is long enough? Strayer et al. (1986) gave 2 rather different definitions of the concept of "long-term." The first considered the length of an investigation with regard to natural processes. It was long-term if it continued through the generation time of the dominant organism, or over a sufficient period to include examples of the important processes that structured the relevant ecosystem. Here, the length of a study is measured against the dynamic speed of the system being examined. Obviously, such a criterion demands that researchers have a good understanding of that system. The authors' second definition involved viewing the length of an investigation in a relative fashion, with long-term studies being those that had continued for a more extensive period than most others. More recently, Lindenmayer et al. (2012) defined "long-term ecological studies" as those that collected field data from a site or set of sites for 10 years or more. Although a 10-year span might seem to be somewhat arbitrary (and it is), they contended that it allowed for repeat sampling over multiple growth or breeding cycles and incorporated a representative degree of bioclimatic variation.

Of course, not all studies need to be long term to provide useful results. Investigations of essentially static patterns (e.g., morphology, genetic characteristics of species), processes at the individual level (e.g., growth, behavior, physiology), or modes of present adaptations do not necessarily require lengthy timeframes. Ecological research conducted at broad spatial scales (e.g., relating presence-absence to vegetation types) is also usually relatively static. The principal disadvantages of long-term studies are practical, rather than ecological. In many cases, observed tendencies cannot be explained without such a perspective. Broad configurations of habitat use frequently are strongly influenced by occasional extreme events. The impetus to quickly publish one's findings tends to favor short- versus long-term investigations. To bridge the gap between the infeasibility of carrying out research that captures multiple rare events and the need to assess the results of these events, workarounds have been developed, although their approaches are far from perfect. There are 4 classes of short-term research that can potentially provide insight into long-term relationships: (1) retrospective studies, (2) a substitution of space for time, (3) the use of systems with quick dynamics as analogues for systems with slow dynamics, and (4) modeling (Strayer et al. 1986). Further, a series of short-term studies can be incorporated into a longer-term research plan. This works especially well in a professor–graduate student relationship. Modeling and an emphasis on space rather than time are 2 common alternatives to longer-term investigations. In substitution studies, sites with dissimilar characteristics (e.g., different developmental stages responding to a common disturbance) are used instead of observing a few sites for an extended period. Such space-for-time sampling is termed a "chronosequence" (Provencher et al. 2001).

An example of a chronosequence is examining the relationship between birds and plant succession over a period of a few years in post-harvest or post-burn sites of different ages (e.g., 1, 5, 15, and 30 or more years). To provide valid results, this design requires all of the areas to have similar histories and characteristics. Naturally, a large number of replicates enhances the reliability of this type of research, but similarity among locations is likely to become compromised as one adds more sites. Chronosequence investigations cannot directly examine the actual historical disturbance events that shaped each locale, as it can only address those effects by averaging data across a large sample size, with the hope that this will yield pertinent results. What is needed instead is an increase in long-term studies of experimental manipulations of vegetation, food, competitors, predators, and other parameters, made in association with demographic research. Lindenmayer et al. (2012:746) noted a clear distinction between these various workarounds and the value of on-the-ground ecological field studies, conducted systematically over many years.

Wiens (2016) suggested that a decision regarding the duration of an investigation ultimately hinged on the characteristics of the system being studied and the questions being asked. Essentially, it requires achieving a balance. A study that is too short may miss important dynamics, and another that averages outcomes over the long term may mask significant relationships. As Wiens (2016:244) summarized, what is needed is a "Goldilocks solution"—not too short, not too long, but just right—with "just right" defined entirely by the research goals. Short-term studies can provide useful information if they have a specific focus that is not likely to be obscured by background variations in habitat relationships. But longer-term investigations are needed if the science of wildlife-habitat relationships is to advance. Graduate students cannot be expected to conduct long-term work in their thesis or dissertation projects. The duty of their advisors is to help direct them toward research where the results are not likely to be swamped by unknown changes over an extended period of time. Long-term investigations done in parallel with carefully matched, short-term substitution studies are needed. Wiens (1989:174–196) provided additional review material on the usefulness of long-term ecological studies.

Where to Measure

As we have noted, organisms may use different areas to acquire resources for their disparate life-history needs. Migratory birds, for example, could choose 1 set of vegetation structures and composition on their breeding grounds and another in their migration pathways or on their wintering grounds. Even within a single site, places where they perch, roost, forage, or nest may be unique locales, selected in various ways. Thus the decision on where to measure rests largely on the research objectives. For example, an investigation of breeding habitat would quantify environmental factors within a group of territories, where measurements could be centered on nests or song perches to describe characteristics specific to those functions.

It is important to recognize that a study conducted at 1 location may not represent patterns of resource use at other sites. Many species exhibit geographic variation in their ecologies (Endler 1977). That is, the types and distribution of resources may vary among locales. The ability of organisms to exploit those resources may be reflected in their behavioral differences, and even by variations in their genotypes and phenotypic traits. Seminal work by Lack (1947) on Darwin's finches (*Geospiza*) illustrated variations in beak depth within a species, given dissimilar conditions among the various islands in the Galápagos archipelago (Fig. 3.7). Often such differences are significant enough to partition a species into subspecies, and dissimilarities may be manifested as a variety of life-history traits.

For example, Bervin and Gill (1983) reported geographic variation in the length of the larval period and size at metamorphosis among populations of wood frogs (*Rana sylvatica* [*Lithobates sylvaticus*]) sampled in the Canadian tundra, Virginia mountains, and Maryland lowlands. These diverse locales may drive a number of biotic and abiotic factors. Unique locations may vary in both the structure and composition of their vegetation, which provide different kinds and arrangements of foods, locales for dens or nests, display sites, or places of concealment from predators. The acquisition of food might require foraging on various substrates, or using alternate foraging modes (Block 1990). Different types and abundances of predators, as well as interspecific competitors, may regulate resource use. Density-dependent factors could influence where conspecifics establish breeding territories; where they forage, rest, and display; and how they move across the landscape.

Consider the Fretwell-Lucas model of habitat use by a species (Fretwell and Lucas 1970). According to their theory, individuals within a species will initially occupy a location until their numbers reach a certain threshold. Thereafter, they move into places where their survival and fecundity often decline. Thus, pooling characteristics across time and space without knowing quality outcomes among higher- and lower-grade locales will most likely result in biased estimates of vegetation structure and composition, topography, and other environmental factors that have little resemblance to those used by the relevant species. Further, their populations might respond differently to climate change, depending on the vegetation type that is occupied. For example, the nest survival of Lewis's woodpeckers (*Melanerpes lewis*) in a period of increasing temperatures rose in aspen (*Populus*) woodlands, whereas it decreased in burned pine (*Pinus*) forests (Towler et al. 2012). Had the study occurred only in deciduous woodlands or conifer forests, the researchers may have reached opposite conclusions. To better understand the effects of climate change on these woodpeckers, research should be conducted across the range of vegetation types that they inhabit.

Sample Size Requirements

Regardless of the care taken in crafting an investigation, all is for naught if an insufficient number of observations are made. To most, this must seem like an

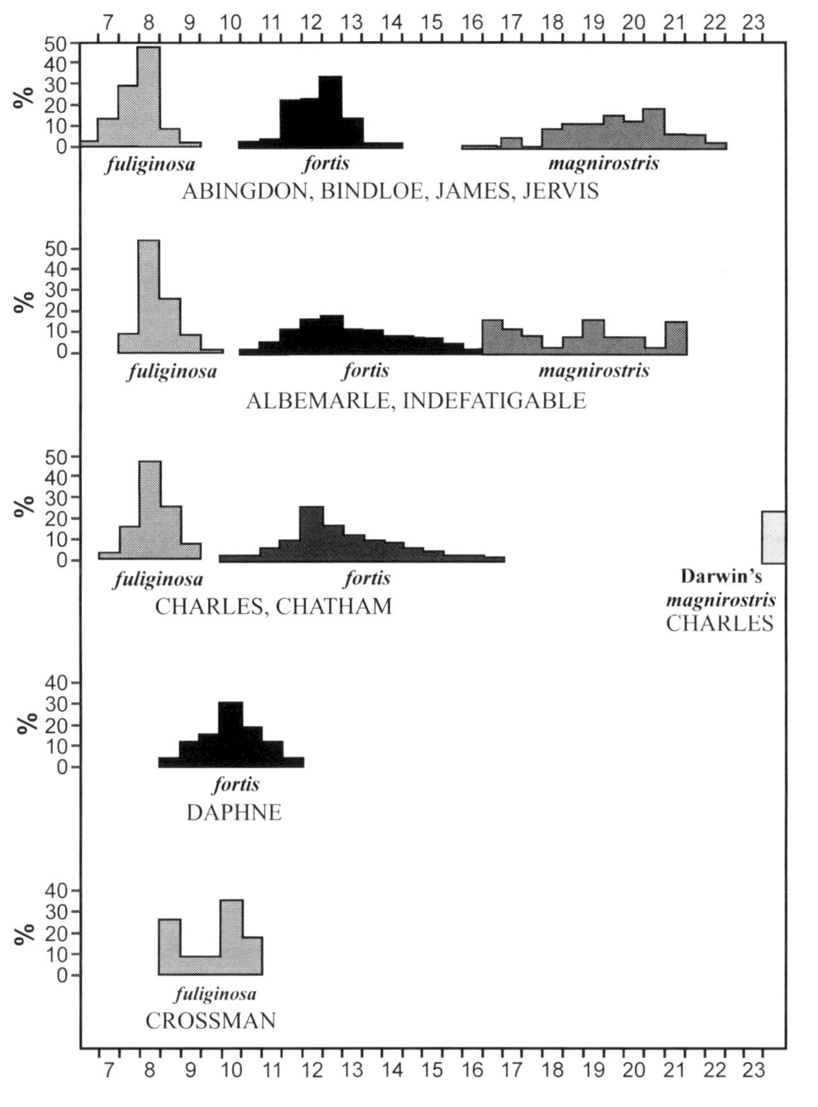

Figure 3.7. Histograms of beak depths in species of Darwin's finches (*Geospiza*) from specific islands (designated in capital letters) within the Galápagos Islands group. (Reproduced from Lack (1947), with the permission of Cambridge University Press)

obvious statement. Yet very little attention has been paid in the scientific literature—both in study designs and in statistical analyses—to this fundamental question. We are not sure why this is so and can only advance the suggestion that researchers have generally tried to collect the largest sample size their budgets and time constraints would allow. All of us are aware of the process, and several of us must admit to having previously followed this "strategy." If investigators must limit their sampling because of time or monetary constraints, then they would be better advised to limit the scope of a study by reducing the number of variables and produce 1 solid result, rather than several that are weak.

Most general statistics texts discuss the determination of sample size as a necessary step in performing certain analyses. The specific methods vary, depending on the type of data being collected. In general, the necessary sample size depends on 4 factors: (1) the amount of variation within the populations; (2) the size of the difference to be detected between sampling units (i.e., effect size); (3) a Type I error

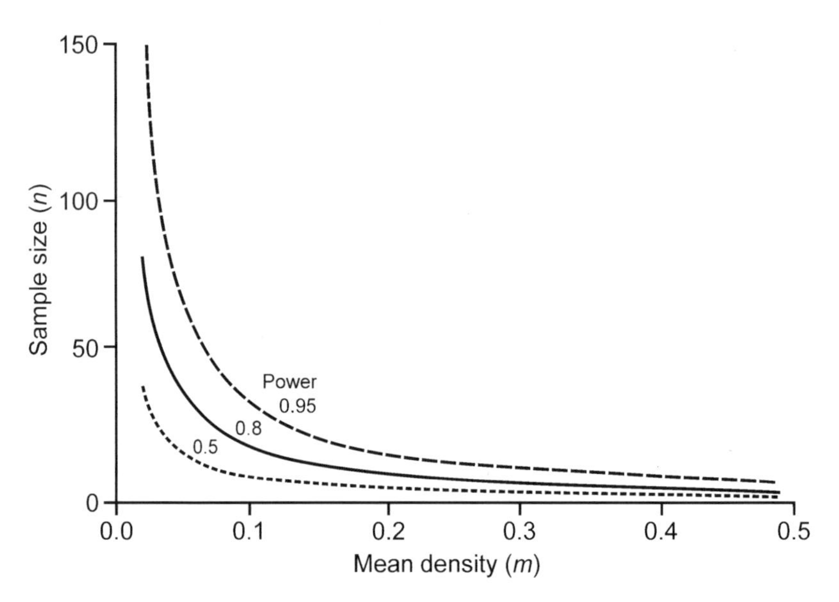

Figure 3.8. A graph of the necessary sample size (*n*), as a function of mean density (*m*), for various degrees of power (1 – β) when sampling a Poisson distribution. (Reproduced from Green and Young (1993:Fig. 3), with the permission of John Wiley & Sons)

(α), which is the probability of incorrectly rejecting a null hypothesis, also called a "false positive"; and (4) a Type II error (β), which is the probability of incorrectly accepting a false null hypothesis, also called a "false negative" (Lemoine et al. 2016).

A key concept often not stated in reported results is "statistical power." It is calculated as 1 – β and is essentially the probability of rejecting a null hypothesis. Once that is established, a researcher can then turn to the other 3 factors, with any 2 of them determining the third (Zar 1999). Note that an initial estimate of population variance is required. This can be derived either from the literature or from data collected during a preliminary analysis of the population of interest, provided that this initial sample is adequate for an unbiased estimate of variance. All of these factors work hand-in-hand in identifying an adequate sample size (Morrison et al. 2008; Fig. 3.8).

Setting a biologically meaningful effect size is also critically important. As noted by Hendry (2019:91), "in eco-evolutionary dynamics—as in many other enterprises—what matters is not whether there is an effect but rather what is the type and magnitude of that effect." An effect size that is too small (e.g., a 2% population decline) may require more samples than it is possible to obtain. Setting it too large (e.g., a 75% population decline) may trivialize the need for the study at all. Establishing β has direct implications for statistical power. Frequently, and more by tradition than by study-specific logic, the default is β = 0.95, resulting in a statistical power of 0.05. For some situations, such as evaluating the effects of management on a rare or listed (i.e., endangered or threatened) species, one might err on the side of caution and use a smaller β to increase the chance of showing an effect. A recent movement in ecology and other fields in the sciences, however, has been toward abolishing the use of statistical significance altogether, in favor of focusing more on effect size, confidence intervals, and meta-analyses (Cumming 2014; Amrhein et al. 2019).

Observer Bias

The scientific method is designed, in part, to help avoid preconceived notions being allowed to determine the outcome of studies. We all, however, carry numerous biases and limitations into the design of any investigation. Our training predisposes us to prefer certain sampling methods; we might believe that animals respond in specific ways to environmental conditions, which tends to narrow our sampling focus; time and monetary constraints limit our ability to sample in the way we would prefer; and so forth. These influence not only how we construct a research experiment, but also increase the potential for additional biases to be inserted into field sampling when multiple observers—most likely all of them with different levels of training and experience—are used to gather data. As noted by Gotfryd and Hansell (1985:224), "Ignoring observer-based variability may lead to conclusions being precariously balanced on artifacts, spurious relationships, or irreproducible trends." This is a strong statement that, unfortunately, is not heeded in the design and implementation of many studies of wildlife and their habitats. Here we will concentrate on the evaluation and subsequent reduction of bias in field sampling.

Gotfryd and Hansell (1985) used 4 observers to independently sample 8 plots located within an oak-maple (*Quercus–Acer*) forest near Toronto, Canada. They followed methods detailed by James and Shugart (1970), measured the variables given in Table 3.1, and found that observers differed significantly in their measurements of 18 of the 20 vegetation variables that were used. As these authors noted, their study addressed only the precision of estimates between observers. No determination was made about the accuracy of the workers' results.

Block et al. (1987) used several univariate and multivariate analyses to test for differences among 3 observers in assessing plant structures and floristics. They found that ocular estimates by these workers differed significantly for 31 of the 49 variables they measured. Their study revealed that perhaps the most confounding aspect of using several observers was the unpredictable nature of the variations among them. Multiple comparisons of estimates for the 31 significant variables resulted in all possible combinations of who reported what. Thus when samples from different observers are pooled, the sampling bias can increase. As we show later on in this chapter, this lack of concordance, especially when combined with low sample sizes, has an especially profound influence on multivariate analyses.

Table 3.1. Vegetation habitat variables

Memnotic	Variable
TRSP	Number of tree species
SHSP	Number of shrub species
SDEN*	Density of woody stems <7.6 cm diameter at breast height (dbh)
CC*	Canopy cover
GC*	Ground cover
BAA	Basal area (BA) of trees 7.6–15.2 cm dbh
BAB	BA of trees 15.2–23 cm dbh
BA1*	BAA + BAB
BA2*	BA of trees 23–53 cm dbh
BA3*	BA of trees >53 cm dbh
CH1	Maximum canopy height
CH4	Maximum canopy height in the quadrant having the lowest canopy
CHAV*	Average of canopy height maxima, by quadrant
CHRNG*	CH1–CH4
CHCV*	CHRNG/CHAV
DTR1	Maximum of the nearest tree distances, by quadrant
DTR4	Distance to the nearest tree >15.2 cm dbh
DAV*	Average of the nearest tree distances, by quadrant
DRNG*	DTR1–DTR4
DCV*	DRNG/DAV

Source: Gotfryd and Hansell (1985:Table 1)

* Variables used in multivariate analysis

Ganey and Block (1994) had 3 observers sample plots for canopy closure, using 2 different estimation devices: a spherical densiometer and a sighting tube (their study is described in Chapter 2). They found significant variations among the workers' assessments of canopy cover with each of these 2 methods. The results, however, were relatively more consistent (i.e., more precise) with the sighting tube.

The biases associated with estimations of animal abundance should also be considered carefully in habitat studies. This is because many of our analytical procedures correlate animal numbers with features of the environment. Obviously, a study that has a low bias among habitat characteristics can be ruined by count data exhibiting a greater bias, and vice versa. For example, Dodd and Murphy (1995) evaluated the accuracy and precision of 9 techniques used to count great blue heron (*Ardea herodias*) nests. Although they found rather high error rates among the techniques, observer bias was low for most methods.

Intriguingly, the greatest disparity was found in their point-counting technique, a result apparently due to the varying choices made by workers concerning the optimum vantage point from which to count nests in the heron colonies.

Researchers can reduce interobserver variability by closely following a set of well-defined criteria for selecting and training such individuals. Although designed for bird censuses, the steps outlined by Kepler and Scott (1981) for counting procedures can generally be applied to most types of sampling. Carefully screen the applicants initially, to eliminate the more obvious visual, aural, and psychological factors that increase observer variability. In addition, organize a rigorous observer training program, which, while it will further reduce inherent variation, still may not entirely eliminate it. In a field experiment, Scott et al. (1981) found that such workers, after training, could estimate the distance to a singing bird within 10–15 percent of a verifiable measurement. Also, periodic training sessions with observers should be conducted to counteract any of the workers' "drift" (i.e., changes in their observation patterns or habits) and recalibrate their records to standard and known values (see Block et al. 1987). Workers can also be trained to develop more-precise ocular estimates of vegetation conditions, although such preparation does not necessarily reduce bias. In an extended study of bird communities, Marcot (1985) first screened his prospective field crews with hearing tests conducted at a health clinic, to ensure that they were capable of detecting frequencies of up to 8 kilohertz, which occur in some bird songs, such as those of golden-crowned kinglets (*Regulus satrapa*) and brown creepers (*Certhia familiaris* [*C. americana*]). He then conducted field tests, where the candidates silently and simultaneously recorded bird species and distances from point counts. Afterward, he compared the results, to determine the degree of concordance or disagreement in species identification and distance estimation; tutored the candidates on any major differences and errors; and iteratively repeated the trials, to reduce any remaining biases among the observers to an acceptable level. Plants ecologists have long recognized disparities among data collection techniques (e.g., Cooper 1957; Lindsey et al. 1958; Schultz et al. 1961; Cook and Stubbendieck 1986; Hatton et al. 1986; Ludwig and Reynolds 1988; Kent and Coker 1994). Morrison (2016) found that 92 percent of the 59 vegetation studies he reviewed resulted in a significant observer error for at least 1 comparison. When he looked at species composition, 5–10 percent of the plant species were misidentified, and 10–30 percent were simply missed by observers. The mean coefficient of variation (CV) among workers in surveys of vegetation cover exceeded 100 percent for taxa with low cover. Observer errors contributed to biases in the estimates of plant community metrics and in the results from multivariate analyses. Although the cost of measuring plant structure and floristics in an adequate number of plots usually is substantial in terms of both time and money, the ramifications of not following a rigorous sampling design are severe. Morrison (2016) concluded that using multiple observers, providing additional training, and continually evaluating and calibrating the results were strategies that could reduce worker errors in vegetation surveys, much as Marcot (1985) did with his bird survey crews. Again, it is better to limit the scope of a study, to ensure that the data are properly collected. Preliminary sampling and analyses of that information can be vital to help identify and correct potential problems with data-collection bias and error.

How to Analyze: Statistical Assessments of Wildlife Habitat

Conceptual Framework

Over the course of our careers, we have witnessed rapid growth and increased sophistication in the analysis of data in animal ecology. For much of the twentieth century, investigators measured a set of variables and, in their publications, displayed the means and associated calculations of dispersion in a table format. They might have recognized that some variables were highly correlated, measured the strength of those associations, and discarded variables that were difficult to interpret or had less biological meaning. Univariate tests—such as analyses of variance (ANOVA), t-tests, chi-square (χ^2) tests, and others—were done with a pad of paper. In the early 1970s, handheld calculators represented a huge advancement in working with basic arithmetic and, later, more-advanced analyses. This approach served us well, and we learned quite a bit about species-habitat relationships.

Then 2 major events facilitated a transition from a univariate to a multivariate approach. The first was a realization that the space used by and the roles of species were multidimensional (Hutchinson 1978). The second was the advent of major advances in computing power, enabling analyses of large, multivariable datasets. Complex statistical methods eventually became fully developed. For example, principal-components analysis was invented in 1901, but nobody could use it yet. After the emergence of mainframe computers in the 1960s, powerful statistical programs followed in short order. By the mid-1970s, a variety of potent software packages became available, including SPSS and SAS. It is important to recognize that without a computer it is not practicable to perform most modern statistical analyses.

Among the first to apply computing power to multivariate statistical analyses in ecology were Green (1971), in his work on the Hutchinsonian niche of bivalves, and James (1971), in her work with breeding bird forests in the Ozark Mountains of Arkansas. Multivariate approaches gained traction with the appearance of a report from a workshop on this topic held in Burlington, Vermont, in 1980 (Capen 1981). Since then, studies using multivariate techniques have been published, evaluated, and improved. Even so, multivariate statistics are not a panacea (James and McCullough 1990). Many multivariate approaches are based on assumptions that are difficult to test, and the resulting patterns can be harder to interpret. Hence biometricians continue to develop new approaches, hoping to improve our understanding of ecological patterns. Below, we first review some of the more widely used multivariate methods and then explore some emerging analytical methods, such as niche modeling, random forests, machine learning, neural networks, and others.

Multivariate Statistics

Multivariate analysis is a branch of statistics used to evaluate multiple measurements that have been made on 1 or more samples of individuals (Stuber et al. 2019). It is distinguishable from other forms of statistical procedures in that multiple variables are considered in combination, as a system of measurements. Because these variables are typically dependent on each other, we cannot separate them and examine each individually (Cooley and Lohnes 1971:3). Biologically, many phenomena only occur if certain combinations of factors are present. For example, an organism might live on north-facing slopes at high elevations. Thus neither elevation nor aspect alone would be particularly informative, but a model containing both probably would be.

In this section we examine the rationale for using multivariate techniques, relating these methods of analysis to our conceptualization of wildlife-habitat studies. In the next, we review the all-important assumptions associated with them. That is followed by a classification and discussion of some of the current techniques and their applications to wildlife studies, concentrating on the frequently used procedures of multiple regression and discriminant analysis. Our intent here is to briefly introduce the employment of multivariate techniques in analyses of wildlife habitat data, concentrating on problems encountered in their use. A formal course in multivariate statistics is not a prerequisite for understanding this chapter. We do, however, strongly recommend making it a part of the academic plan for all graduate students, as it will enhance their ability to evaluate the literature, even if they never conduct a multivariate analysis of their own data. Many good multivariate texts are available and should be consulted for details not in-

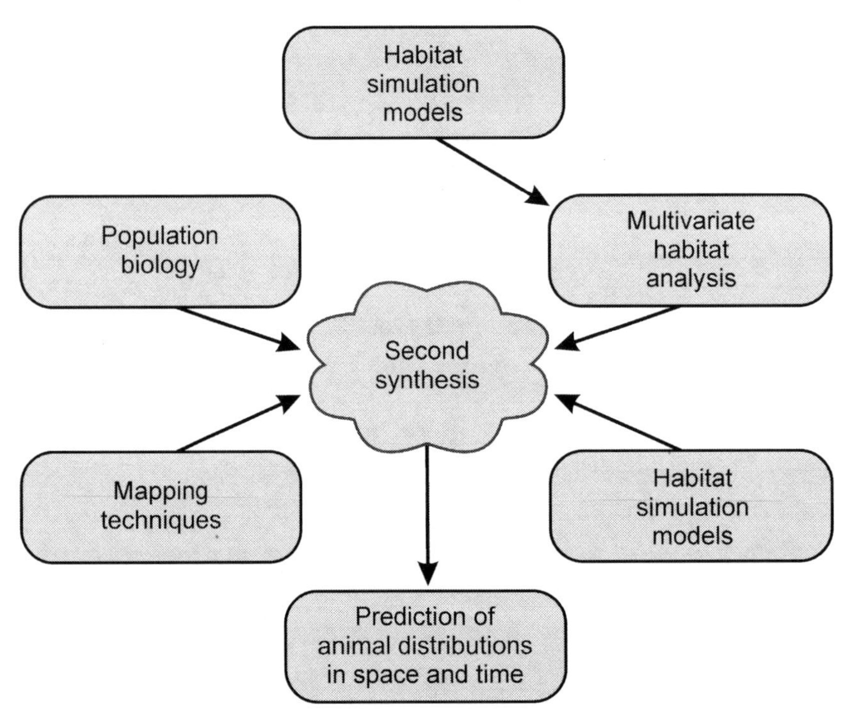

Figure 3.9. A schematic diagram of the scientific research elements that are combined in a synthesis to produce a multivariate habitat analysis. (Reproduced from Shugart (1981:Fig. 1), courtesy of the US Department of Agriculture, Forest Service)

cluded herein (e.g., Cooley and Lohnes 1971; Dillon and Goldstein 1984; Johnson 1992; McGarigal et al. 2000; Timm 2002; Shaw 2003; Manley and Navarro Alberto 2016). The publications by McGarigal et al. (2000) and Manley and Navarro Alberto (2016) were written by wildlife scientists and provide many examples of how these methods can be applied to wildlife questions.

Multivariate statistical techniques were not originally designed for the analysis of wildlife habitat and behavioral data. Indeed, they have been used for other purposes since the late 1880s, with a progression of new, more sophisticated methods following throughout the 1900s (Cooley and Lohnes 1971:4). Only since the 1960s have animal ecologists placed an emphasis on quantitative analyses of habitat. The application of multivariate analyses to wildlife data is the product of a synthesis, beginning in the early 1970s, that united several lines of ecological research. As outlined by Shugart (1981), this involved linking 2 analytical tools with 3 ecological concepts—niche theory, microhabitat, and individual response (Fig. 3.9). Hutchinson's reformulation of the niche concept (1957) in terms of an n-dimensional hypervolume caused ecologists to alter their view of wildlife habitat and the ways in which they analyzed data (Stauffer 2002).

Assumptions

There are 4 common assumptions associated with parametric multivariate analyses: multivariate normality, equality of the variance-covariance matrices (i.e., group dispersions), linearity, and independence of the error terms (i.e., residuals). Violation of any of these assumptions can bias or taint the results of an analysis and the conclusions derived from them. Unfortunately, many published papers in the wildlife literature have failed to discuss these assumptions or their ramifications on results (but see Williams 1983 for a diagnosis of discriminant analysis, a widely used multivariate method). As we discuss beyond, low sample sizes are often a major factor in such violations, as is the large number of variables associated with low sample sizes.

Normality The assumption of multivariate normality is more than simply a presumption that each variable is itself normally distributed, in the univariate sense. Unfortunately, tests of this premise are cumbersome and yield only approximations. A variable-by-variable examination of normality, however, will certainly help identify those that greatly depart from normality. Korkmaz et al. (2014) suggested that if data do have a multivariate normal distribution, then each of the variables has a univariate normal distribution, but the opposite does not have to be true. Standard univariate transformations (e.g., logs, square-root transformations) can also be applied as appropriate. These authors created an R package, containing 3 frequently employed tests (Mardia's, Henze-Kirkler's, and Royston's), to assess normality. Each has its merits and limitations, but it is beyond the scope of this volume to evaluate them.

Variance and Dispersion In ecology, it is well known that the distribution and behavior of animals change along gradients of environmental variables (e.g., soil moisture, canopy cover, air temperature). As noted by Pimentel (1979:177), however, biologists seem too concerned with the mean responses of animals to environmental gradients, rather than the distribution of animals along such gradients. If populations have unequal dispersions, then they are different, even if their central tendencies (i.e., means) are the same. Thus trying to force normality to meet the formal assumption of the equality of dispersions is biologically unsound. The magnitude of this dispersion, however, can be used as a multivariate measure of niche breadth (see Carnes and Slade 1982). Nothing prevents one from comparing populations having unequal dispersions. Fortunately, research indicates that tests of equality of group centroids are rather insensitive to moderate departures from multivariate normality and homoscedasticity. If sample sizes are large and are equal between groups, then the inequality of the dispersions has no real effect on an interpretation of the results. The application of this rationale requires careful planning, to ensure that large, equal samples are collected for all groups under investigation—a caveat that has been ignored in most research on wildlife-habitat relationships. As noted by Wiens (1989:66), a considerable portion of the variance that is so easily discarded in statistical analyses may contain important insights into the dynamics of the system being studied.

Linearity Linearity in parametric multivariate analyses is important in 2 main ways. First, most models are based on assumptions of linear relationships. Second, the correlation coefficient—which forms the basis for most multivariate calculations—is sensitive only to the linear component of the relationships between 2 variables. Assuming such linearity is a more parsimonious approach. Therefore, non-linear multivariate models should be used if a non-linear connection is anticipated. Fortunately, many such associations can be approximated using linear models, even though non-linear components may exist. The data transformations noted in the section on "Normality" may help linearize a non-linear relationship. Non-linearity, however, changes the probabilities in tests of significance. For example, an investigator is likely to fail to reject null hypotheses of equality of group centroids, yet reject null hypotheses of the equality of group dispersions (Pimentel 1979:178–179). Researchers would be well served by first examining the linearity of their data, variable by variable, before plunging those data into the "black box" of a canned multivariate statistical package. In biology, a common form of non linearity is a sudden shift from 1 state to another. In many analyses, this type of variation can often be conceived of as a categorical variable. For example, vegetation tends to remain fairly similar in areas for a while and then shift rather abruptly. Viewed as a continuous variable, vegetation is strongly non-linear in space. Vegetation descriptions, therefore, tend to break these non-linear functions into categorical groups (e.g., forest, non-forest), with boundaries drawn along the areas of rapid change.

As noted by McGarigal et al. (2000), resource utilization by a species is often assumed to approximate a Gaussian distribution. If that is so, linearity may only apply toward the tails of the distribution. Thus the selection of linear models is usually more a matter of past practice and statistical convenience than a decision based on ecological reality. As noted above, we should expect biological entities to often have non-linear and non-normal distributions. Linear analyses, however, are relatively straightforward, and their statistical properties are well understood. Further, a model has little utility if other researchers—and especially managers—find it difficult to understand. Thus it can be argued that linear models should be used unless the data are *highly* non-linear.

There are 2 basic methods for applying non-linear methods to statistical models. The first is to include transformed variables in a linear model. A common approach is to transform such variables polynomially. Linear models are first-order polynomial models. Second-order, or quadratic, models are obtained by squaring a variable (X^2), resulting in a U-shaped relationship with a single inflection point. Third-order, or cubic, models are obtained by cubing a variable (X^3), producing a curved relationship with 2 inflection points. Other common transformations are used to capture non-linear relationships, include log functions. We believe that these sorts of transformations should be driven by the expectation that a quadratic or cubic relationship is appropriate, rather than in an attempt to fit a function into a single model. Simple examples include the measurement of 1 dimension from a multidimensional entity (e.g., tree volume is expected to be cubically related to tree height), or where a well-understood non-linear pattern is expected (e.g., wind speed decreases exponentially as you approach a surface).

Higher-order models are possible, but they become increasingly difficult to visualize and are unlikely to coherently relate to expected patterns. Thus relationships based on highly transformed variables are difficult to interpret and evaluate. An informative example is provided by Burger et al. (1994), who regressed arcsine-transformed percentages of nest pre-

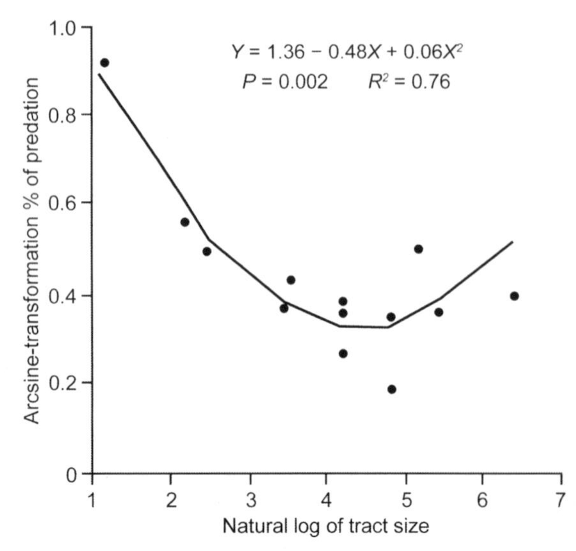

Figure 3.10. A graph of arcsine-transformed predation on artificial nests (n = 540) in 15 prairie fragments, regressed on natural log-transformed tract areas (ln[size]) and the square of such areas (ln[size]^2). (Reproduced from Burger et al. (1994:Fig. 1), with the permission of The Wildlife Society)

dation onto natural log-transformed areas (ln[size]) and the square of these areas (ln[size]^2). This regression explained 77 percent of the variation in predation rates among study tracts of differing sizes (Fig. 3.10). It is exceedingly difficult, however, to interpret these transformations in terms of real-world natural history. Rather than creating highly transformed models, you should seek to limit transformations. If a linear model employing untransformed variables can meet your objectives, then use it.

Meents et al. (1983) evaluated the use of first-, second-, and third-order independent variables on predictions of bird abundance that had employed multiple regressions. They found that polynomial variables resulted in significant regressions in many cases, whereas linear variables did not. They also showed that while linear relationships dominated during certain parts of the year, non-linear ones were important during other periods. This explains, in part, why models often must be season specific. In addition, the authors showed why non-linear relationships can have important management implications. For instance, woodpecker abundance might be related to snag density in a curvilinear fashion, where relatively low or high densities of snags results in a smaller number of birds than those found in moderate snag densities. Thus woodpeckers may disappear when the quantity of snags declines below some critical non-zero threshold. In this simple case, linear models could mask important biological relationships, which could result in faulty management decisions. Some examples of non-linearity are readily apparent, such as soil-moisture tolerance in salamanders, the micronutrient content of herbivore forage, and grass density for rodents. Other examples of non-linear analyses of wildlife-habitat data are included below, along with their specific analytical techniques (e.g., logistic regression).

To this point we have discussed using variable transformations to impart non-linear behavior onto linear models, but there are also a variety of non-linear models. Wester (2019) provided some linear and non-linear analyses of animal ecology data using different regression techniques: single and multiple variables, and non-parametric approaches. Draper and Smith (1981:Chapter 10) gave examples of such non-linear models, and Seber (1989) and Ratkowsky (1990) presented a thorough development of non-linear modeling as it applied to regression analysis. On computers, non-linear models are solved as approximations, through relatively complicated iterative numerical calculations. (The form of the equations and the iterative methods are beyond the scope of this book.) Non-linear models have, however, been incorporated into most of the larger statistical software packages.

Random Samples and Independence Independence and random sampling are cornerstones of many biological investigations, regardless of how the data are analyzed. In studies of biological populations, truly random sampling is often difficult to achieve. As discussed elsewhere, an inability to randomly sample from a biological population precludes an understanding of how the attributes of the sampled group are consistent with the biological population. This is a fundamental problem, which modern statistics cannot alleviate. Requirements for independence between samples, however, have been greatly reduced in recent years through the development of mixed models, in which some variables are assumed to be correlated. For example, many studies that use radio telemetry often violate assumptions of independence, because numerous fixes on the same individual are not independent events, regardless of the time that had elapsed between them. Similarly, multiple samples from flocks, herds, coveys, or similar groupings of animals are probably not independent. Historically, these sorts of correlated data were problematic, because the information content of correlated samples is less than that of independent samples. Therefore, the sample count inflated the true statistical *n*. It is important to recognize correlated data streams and both sample and model them appropriately. It is also critical to understand the nature of the population that you are sampling from, as well as its relationship to the larger biological population in which it is embedded. Thus we should tightly restrict the definition of the population that is sampled (here, "population" refers to a sampled group, not a biological population). The biological population will have attributes associated with sex, age, time, and location, which may differ from those of your sample population. When these elements are ignored, unequal representation of the factors between samples might imply differences that do not exist between populations (Pimentel 1979:176). Unfortunately, few published studies specifically or adequately define their statistically sampled population (Morrison et al. 2020:Chapter 1). Failure to do so has clear and adverse management implications, as users will be unable to apply these results at the appropriate time and place, thus overextending them beyond the target population (Tacha et al. 1982).

Classification of Multivariate Techniques

We can divide multivariate techniques into 3 general categories: ordination, classification, and model selection. Whereas all of them can be analyzed using parametric procedures, non-parametric and Bayesian approaches are becoming more widely used and accepted. James and McCullough (1990) summarized many of these multivariate measures, noting both their objectives and their limitations (Table 3.2). Methods are broadly classified by the number of dependent and independent variables of interest and the goal of the researcher in analyzing the data. That is, does the investigator wish to examine the structure or interdependence of the variables (principal components analysis); determine the relationship (i.e., correlation) among variables (multiple regression); separate groups (multivariate analysis of variance); or develop predictive equations (discriminant analysis)? Dillon and Goldstein (1984:Fig. 1.5–1) and Harris (1985:Table 1.1) gave similar classifications of multivariate techniques. Below, we briefly outline some of the specific methods found within these categories and provide examples of their application to wildlife-habitat data. Because most studies of wildlife-habitat relationships have concentrated on principle components analysis (PCA), multiple regression (MR), multivariate analysis of variance (MANOVA), discriminant analysis (DA), and logistic regression (LR), we will concentrate our discussion on how these methods are used and interpreted. There are many additional analyses—such as detrended correspondence analysis (DCA), non-metric multidimensional scaling (NMDS), and reciprocal averaging (RA)—that could be applied to habitat analysis. Capen (1981), Verner et al. (1986), McGarigal et al. (2000), and Scott et al. (2002) should be consulted for many more specific examples.

Data Structure: Ordination and Clustering Methods within this broad category of analyses seek to reduce a complex dataset (i.e., many variables) to a small number of dimensions (i.e., axes) that are internally correlated but are unique (i.e., not correlated) with regard to other derived dimensions. Ordination and clustering are similar in that no groups are assumed to exist prior to the analysis. Because wildlife biologists usually collect data on numerous, usually intercorrelated variables, they need a way to reduce the number of active variables to a manageable level. A common technique involves a 2-step procedure. First, variables that are highly intercorrelated are identified. The degree of correlation between such variables is somewhat arbitrary and user defined, but $r > 0.7$ is frequently employed. Second, for each correlated pair that is identified, the 1 having less biological relevance (as interpreted by the investigator) or the least power to separate groups (as identified through a *t*- or *F*-test) is removed from the analysis. Unfortunately, unless the excluded variable had a perfect correlation with the included variable, this method results in some information being lost from the dataset. The lower the correlation (r), the greater the loss. Thus data-reduction techniques that retain most of the information from the variables are desirable.

Principal components analysis is a method that identifies new sets of orthogonal (i.e., mutually perpendicular and thus not correlated) axes to encompass the greatest variance among observations (Stuber et al. 2019). The first axis is the line that goes through the observations—in other words, it is oriented so that the projections of the observations onto the axis have maximum variance. The second axis should be in the direction of the greatest remaining variance, perpendicular (i.e., orthogonal) to the first axis—that is, it does not duplicate the variance explained by the first axis. Additional axes are derived until all of the explainable variances are accounted for. A general rule is to retain axes with an eigenvalue greater than 1, as these axes elucidate the largest amount of variations in the dataset.

Shew et al. (2012) used principal components analysis to better understand habitat use by western foxsnakes (*Pantherophis vulpinus* [*P. ramspotti*]) in Missouri. They tracked radio-marked snakes and recorded habitat information on each snake 1–4 times each week. The number of locations where habitat

Table 3.2. Objectives and limitations of 12 multivariate procedures commonly used in ecology

Procedure	Objectives	Limitations
Multiple regression	• Predict response of 1 variable, *Y*, from a set of explanatory variables, Xs. • Investigate the association of an X variable with the Y variable in the presence of other variables. • With experiments, investigate cause and effect relationships.	• Strong correlations do not allow inferences of cause and effect. • Predictions apply only to situations similar to those in which the model was developed. • Stepwise regression is not encouraged. • It only considers linear relationships. • It only is applicable to situations where *Y* is a continuous variable. • Errors should be normal, and sampling random.
Multivariate analysis of variance	• Test for a difference between/among 2 or more groups of objects with multiple Xs.	• The procedure is intended for use with continuous multivariate data. • Each vector of observations is assumed to be independent.
Discriminant function analysis	• Develop linear combinations of Xs with the maximal ability to discriminate groups of objects. • Refer to canonical variates analysis when used to reduce dimensions of the data. • Can use the discriminant function to classify current or future observations to groups.	• It is intended primarily for continuous data. • It assumes multivariate normality. • It assumes homogeneous dispersion matrices between/among groups. • Only linear combinations are considered. • Groups must be defined a priori.
Principal components analysis	• Reduce the dimensionality of a dataset consisting of objects and attributes of observations. • Develop linear combinations of the variables that encompass maximal variance. • Suggest new combinations of variables for future study.	• It is intended for continuous data. • Only linear combinations are considered.
Principal coordinates analysis	• Describe data by reducing the dimensions of a distance matrix among objects. • Generalize a principal components analysis in which non-Euclidean distances may be used.	• Results vary with the distance measure used. • It cannot indicate the combination of variables, because it relies on the distance matrix only.
Factor analysis	• Reproduce a correlation matrix among original variables by hypothesizing the existence of 1 or more underlying factors. • Discover the underlying data structure by interpreting the factors.	• Exploratory methods are unstructured, thus interpretations are subjective. • It is inefficient for data not summarized well by correlations. • It is not useful for non-linear relationships or categorical data.
Canonical correlation	• Analyze the correlation between 2 groups of variables about the same set of objects simultaneously, rather than calculating pairwise correlations.	• It is inefficient for data not summarized well by correlations or linear combinations. • It is not ideal for non-linear relationships or categorical data.
Logistic regression	• Model a dichotomous or multinomial *Y* variable as a functions of multiple explanatory, *X*, variables. • Use with X variables, which can be categorical or continuous. • Investigate the association of an *X* variable with the *Y* variable in the presence of other variables. • With experiments, investigate cause and effect relationships. • Is an alternative to discriminant function analysis.	• Good predictability does not infer causation. • Stepwise procedures are inappropriate. • The procedure considers only linear combinations of Xs. • Predictions apply only to situations similar to those in which the model was developed.
Log-linear models	• Investigate joint relationships among categorical variables.	• Variables must be categorical. • When there are both Y and X variables, logistic regression may be more appropriate.

(*continued*)

Table 3.2. continued

Procedure	Objectives	Limitations
Correspondence analysis	• Describe data consisting of counts by reducing the number of dimensions. • Usually used for graphical display. • Suggest new variables for future study.	• It is inefficient for data that are not counts, because they will not be described by chi-square distances. • It is not suitable for non-linear data. • It will not discover non-linear relationships.
Non-metric multidimensional scaling	• Describe data by reducing the number of dimensions. • Usually used for graphical display. • Discover non-linear relationships.	• The procedure uses rank-order information only.
Cluster analysis	• Classify or group objects based on a similarity measure. • Reduce a set of objects to a smaller set of objects.	• Results vary with the clustering algorithm used. • Results depend on the distance measure used.

Source: Modified from James and McCollough (1990)

was measured ranged from 26 to 42 for each snake. They also measured habitat variables at random plots within the same general area where snakes were captured. Prior to conducting a PCA, they examined intervariable correlations, and if $r > 0.7$, they retained the variable likely to have the broadest influence. Prior to conducting the analysis, they examined each variable for normality and used transformations to improve normality where needed. Because the resulting variables differed in scale, they employed a correlation matrix, rather than a variance-covariance matrix, in their analysis. They retained the 3 principal components with eigenvalues >1, which explained more than 56 percent of the variation in the dataset. The first principal component (PC1) represented a gradient from live to dead herbaceous vegetation; PC2 was associated with the density and height of live vegetation. On both components, male and female snake locations were separated from random plots by using areas of denser yet dead herbaceous vegetation and denser plus taller live vegetation (Fig. 3.11). These results suggested that mowing dead herbaceous material should be delayed during the breeding season, as this type of vegetation was preferentially selected by foxsnakes.

We often forget that humans are part of the systems where wild animals occur. As such, human activities and our ideas of wild animals have both direct and indirect effects on species' ecologies. Multivariate analyses are widely used to evaluate social perceptions of wildlife, in addition to describing ecological patterns. For example, Hazzah et al. (2013) studied the conflict between protecting fauna and flora and accommodating the needs of the local communities within and bordering Tsavo and Nairobi National Parks in Kenya. Allowing local ranchers access to the parks and their resources is a contentious issue, because the retaliatory killing of African lions (*Panthera leo*) has caused a steep decline in the species' population, threatening local extinction. The authors interviewed 206 men who headed households with livestock holdings. They conducted 2 separate factor analyses: the first to understand attitudes toward wildlife, and the other to evaluate factors that predisposed these men to kill lions. The first analysis (Table 3.3) revealed that respondents thought lions were important to conserve now and into the future, for a variety of reasons. The second (Table 3.4) depicted social elements of lion-killing behavior (e.g., to gain prestige within the community, or for entertainment) and illustrated that lion killing was not always provoked by livestock depredation.

Recently, we have seen an increase in the use of non-metric dimensional scaling (NMDS) as an ordination technique. NMDS differs from principal components analysis by using measures of similarity, instead of correlations, to order species along environmental gradients. Whitney et al. (2015) studied the effects of uncharacteristically large wildfire on colonization by native and non-native fishes in

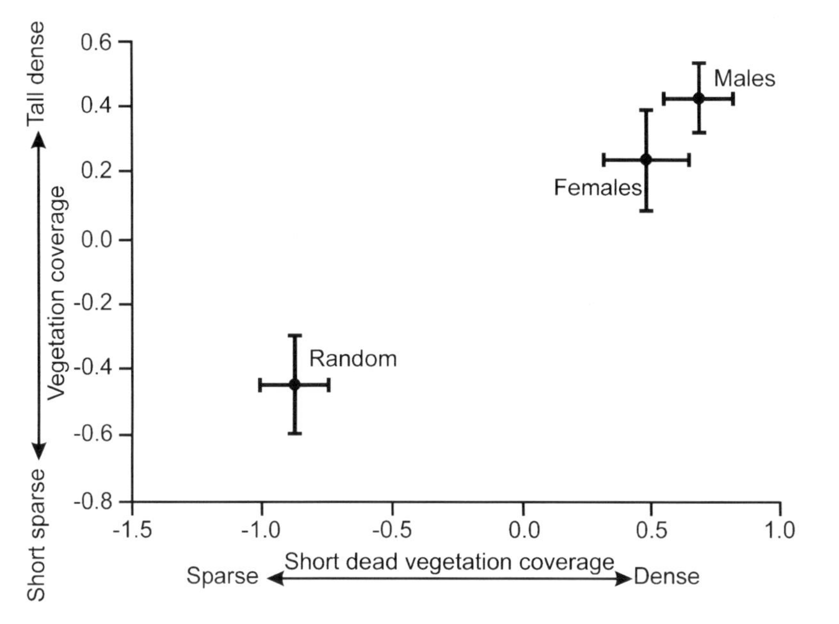

Figure 3.11. A graph of mean principle components scores (with standard errors) for males, females, and random locations of western foxsnakes (*Pantherophis vulpinus*) at Squaw Creek National Wildlife Refuge in Missouri in 2013. (Reproduced from Shew et al. (2012), courtesy of the Society for the Study of Amphibians and Reptiles)

Table 3.3. A pattern matrix of wildlife attitudes

Statement item	Factor 1	Factor 2
Lions have a right to exist.	0.929	−0.176
God would want me to protect all wildlife.	0.806	−0.018
It is important to me that my grandchildren see lions.	0.741	−0.069
I feel that lions are beautiful animals.	0.724	0.077
Lions deserve protection.	0.692	−0.004
The wildlife in the ecosystem is a national treasure.	0.643	−0.026
I like to watch wildlife in their natural environment.	0.623	0.098
I feel lions have the same rights as livestock.	0.556	−0.008
I would like to communicate more with scientists.	0.483	0.184
I appreciate the role that wildlife plays in the environment.	0.455	0.321
I am concerned about over-hunting wildlife.	−0.054	0.753
I am concerned about the future of wildlife.	−0.009	0.589

Source: Hazzah et al. (2013)

Note: This is a 2-factor model, using principal-axes factoring with Promax rotation ($n = 206$)

Table 3.4. A factor-loading matrix of lion-killing propensities

Statement item	Factor 1	Factor 2
Killing a lion for prestige/status is acceptable.	0.894	−0.074
I will kill a lion just for fun.	0.865	−0.110
Traditional hunts are acceptable.	0.792	−0.025
It is acceptable for young boys to kill a lion for practice.	0.718	0.085
When I see a lion, it is acceptable to kill it.	0.673	0.110
If a lion entered my *boma* I would kill it.	−0.045	0.740
I will kill a lion to defend my property.	−0.117	0.693
If a family member was injured by a lion, I would kill it.	−0.160	0.633
If my cow was killed by a lion, it would be acceptable to kill it.	0.134	0.628
Snaring a problem lion is acceptable.	0.191	0.523
If my father asked me to kill a lion in revenge for our cows . . .	0.127	0.443

Source: Hazzah et al. (2013)

Note: This is a 2-factor model, using principal-axes factoring with Promax rotation ($n = 206$)

the American Southwest. They used NMDS with a Bray-Curtis distance matrix of occupancy probabilities across 16 sites to identify those that maintained a high degree of occupancy for each species and could serve as potential sources for recolonization (Fig. 3.12). Ordination of the species' occupancy across the 16 sites sampled during wildfire—looking at both among-site (Fig. 3.12*A*) and interspecific (Fig. 3.12*B*) variations—indicated that native occupancy was not necessarily tied to sites that experienced minimal

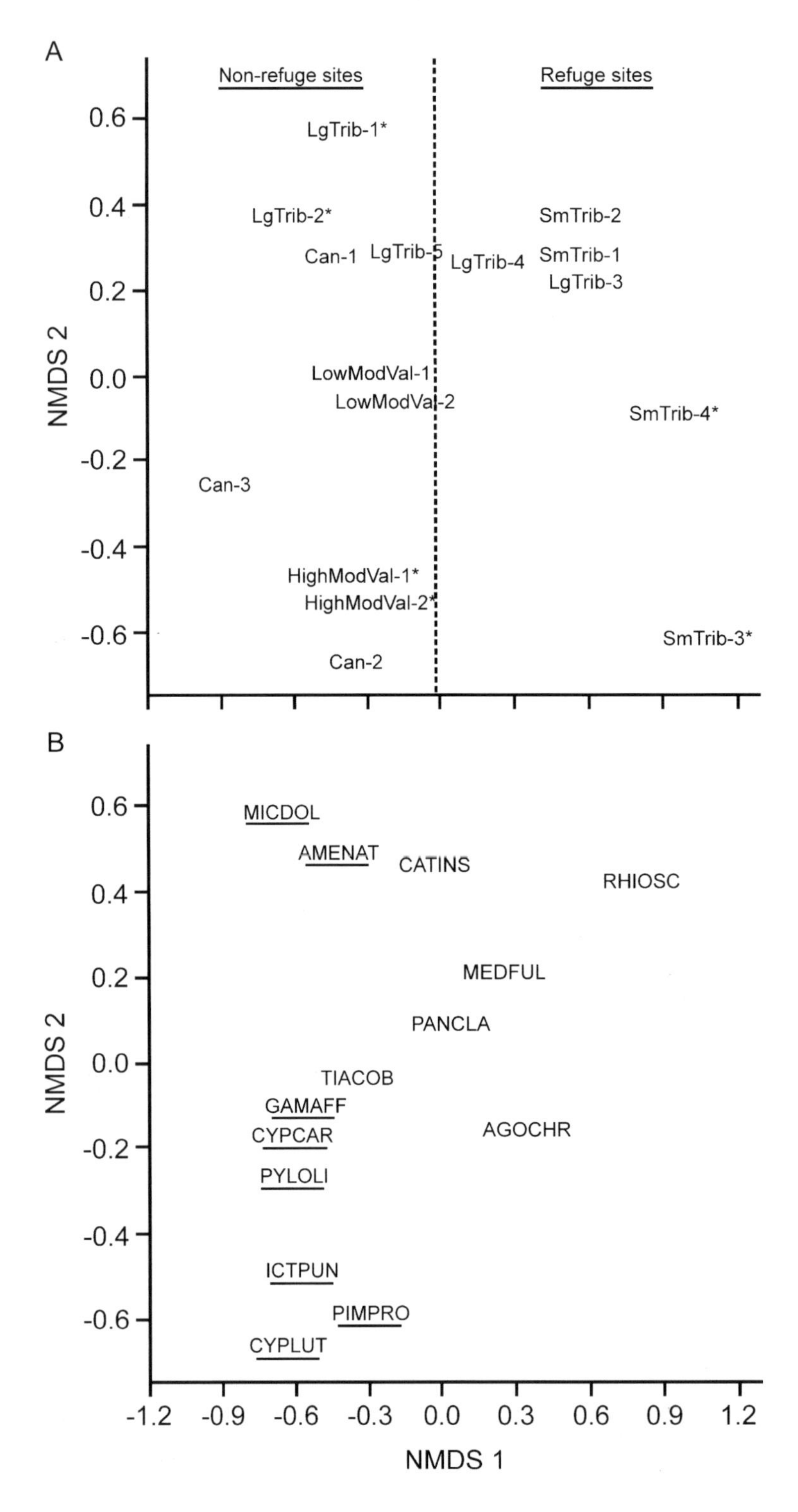

Figure 3.12. Graphs of non-metric multidimensional scaling biplots summarizing among-site (*A*) and interspecific (*B*, with standard abbreviations used for each species) variation in occupancy probabilities for native and non-native fishes in the American Southwest. (Reproduced from Whitney et al. (2015), courtesy of Blackwell Scientific)

disturbance effects of wildfire. The first axis of the NMDS separated sites with high native occupancy (positive NMDS1 scores) from those with high non-native occupancy (negative NMDS1 scores). Because of this, the authors reasoned that sites with positive NMDS1 scores were refugia from disturbances for natives, whereas those with negative scores were refugia for non-natives. The second axis tended to separate large tributaries (positive NMDS2 scores) from main channels (negative NMDS2 scores). NMDS illustrated that the chance that a site would provide refuge during disturbance was species-dependent.

Various other parametric and non-parametric methods fall within this category of multivariate techniques, including factor, principal coordinates, and correspondence analyses; non-metric multidimensional scaling; cluster analysis; and their relatives. Miles (1990), for example, compared results using several of these methods. The non-parametric techniques are designed for situations where data are highly non-linear, or when sample sizes are too low for normality-linearity to be adequately determined. No method is without its drawbacks, however, so we recommend that students proceed cautiously when selecting an analytical technique. Because of the qualitative nature of interpretations of the outputs from all multivariate methods, one must fully understand how a technique operates to assign a meaningful biological explication to the results. The greatest problem is less about the specific method used, but rather about the absence of any follow-up confirmatory or validation studies (e.g., Marcot et al. 1983; Raphael and Marcot 1986; Fielding and Haworth 1995). Methods for validating results include bootstrapping, jackknifing, and the use of independent datasets. Even if multiple techniques are used to confirm study results, this does not justify inadequate sample sizes, biased sampling methods, or gross violations of statistical assumptions. Remember that the components derived from these techniques cannot be properly considered to be niche dimensions or habitat dimensions, as they reflect an arbitrary, operational decision by the researcher (Wiens 1989:65).

Assessing Relationships: Regression Analysis Regression analysis is probably the most widely used method of data analysis in animal ecology. Regression provides 3 general types of results (Wester 2019). First, it can predict or estimate a response variable, such as abundance or occupancy rates, from 1 or more independent (i.e., predictor) variables. Second, it produces the best combination of variables (based on available data) for anticipating a certain relationship. Third, the success (i.e., precision) of a regression analysis can be ascertained, usually through the use of correlation coefficients (Pimental 1979:33).

Although popular, regression analysis has many problems associated with its use that can substantially bias biological interpretations of the data. Pimentel (1979) provided an especially sobering review of regression analysis in general. In James and McCollough's (1990) review of multivariate methods, they lamented that they could not find even 1 good example of regression in their literature search. As in all multivariate techniques, the multivariate extension of simple linear regression magnifies these problems. Draper and Smith (1981:Chapter 8) supplied a flow diagram of the steps necessary to ensure the proper development of predictive models, a frequent goal of biologists using multiple regression, or MR (Fig. 3.13). Their diagram identified 3 primary stages: planning the analysis, developing the models, and verifying (i.e., validating or testing) the initial model outputs.

Regression analysis allows us to identify how much of the observed changes in the dependent variable is explained by the independent variables, and how much is not (i.e., the error term, *e*) (Wester 2019). A measure of the relative importance of each of these sources of variation (i.e., the independent variables) is termed the "coefficient of multiple determination," or R^2. R^2 ranges from 0 (for no linear relationship) to 1 (for a perfectly linear relationship). The value of R^2 is thus a measure of the explanatory value of the linear relationship (Wesolowsky 1976:43).

When sample sizes are small in relation to the

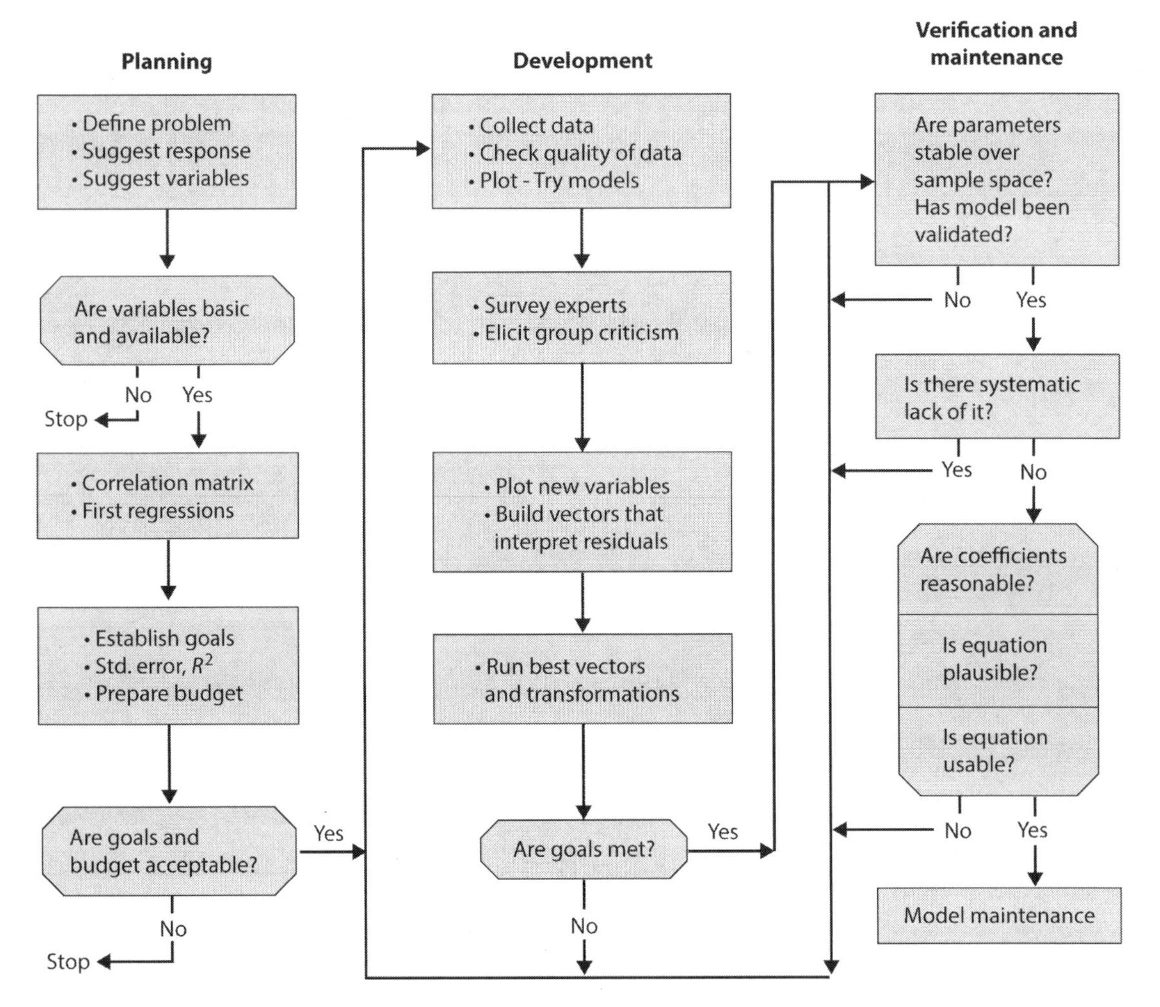

Figure 3.13. A summary diagram of predictive model-building procedures. (Reproduced from Draper and Smith (1981), with the permission of John Wiley and Sons)

number of parameters that are fitted (i.e., the number of independent variables), it is possible to get a large R^2, even when no linear relationship exists. Conversely, a low R^2 value does not necessarily indicate a bad relationship. Here, 1 or more individual coefficients may be significant, and the corresponding parameters, rather than the overall regression model, may be of primary interest. In addition, a low R^2 may simply show insufficient variation in the mean values. Again, spatial scale plays an overriding role in determining the results of our studies. In Figure 3.14*A*, we see that fitting a line through all of the points results in a good model for predicting the abundance of a bird species in relation to tree density. When our interest becomes more site specific, however, we see that tree density alone is a poor predictor of bird abundance (Fig. 3.14*B*). This is because we have moved further into the realm in which increasingly finer, site-specific indicators of an animal's habitat must be measured to explain its abundance. This is a simple, univariate example, but it illustrates the relationship between sampling scale and the results of a regression analysis. You can almost always find a significant regression result by sampling across a wide enough range of conditions, such as including young-growth forest in an analysis of the relationship

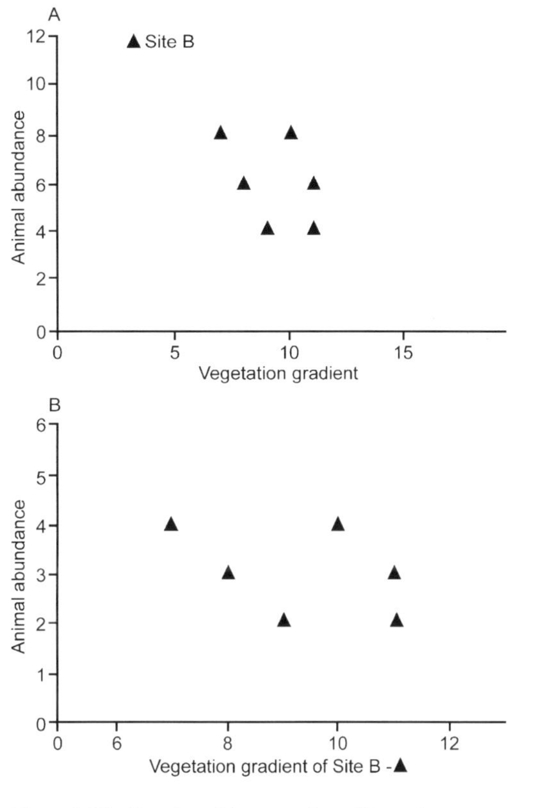

Figure 3.14. Graphs with examples of how the extent, or range, of an environmental gradient from which samples are taken can influence conclusions. (*A*) Sampling from an arbitrary, relatively wide gradient shows a positive relationship between animal abundance and vegetation. (*B*) Sampling from only a subset of the gradient (*Site B*) results in a negative abundance-vegatation relationship. (Reproduced from Morrison et al. (2006), courtesy of Island Press)

between numbers of canopy-dwelling birds and forest structure.

There are many specific ways in which your data can be processed and statistically significant variables identified within the broad category of multiple regression. The value of these different variable selection procedures (e.g., forward inclusion or backward elimination, stepwise regression, and all-possible-subsets regression), however, is beyond the scope of this chapter. Many have questioned employing stepwise and all-possible-subsets approaches (as reviewed in James and McCollough 1990), so use them cautiously, if at all. Additionally, to us the way in which variables are entered into a multiple regression are far less important (from the standpoint of ecology) than the way in which the data were collected in the first place. Crafting a priori models—which combine subsets of variables from a variable set, based on previous ecological knowledge—may be a more reasonable approach (see the section on "Model Selection" later in this chapter).

Separation: Multivariate Analysis of Variance

Multivariate analysis of variance (MANOVA) is the multivariate analog to univariate analysis of variance (ANOVA). MANOVA is particularly useful when we wish to examine the combined effects of 1 or more factors (e.g., habitat variables) on distinguishing a priori groupings (e.g., species, study areas). Thus, where we have more than 1 variable for each individual or grouping, we can apply MANOVA. It can be used to test the difference(s) between 2 or more groups, and many of the ANOVA models (1-way, 2-way, repeated measures, etc.) can be applied to MANOVA. The most commonly encountered MANOVA statistic is termed "Wilks' lambda (λ)." It tests whether the group's centroids are significantly separated. Wilks' lambda is calculated as the ratio of the determinants of the pooled within-groups, sums-of-squares, cross-product matrix divided by the total sample's sums-of-squares, cross-product matrix. Lambda ranges from 0 to 1, and the smaller the value, the greater the separation between (or among) groups. A measure of the variability among the data that is explained by the experimental factor is $1 - \lambda$. Computer programs usually present λ as a statistic that has been transformed into a value for F or chi-square, with an associated P. In such cases, a large F or chi-square is associated with a small P. Wilks' lambda is the most commonly used MANOVA test statistic, although other measures are also available (Zar 1999:Chapter 16).

In wildlife ecology, MANOVA is employed when there is a specific experimental design that has been developed to test a stated hypothesis. For example, Schmid et al. (2003) used MANOVA to test whether differences in the use and availability of proportions of the habitat (i.e., benthic substrates) of Kemp's ridley turtles (*Lepidochelys kempii*) were significantly different from zero. Bond et al. (2002) utilized MANOVA on data matrices, with land-cover types (e.g., woody, rowcrop, or grass) as dependent variables, and with the group (used or available land-cover types) and individual cottontail rabbits (*Sylvilagus floridanus*) as independent variables.

A non-parametric form of MANOVA, permutation MANOVA (PERMANOVA) can be applied to situations where the assumptions of MANOVA are violated (Anderson 2001; Anderson and Walsh 2013). Anderson and Walsh (2013) found that PERMANOVA was largely unaffected by heterogeneity in the dispersion matrices but performed poorly with unbalanced designs (i.e., groups with different sample sizes.). Crisol-Martínez et al. (2017) used PERMANOVA to evaluate the species richness, activity, and guild activity of several species of insectivorous bats found within an agricultural landscape of macadamia nut (*Macadamia*) trees in eastern Australia. They analyzed a mixed model containing "site" as a fixed factor and "area" (i.e., north, central, or south) as a random factor. Whenever significant values were found, PERMANOVA pair-wise tests were used to check for differences across sites and areas. The authors discovered that bat communities were dissimilar between areas (e.g., north vs. south) but not among sites within areas. They also reported no difference in the bat communities between riparian and upland sites.

Classification Discriminant analysis (DA) is widely applied throughout the scientific disciplines, including animal ecology. DA refers to a general group of methods, each of which has slightly different objectives. The overall goal of DA, however, is the classification of individuals into specific groups (e.g., species, vegetation types). Researchers can use DA methodology to order sites or individual samples along environmental gradients, so it resembles principal components analysis in its ordination capabilities. Unlike PCA, however, DA starts with sets of groups (2 or more) and a sample from each group. Thus, while the goals of PCA and some applications of DA are similar, the experimental design for collecting data and the underlying objectives for their analyses differ markedly.

Dillon and Goldstein (1984:Chapter 10) provided an excellent description of how discriminant analysis works, which we will summarize here, along with material from Pimentel (1979:Chapter 10) and Neff and Marcus (1980). In DA, researchers evaluate 1 categorical dependent variable and a set of independent variables. Although there is no requirement that these independent variables be continuous in nature, DA often performs poorly when independent variables are categorical. A categorical dependent variable is a grouping factor that places each observation into 1, and only 1, predefined group. For example, a researcher might be interested in examining differences among species or study sites, based on a series of environmental characteristics. After all individuals are assigned to these groups, the investigator might further wish to discriminate among the groups, based on the values of the independent, or predictor, variables (i.e., the habitat characteristics). DA is thus a method of separating groups—based on measured characteristics—and determining the degree of dissimilarity in observations and in groups, as well as the specific contribution of each independent variable to this dissimilarity (as in the variable loadings described for PCA earlier in this chapter).

After developing the linear discriminant functions, a researcher can then use these functions to identify or classify unknowns into the group predicted by discriminant analysis. Such a classification analysis is often used to determine how well DA can identify members of the groups used. Table 3.5 presents a simple example. In it, species that are not well separated from each other by DA will show high classifications (i.e., misclassifications) for a different species. In this table, many orange-crowned warblers (*Vermivora* [*Leiothlypis*] *celata*) were misclassified as MacGillivray's warblers (*Oporornis* [*Geothlypis*] *tol-*

Table 3.5 A classification matrix derived from a discriminant analysis program

Actual group	Predicted group membership (%) Orange-crowned	MacGillivray's	Wilson's
Deciduous tree sites			
Orange-crowned warblers	*13*	48	39
MacGillivray's warblers	9	*88*	3
Wilson's warblers	26	15	*59*
Non-deciduous tree sites			
Orange-crowned warblers	*16*	58	26
MacGillivray's warblers	16	*74*	10
Wilson's warblers	26	13	*61*

Source: Morrison (1981:Table 5)

Note: The table shows the actual and predicted group memberships for species of male singing warblers, based on their habitat use on deciduous and non-deciduous tree sites. Italicized numbers denote the percentage correctly classified.

miei), indicating a large overlap between the habitats used by both species in the study.

Findley and Black (1983; see also Findley 1993) summarized their conception of how a Zambian insectivorous bat community would appear in multivariate space. Their drawing (Fig. 3.15) provides a good example of how projections of species can be interpreted, and it applies to most methods. Each sphere represents the morpho- or ecospace occupied by 1 species. The volume of the sphere equals the total niche volume, and the diameter indicates the amount of intraspecific variation, while the center is the species' centroid (i.e., the mean value of all its morphological or ecological variables). Overlap—a function of both interspecific distance and intraspecific variability—is shown if the spheres intersect each other. In Figure 3.15, the region of the community's centroid is occupied by a number of closely packed species with low intraspecific variability (i.e., narrow niches).

Rexstad et al. (1988) conducted a simulation study to examine, in detail, the ramifications of violating the assumptions of discriminant analysis. Although some of their conclusions should be tempered (see Taylor 1990; also see Rexstad et al. 1990), the message contained in their paper remains. There is little published evidence to suggest the widely held belief that discriminant analysis is robust for violations of the variance-covariance matrix, or that using DA for descriptive purposes only results in ecologically meaningful analyses. Because of these shortcomings, many people are now using logistic regression analysis as a replacement for discriminant analysis (see the next section).

Sollmann et al. (2016) employed DA to study habitat relationships of a recolonizing black bear population in the Ozark Mountains of Missouri. Bears were introduced in Arkansas to repopulate their historical range, and they expanded into Missouri. Thus information on the population status and habitat relationships of these bears was needed to manage between-bear and human-bear interactions. The researchers sampled bears using a grid of hair snares and estimated density by employing a spatial capture-recapture model. DA was applied to their data to distinguish environmental characteristics among 5 sampling areas with varying bear densities. The authors considered discriminant functions that, when combined, explained at least 95 percent of the between-group variance. To determine which covariates differentiated the groups, they evaluated the correlation of covariates with the discriminant functions retained in their analysis. The first discriminant function explained 86.86 percent of between-group variance among the 5 sampling grids. It was strongly positively correlated with the percentage of forest and strongly negatively correlated with elevation. The second function explained an additional 8.33 percent of the variation and had a strongly negative correlation with the distance to human settlements and average slopes. The 3 grids with highest bear densities separated along the first discriminant axis, indicating that higher densities occurred in grids with a lower percentage of forest and a shorter distance to rural settlements, which may have represented food sources (Fig. 3.16).

Logistic regression analysis (LR) has seen increased use in recent years as a non-parametric, non-

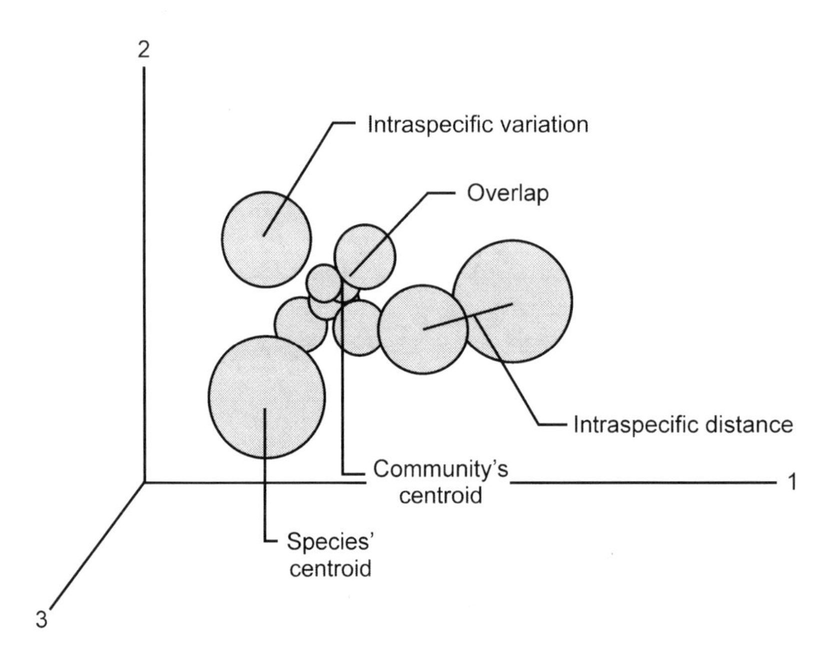

Figure 3.15. A tentative model of a community in attribute space. The attributes could be morphology or diet, or perhaps other physiological, behavioral, or ecological parameters. The community is composed of some closely packed, rather than invariable, species, as well as some more distant, more variable kinds. (Reproduced from Findley and Black (1983:Fig. 1), with the permission of John Wiley & Sons)

linear alternative to 2-group discriminant analysis (Keating and Cherry 2004). Recent advances in programming and computer power have enabled LR to be extended to situations with more than 2 groups (Harrell 2015). LR can be used to analyze independent variables that are true categorical data (e.g., coat color) and those that have been summarized into categories (e.g., height intervals). It can also analyze continuous data. In addition, LR can be employed to develop predictive models.

An early application of LR in wildlife research was by Martinka (1972), who compared used and unused sites for male blue (or dusky) grouse (*Dendragapus obscurus*) in Montana. Brennan et al. (1986) assessed the predictive abilities of discriminant analysis and logistic regression to distinguish mountain quail (*Oreortyx pictus*) habitat at 4 study areas in northern California. LR consistently predicted quail habitat better than DA. Nadeau et al. (1995) used logistic regression to create a habitat model based on muskrat (*Ondatra zibethicus*) presence along wetlands near James Bay in Québec. They developed 2 models, with 1 based only on the presence of burrows, and the other using the presence of muskrat feeding signs and droppings as dependent variables. Independent variables were a mix of continuous (e.g., water depth, floating plant cover) and categorical (e.g., bank slope, dominance of peat soil) variables. Chandler et al. (1995) used LR to predict the probability of bald eagle (*Haliaeetus leucocephalus*) use of a shoreline segment on northern Chesapeake Bay, based on a series of habitat variables. LR can be further improved by incorporating mixed effects into conditional models to evaluate habitat selection when resource availability changes or when a species exhibits interindividual heterogeneity in its habitat use (Duchesne et al. 2010).

LR is potentially more robust than similar parametric procedures with regard to deviations from multivariate normality and equal covariation. Further, various studies have found that LR provides better between-group separations and classification success than discriminant analysis allowed. Further, the non-linear tendencies of ecological data are usually distorted by linear analyses. Thus LR's nonparametric procedures should provide a more meaningful interpretation of ecological phenomena when linear models fail to adequately explain variation (see Efron 1975; Press and Wilson 1978; Brennan et al. 1986).

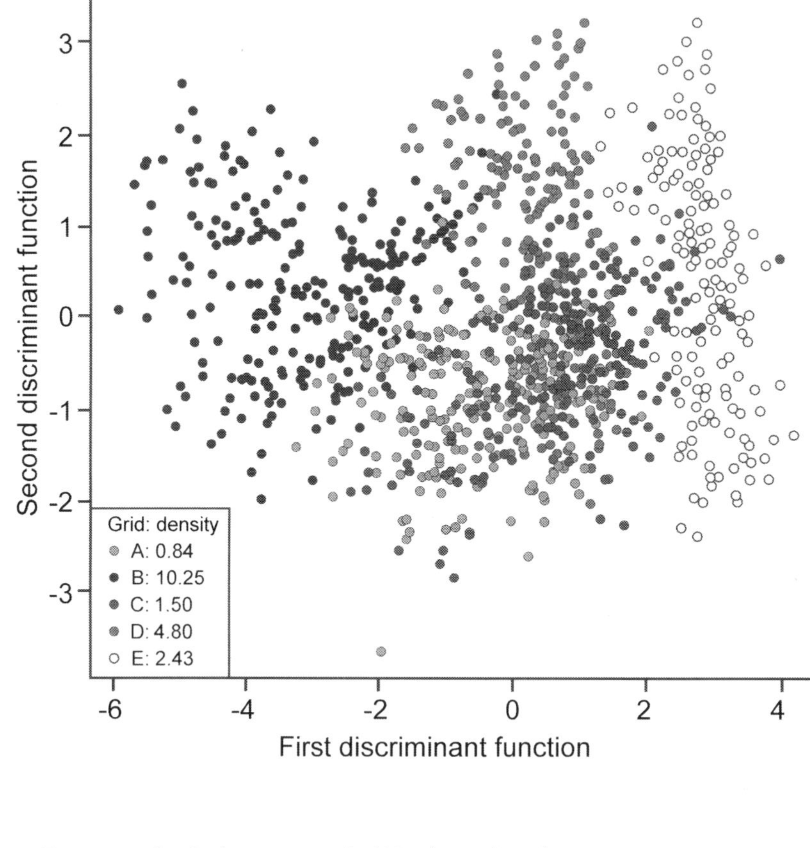

Figure 3.16. Two primary axes from a discriminant analysis, displaying the distribution of habitat features from 5 study sites representing varying population densities of black bears (*Ursus americanus*) in the Ozark Mountains of Missouri. (Reproduced from Sollmann et al. (2016), courtesy of the Ecological Society of America)

For example, Jochimsen et al. (2014) used multinomial LR to distinguish environmental variables along roads where western rattlesnakes (*Crotalus oreganus*) and gophersnakes (*Pituophis catenifer*) were encountered in the shrubsteppe region of southeastern Idaho. They compared random points ($n = 251$) along the road with snake-crossing points, grouped by species ($n = 233$). Explanatory variables were included in the LR, all with pairwise correlation coefficients (0.6, transformed for normality when appropriate. LR identified 9 environmental variables associated with random points and snake localities (their Table 1). Random points were negatively associated with the percentage of roadside cover, areas dominated by grass cover, the percent of juniper (*Juniperus*) cover within 500 m, and the presence of basalt within 100 m. These points were positively associated with areas dominated by shrub cover and distance to the nearest den. Rattlesnake crossings were associated with a high percentage of roadside cover, juniper cover within 500 m, grass cover, the percent of urban cover within 100 m, and areas in close proximity to dens. Gophersnake crossings included the dominance of grass cover along roadsides, the presence of basalt within 100 m, areas lacking shrub cover, low burrow densities, and the percentage of juniper cover within 500 m.

Cautions and Solutions As discussed in our opening comments in this chapter, and in more detail in Chapter 2, there are important limitations to the pattern-seeking, descriptive methods of habitat use typically found in multivariate analyses to which the researcher need attend. Kaufman and Kaufman (1989) aptly summarized the cautions that should be applied to results of descriptive studies:

1. Statistical separation of species on a habitat axis does not prove that the animals recognize and respond to the characteristics measured. At best,

this provides testable hypotheses to evaluate with experimentation.

2. Even though average positions along habitat axes occupied by a series of species are different, interspecific overlap may be considerable and must be explained (would also apply to intraspecific, sex-age comparisons).
3. Most studies are not replicated in time or space, so that the generality of most studies is unknown.
4. Differences in resource use do not necessarily elucidate the mechanisms that are ultimately responsible for the patterns observed.
5. Understanding cause and effect of observed patterns of habitat use will require experimental manipulations and not just additional studies using refined descriptions of habitat use (Chapters 4 and 5).

It is critical that researchers carefully examine the magnitude of differences for each independent variable between groups in a multivariate analysis. Remember that the variable(s) entered and the order of entry are primarily a statistical decision based on some selection criterion (e.g., *F*-value). A variable with a small difference in means and small variance could have a larger *F*-value and be entered into an analysis before a variable with a large difference in means but also a relatively large variance. As discussed earlier, biologists tend to concentrate on statistical significance rather than on an ecological interpretation of means *and* variances of variables prior to entry into an analysis. Thus the ecological interpretation made after an analysis in such situations can be misleading. Multivariate methods worsen this tendency because of the black box that field data disappear into and the myriad of statistical parameters that return on the printout. The high natural variance usually seen in biological populations makes application of models to other places and times difficult (see Fielding and Haworth 1995 for a detailed analysis and discussion).

Another concern in wildlife analyses in general, and habitat studies in particular, is spatial autocorrelation. That is, most habitat features do not occur independently of surrounding features. This means that where an animal occurs may be based on the features at that point in space as well as features that are nearby. Recall our earlier discussions about how habitat is usually measured in observer-defined areas around an animal. Marzluff et al. (2004) reviewed this issue and presented a new method for analyzing resource selection that incorporates a probabilistic measure of use, called the utilization distribution. Rather than assuming that resource use is uniform within a home-range boundary, differential use can be quantified using the utilization distribution.

Multicollinearity is another issue that confounds habitat analyses. Multicollinearity occurs when explanatory variables are correlated among each other. Multicollinear explanatory variables are difficult to analyze because their effects on the response variable can be due to either true synergistic relationships among the variables or spurious correlations. Graham (2003) reviews this subject in detail and offers several relatively simple solutions.

Different sampling methods may provide different results. Block et al. (1998) used classification success as an ad hoc index of model performance. They sampled small mammals in California oak woodlands using both pitfall and live traps and developed classification functions using both DA and LR. Variables included in the live trap and pitfall models overlapped but were not identical. For example, the DA model for brush mouse (*Peromyscus boylii*) using live trap data included slope, litter depth, dead woody debris (1–10 cm diameter), height of herbaceous vegetation, low canopy cover (1–2 m height interval), blue oak (*Quercus douglassii*) cover (2–5 m height interval) and grass cover, whereas the model using pitfall trap data included herbaceous vegetation height, number of burrows, California black oak (*Q. kelloggii*) cover (2–5 m height interval), forb cover, and blue oak cover (2–5 m height interval). Classification success was comparable as DA correctly classified 65% of the pitfall trap stations and 69% of the live trap stations as used or unused. Block et al. (1998) attempted to cross-validate the models by using pitfall data to test the live trap model and vice versa. Results were not encouraging in that the models correctly classified roughly half the stations as used or unused (i.e., basically a flip of the coin). This serves as an important reminder to constrain extension of your results to the types of methods used to sample a population.

Model Selection

Much of the work conducted by ecologists is exploratory in nature and results in extremely large datasets composed of numerous variables. Historically, ecologists have subjected these datasets to multiple statistical procedures, often beginning with various data reduction techniques (discussed above, in the "Cautions and Solutions" section). Further, the multivariate methods outlined in this chapter are often used to create multiple habitat models and descriptions. Thus we need unbiased methods for selecting the best of what is often a host of potential results. Burnham and Anderson (2002:35–37) presented a summary of the various model selection techniques. Available methods included stepwise procedures, cross-validation, goodness-of-fit tests, and an adjusted R^2. All of these approaches have strengths and weaknesses, and all are applicable, depending on the form of the data.

Many traditional statistical methods and modeling approaches rely on *P*-values to assess statistical significance. This choice has come under extensive criticism over the past decade. The American Statistical Association published an assessment of *P*-values and summarized a set of principles and warnings (Wasserstein and Lazar 2016:131–132). Among these principles were: (1) scientific conclusions and business or policy decisions should not be based only on whether a *P*-value passes a specific threshold; (2) a *P*-value, or statistical significance, does not measure the size of an effect or the importance of a result; and (3) by itself, a *P*-value does not provide a good measure of evidence regarding a model or a hypothesis. They concluded by recommending methods that emphasized estimation over testing, such as decision-theoretic modeling (aka "model selection"). Model selection forgoes the use of *P*-values and applies Akaike's Information Criterion, or AIC (Akaike 1973), to assess a model's fit. Briefly, AIC selects the best-fit model without overfitting the data with too many parameters (i.e., an assumption that the model can be slightly improved by adding yet 1 more parameter). The model with the smallest AIC is the best approximating model for the data, although AIC does not determine if the model is good ecologically. It only helps select among models that were developed by a particular scientist (Burnham and Anderson 2002:62). (Note that this statement applies to all model selection procedures.) These authors also discussed many of the misuses of AIC and provided several detailed examples. A few other relatively recent papers using AIC and related terms include Woodward et al. (2001), Kuehl and Clark (2002), and Otis (2002).

As summarized by Burnham and Anderson (2002:96), the best scenario would be for an investigator to develop a set of multiple working hypotheses that are based on a thorough understanding of the available literature and experience with the relevant system (i.e., natural history). Next, scientific knowledge, experience, and expertise should be employed to define a set of a priori candidate models that represent each of the working hypotheses. This allows a clear development of the research problem, followed by careful planning in the design of the study, including gathering adequate sample sizes. Third, appropriate statistical techniques should be chosen, including model selection procedures, to compare the various hypotheses. We follow Arnold (2010) in recommending that scientists concentrate more on estimations of effect size (i.e., the magnitude of difference) and associated confidence intervals, rather than placing an overreliance on *P*-values.

Pitt et al. (2017) applied logistic regression within a model-selection framework to evaluate the habitat relationships of 3 anuran species—wood frogs (*Litho-*

bates sylvaticus), southern leopard frogs (*L. sphenocephalus*), and green frogs (*L. clamitans*)—within the Piedmont ecoregion of South Carolina (Table 3.6). They first calculated Spearman's rank correlation coefficients to identify highly correlated ($P \geq 0.6$) variables and removed those with limited representation in the dataset. They then developed a priori models for each species, based on the natural history of the species, as well as on published literature, and selected the best models from those with the lowest AIC scores and highest model weights. They analyzed habitat within both 2 m and 20 m of frog locations, but, for illustration, we discuss their results only from the 2 m scale. The model that best described wood frog habitat included cooler temperatures; increased canopy openness, deciduous leaf litter, and relative humidity; and slightly less soil moisture. The best model for southern leopard frogs involved fern cover; increased grass and herbaceous cover and relative humidity; and a slightly denser understory. The model that best described green frog habitat contained deciduous leaf litter; increased soil moisture and fine woody debris; and fewer trees. The weights for other competing models were substantially less.

Although the use of model selection approaches has seen tremendous growth over the past 15 years, you should be aware of their limitations. Guthery et al. (2005:457) concluded that "it is largely an inductive approach to knowledge accrual and, therefore, subject to the pitfalls of induction." The algorithm tends to overfit data (i.e., use too many variables), resulting in models that contain useless variables and generalize poorly. Errors of commis-

Table 3.6. **Paired logistic regression models of habitat selection by wood frogs (*Lithobates sylvaticus*), southern leopard frogs (*L. sphenocephalus*), and green frogs (*L. clamitans*)**

Species	Model	Variables	AICs value	Rank	Model weight
Wood frogs	Forest with deciduous leaf ground cover	Temperature + canopy openness + relative humidity + soil moisture + deciduous leaf cover	157.57	1	0.84
	Global model	All variables	161.01	2	0.15
	Forest with fine woody debris ground cover	Temperature + canopy openness + relative humidity + soil moisture + fine woody debris	168.06	3	<0.01
	Forest with herbaceous ground cover	Temperature + canopy openness + relative humidity + soil moisture + herbaceous cover + grasses	172.64	4	<0.01
Southern leopard frogs	Wet meadow 1	Herbaceous cover + grasses + ferns + understory openness + relative humidity	116.95	1	0.75
	Wet meadow 2	Herbaceous cover + grasses + reeds	121.16	2	0.09
	Floodplain forest	Light + grasses + shrubs + herbaceous cover	122.64	3	0.04
	Global model	All variables	122.91	4	0.04
	Marsh	Water + reeds + relative humidity	123.05	5	0.04
	Forest with deciduous leaf ground cover	Light + deciduous leaf cover	124.39	6	0.02
	Forest wetland	Light + water + shrubs + soil	124.45	7	0.02
Green frogs	Forest with deciduous leaf ground cover	Soil moisture + trees + fine woody debris + deciduous leaf litter	64.35	1	0.94
	Marsh grass	Water + grasses	71.94	2	0.02
	Global model	All variables	72.03	3	0.02
	Swamp	Soil moisture + trees + herbaceous cover + shrubs	72.98	4	0.01
	Forest with herbaceous ground cover	Soil moisture + trees + herbaceous cover + shrubs	74.61	5	<0.01
	Shrubby shoreline	Water + shrubs	77.46	6	<0.01

Source: Pitt et al. (2017)

Note: Data were collected at the 2 m scale in 2 study landscapes within the Piedmont ecoregion of South Carolina

sion in information theory–AIC-based papers included hopelessly uninformative lists of encrypted models and an imposition of the model selection approach on studies that would have been better executed in a simple, descriptive format. Arnold (2010) noted that a problem plaguing model selection was the inclusion of variables that were considered to be uninformative. He added that AIC-based model selection had about a 1-in-6 chance of admitting a spurious variable, compared with a 1-in-20 chance for a model based on traditional hypothesis testing, with $\alpha = 0.05$. Guthery et al. (2005) stated that model selection represented 1 tool in the toolbox, but they recognized that traditional hypothetico-deductive approaches remained a valid method to use in addressing ecological questions. Additionally, these authors predicted that model selection would become a ritualistic replacement for the widespread use of *P*-values and frequentist statistics in animal ecology. A decade and a half later, this prediction has apparently proven to be true.

Ecological Niche Modeling

Many equate ecological niche modeling (ENM) with species distribution models (SDM). As reviewed by Peterson and Soberón (2012), subtle differences distinguish them. The authors asserted that those who used SDMs preferred not to overinterpret the ecological significance of the model and had no need to characterize all of the niche dimensions. In contrast, those who applied ENMs focused on a subset of ecological niche dimensions (defined by coarse-resolution dimensions) that were not necessarily affected by the population processes of the species in question. Thus, for SDMs, one might model species' distributions without considering a niche definition or even referring to any environmental variables. In contrast, ENM permits modeling the processes that produce and shape the area of distribution, either transferring causal factors to time or geological space, or biologically interpreting the obtained pattern. These areas of inquiry require some type of hypothesis about the ecology of the species, which is clearly niche-related.

For example, southwestern willow flycatchers (*Empidonax traillii extimus*) are currently listed as an endangered subspecies under the US Endangered Species Act. Their designation as a subspecies was based largely on observational studies that distinguished song and plumage characteristics. Zink (2015) tested the subspecies status of willow flycatchers by using a combination of genetics, morphology, and ENM. He constructed correlative ecological niche models using breeding records that were input into Maxent (version 3.2.2) and partitioned locations into those representing the endangered southwestern willow flycatchers and the neighboring subspecies (*E. t. adastus*). He obtained 19 layers of climate data from the Worldclim global bioclimatic database (Hijmans et al. 2005) and found that southwestern willow flycatchers did not have a significantly different climatic niche from that nearest geographic neighbor (i.e., in the Great Basin and the Rocky Mountains). He concluded that willow flycatchers in the American Southwest were probably peripheral populations of an otherwise widespread species that did not merit subspecific recognition. These results may have a bearing on continuing to recognize southwestern willow flycatchers as an endangered species.

Various statistical approaches used in ENM include the maximum entropy algorithm, general linear models, general additive models, and others. Qiao et al. (2015) conducted a simulation study to evaluate alternative procedures. They developed virtual species with known niche and dispersal properties to test the ability of various algorithms to estimate potential areas of distribution. The authors found that different ones performed in dissimilar ways, depending on the species. Thus they opined that researchers planning to use ENM studies should first evaluate a set of algorithms and then choose the algorithm that performed the best. In practice, however, this is rarely done, with investigators perhaps using a suboptimal algorithm in their research system.

Machine Learning, Neural Networks, and Random Forests

Machine learning (ML) is a discipline in computer sciences that develops algorithms capable of data-driven decisions (Thessen 2016). Part of what makes ML appealing to many is the variety of its algorithms, which may be capable of making sense out of messy datasets. Most ML techniques can be employed to perform multiple tasks, and several of them can be used in combination to address the same problem. General tasks done with machine learning include function approximation, classification, clustering, and rule induction. Many of them are somewhat similar to tasks carried out using more traditional statistical analyses. For example, functional approximation is analogous to linear regression; classification is comparable to discriminant analysis; clustering resembles cluster analysis; and rule induction parallels decision trees (Thessen 2016).

Both artificial neural networks and random forests are tools employed within the machine learning environment to address various questions related to animal ecology. Artificial neural networks have been used to model species distributions resulting from climate change (Thuiller 2003), identify birds and bats by vocalizations (Kasten et al. 2010; Jennings et al. 2008), and classify bird communities (Lee et al. 2007). Random forests is a method that fits a user-selected number of decision trees to a dataset and then combines the predictions from all of these trees (Breiman 2001). The random forests algorithm initially creates a tree for a subsample of the dataset. At every decision point, only a randomly selected subset of variables is used for the partitioning. The predicted class of an observation in the final decision tree is calculated by a majority vote of the predictions for that observation in all trees, with any ties split randomly. The advantages of random forests are that it has high classification accuracy, can assess the importance of variables, is able to model complex interactions among predictor variables, has the flexibility to perform several types of statistical data analyses (e.g., regression, classification, survival analysis), and can impute missing values (Cutler et al. 2007).

Bayesian Methods

Bayesian methods are important statistical tools that are being used more frequently by ecologists. Although Bayes' theorem was established long ago, it has only been within the past few decades that Bayesian analysis has become more widely applied. In these analyses, the user develops a prior probability distribution, which is effectively a quantitative model or hypothesis. Bayes' theorem uses this prior probability distribution and the likelihood of the data to generate a posterior probability distribution, which provides a direct measure of the certainty that can be placed on models, hypotheses, or parameter estimates. Bayesian methods can be applied to model selection to rank alternative models, and multiple models can be averaged into a single model, given the uncertainty in plausible models (Hooten and Hobbs 2015). Bayesian inference differs from classical, frequentist inference in 4 ways (Ellison 2004:509):

1. Frequentist inference estimates the probability of the data having occurred, given a particular hypothesis, or $P(Y|\mathrm{H})$, whereas Bayesian inference provides a quantitative measure of the probability of a hypothesis being true in light of the available data, or $P(\mathrm{H}|Y)$.
2. Their definitions of probability differ, as frequentist inference describes it in terms of the long-run (i.e., infinite) relative frequencies of events, whereas Bayesian inference characterizes it as a individual's degree of belief in the likelihood of an event.
3. Bayesian inference uses prior knowledge along with the sample data, whereas frequentist inference employs only the sample data.
4. Bayesian inference treats model parameters as random variables, whereas frequentist inference considers them to be estimates of fixed, true quantities.

Ecologists have used Bayesian methods for a variety of studies, ranging from predicting single-species population dynamics to understanding ecosystem processes. Sanderlin et al. (2016) employed Bayesian hierarchical models to evaluate avian species richness, utilizing a 10-year dataset that was compiled after wildfires had occurred in ponderosa pine (*Pinus ponderosa*) forests in northern Arizona. Birds were sampled, using point counts, at 149 stations that were stratified among severely burned, moderately burned, and unburned areas. The authors used species, time since the wildfires, and burn severity to model the probability of occupancy, and species and sampling effort to model the detection probability. They employed independent, non-informative priors, because there was no previous knowledge about the parameters. Avian species richness, on average, increased over the 10-year time period, although 95-percent Bayesian credible intervals (BCIs) overlapped among the transects (Fig. 3.17). While patterns were not significant (because the 95-percent BCIs overlapped), the average posterior median species richness estimate of the moderately burned transects (median = 64 species) was greater than both the severely burned ones (55 species) and the unburned ones (50 species).

Sample Size Analyses

Inadequate sample size is a chronic problem found in many papers using multivariate methods. Regardless of the refinement of the study design and the care taken in recording data, a proper ecological interpretation of multivariate analyses is difficult, at best. Interpretations based on inadequate samples are a waste of time, as well as being potentially misleading. Earlier in this chapter we identified methods of determining proper sample sizes for univariate analyses. Those methods also apply here, on a variable-by-variable basis. In multivariate analyses, however, our problems are increased by orders of magnitude, given our desire to simultaneously interpret ecological phenomena in many dimensions for multiple species.

Many published studies have only slightly more observations than variables, and some have even fewer. Johnson (1981) outlined general guidelines for determining adequate sample sizes in multivariate investigations, noting that more observations were needed when the number of independent variables was large. He thought that an appropriate minimum sample size for multivariate analyses was 20 observations, plus 3–5 additional ones for each variable in the analysis. Tabachnick and Fidell (1983:92) suggested a minimum ratio of 4 to 5 samples per variable for multiple regression. We note that an additional 5–10 observations for each variable would give a more conservative target for the sample size. Larger sample sizes do not, however, provide an answer for poorly designed studies or biased data. As Johnson (1981:16) noted, calls for larger samples are the "knee-jerk" reaction when variability is excessive.

In a study of habitat use by birds in Oregon, Morrison (1984b) found that a minimum of 35 plots was necessary to obtain stable results. Stability was determined when the means and variances did not change with an increase in the sample size. His review of the animal ecology literature showed, however, that very few studies met the minimum criteria established by Johnson (1981). A similar study by Block et al. (1987) found that even larger numbers of plots—up to 75 or more—were needed in their bird-habitat analysis. Few researchers had even discussed the issue of sample size. Collecting "all the data that I could" is no excuse for publishing results based on inadequate sample sizes.

The minimum number of sample sizes required is a study-specific question, although the papers mentioned above give approximations of the range of samples you can expect to need. Remember, however, that these minimum sample sizes apply to each biological period being considered (e.g., fall, winter), and the appropriate n might vary among them. All investigations should include a justification for the amount of data used in the analyses.

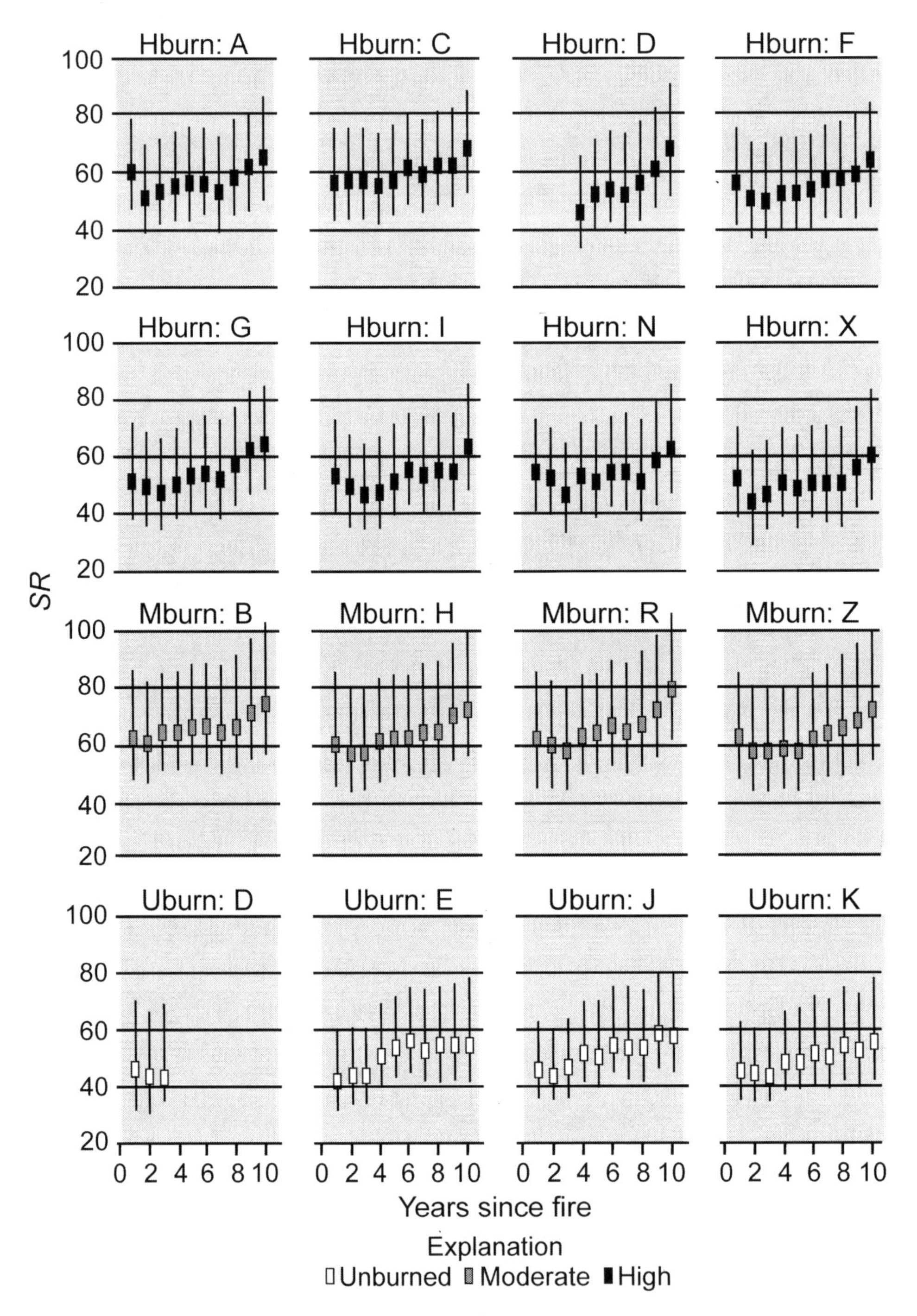

Figure 3.17. A set of graphs noting trends in avian community richness (i.e., species richness, or SR) following wildfires in 1996 in ponderosa pine (*Pinus ponderosa*) forests of northern Arizona. They summarized data from 1997 to 2006, showing Bayesian credible intervals over 3 different burn severities: high, moderate, and unburned. (Reproduced from Sanderlin et al. (2016), courtesy of the US Geological Survey)

Computer Statistical Packages

As noted earlier, advances in computing power and in the development and application of new statistical methods have greatly improved our ability to analyze large, multivariate datasets. We have witnessed the transition from mainframe computers that would fill an entire room to desktop computers with faster co-processors, a larger storage capacity, and greater memory. Pre-programmed statistical packages (e.g., SPSS, BMDP, and SAS) have improved over time to incorporate new methods, algorithms, and statistical tests. These revisions provide more options for analysis and greater flexibility in customizing an analysis to address specific study objectives. Unfortunately, the availability and ease of use of these packages probably has led to an increasing number of misuses of statistical methods. Novices employing them should be aware that the default settings for a particular analytical method often must be adjusted for every application. Further, each package has a set of options and statistics that the user must specifically request.

A necessary prerequisite to using any statistical package, however, is to be well versed in statistics. Touchon and McCoy (2016) surveyed 154 universities for programs requiring doctoral students to take courses in this discipline. They found that only one-fourth of the programs did so, but most of those requirements were not courses in contemporary statistics or advanced techniques. Many university computer centers offer short courses in the use of statistical packages, and professional societies frequently hold workshops on sampling methods and statistics at their annual meetings. In addition, an increasing number of statistics departments are offering more in-depth courses on the use and interpretation of such software. Stauffer (2002) noted that a problem with multivariate statistics is that anyone capable of entering data in a format that could be analyzed by a computer package could conduct a multivariate analysis, whether or not the data were appropriate for such an analysis. These problems have lessened in the literature over time, as referees and editors have become more sophisticated in their knowledge of what are appropriate multivariate analyses (e.g., addressing assumptions, evaluating sample sizes).

Touchon and McCoy (2016) surveyed statistical methods and analysis software used in 19,526 papers published in major ecology journals from 1990 to 2013, with 2 informative and related trends emerging from their analysis. First, they found that the use of many traditional statistical tests (e.g., ANOVA, t-tests, Mann-Whitney U tests) had declined, whereas the employment of more-recent types of analyses (e.g., generalized linear model [GLM], maximum likelihood, model selection, mixed-effects models, Bayesian) had increased considerably. Second, concomitant with the rise of contemporary statistics, these authors noted a major shift among the use of programs to analyze data. From 1990 through 2008, SAS was the dominant program cited in the papers they surveyed. Whereas recent use of SAS, JMP, and SPSS has been flat or decreasing, citations of the open-source program R have gone up dramatically. R was first cited in 2003; by 2013 it was referenced more than 3 times as often as any other program. R offers quite a bit of flexibility in both statistical analysis and the development of publication-ready graphics. It is essentially an environment within which statistical techniques are implemented and can be extended via packages. There are about 8 packages initially supplied with R, and many more are available. Its applications run the gamut from traditional to contemporary statistical analyses (e.g., linear and non-linear modeling, classical statistical tests, time-series analysis, classification, clustering, Bayesian) and graphical techniques. R is designed around a true computer language, and it allows users to add to its capabilities by defining new functions. Much of the system is written in the R dialect of S, which makes it easy for users to follow the algorithmic choices being made. For computationally intensive tasks, C, C++, and Fortran code can be linked into R. Advanced users can write C code to manipulate R objects directly. Bocard et al. (2018) provided useful guidance and suggested applications of R for various multivariate analyses.

Summary

Reporting the results of animal ecology research hinges on placing an investigation in the proper context. By this, we mean that the application and extrapolation of the results largely depend on the study design, in terms the spatial and temporal extent of the data. Certainly, much of this is based on the research objectives and the available resources. For example, an investigation designed to understand forage availability for elk (*Cervus canadensis* [*C. elaphus canadensis*]) on their wintering range would be conducted during that season, at locales where elk are found. Available funding and personnel, however, may constrain the number of study sites, the types and number of variables collected, and sample size. Acquiring an adequate amount of samples may mean reducing the quantity of variables collected and the number of ranges sampled. More often than not, we see studies conducted at a particular location (e.g., an experimental forest, a conservation district) because of convenience and the interest(s) of the funding source (Morrison 2016). This, of course, does not negate the value of the investigation, but it does limit the inferences that can be drawn for other locales.

As we have discussed, conducting research at 1 location or during 1 season tends to be more the norm than the exception. We contend that such practices limit our ability to better understand variations in resource use and factors that limit populations. If the science of animal ecology is to advance, researchers should strive to design investigations that can be replicated across space and conducted over sufficient durations of time—that is, seasons and years. This does not require any 1 study to address overarching questions. Rather, researchers could pursue work-arounds by having multiple investigations conducted at replicate sites (e.g., the spotted owl demography studies in the Pacific Northwest, mentioned earlier in this chapter) and shorter side-studies done in a piecemeal fashion, while preserving the continuity of long-term data collection. This strategy has worked well in many academic settings, where graduate students maintain long-term data streams while selecting subsets of the data to address short-term objectives.

Researchers have multiple options for analyzing data. Our intent in this chapter has not been to recommend 1 particular approach over another, but to provide a taste of past and present techniques. Indeed, advances in computer capabilities have increased our options for analyzing large, complicated, multivariate datasets. The key point to keep in mind, however, is that analysis is a tool to help us better understand the patterns and processes that we observe. Remember that 1 size (of analysis) does not fit all. In many cases, alternative approaches to an analysis may be valid, and multiple approaches could be explored. It is incumbent upon researchers to first understand what the expected outputs from a given method are, and then to consider the various assumptions inherent in each approach, assess whether the dataset will meet the assumptions, and evaluate the ramifications of violating those assumptions. A number of statistical packages and programming languages are available to conduct analyses, and these tools are always evolving.

LITERATURE CITED

Afifi, A.A., and V. Clark. 1984. Computer-aided multivariate analysis. Lifetime Learning, Belmont, CA.

Akaike, H. 1973. Information theory and an extension of the maximum likelihood principle. Pp. 267–281 *in* B. N. Petrov and B. F. Csaki, eds. Second International Symposium on Information Theory. Académiai Kiadó, Budapest, Hungary.

Amrhein, V., S. Greenland, and B. McShane. 2019. Retire statistical significance. Nature 567:305–307.

Anderson, D. R., and K. P. Burnham. 2002. Avoiding pitfalls when using information-theoretic methods. Journal of Wildlife Management 66:912–918.

Anderson, M. J. 2001. A new method for non-parametric multivariate analysis of variance. Austral Ecology 26:32–46.

Anderson, M. J., and D. C. I. Walsh. 2013. PERMANOVA, ANOSIM, and the Mantel test in the face of heterogeneous dispersions: What null hypothesis are you testing? Ecological Monographs 83:557–574.

Arnold, T. W. 2010. Uninformative parameters and model selection using Akaike's information criterion. Journal of Wildlife Management 74:1175–1178.

Bervin, K. A., and D. E. Gill. 1983. Interpreting geographic variation in life-history traits. American Zoologist 23:85–97.

Birkhead, T. M., J. O. Wimpenny, and B. Montgomerie. 2016. Ten thousand birds: Ornithology since Darwin. Princeton University Press, Princeton, NJ.

Block, W. M. 1990. Geographic variation in foraging ecologies of breeding and nonbreeding birds in oak woodlands. Studies in Avian Biology 13:264–269.

Block, W. M., M. L. Morrison, and P. E. Scott. 1998. Development and evaluation of habitat models for herpetofauna and small mammals. Forest Science 44:430–437.

Block, W. M., K. A. With, and M. L. Morrison. 1987. On measuring bird habitat: Influence of observer variability and sample size. Condor 72:182–189.

Bock, C. E., and Z. F. Jones. 2004. Avian habitat evaluation: Should counting birds count? Frontiers in Ecology and the Environment 2:403–410.

Bond, B. T., L. W. Burget Jr., B. D. Leopold, J. C. Jones, and K. D. Godwin. 2002. Habitat use by cottontail rabbits across multiple spatial scales in Mississippi. Journal of Wildlife Management 66:1171–1178.

Borcard, D., F. Gillet, and P. Legendre. 2018. Numerical ecology with R, 2nd edition. Springer, New York.

Breiman, L. 2001. Random forests. Machine Learning 45:5–32.

Brennan, L. A., W. M. Block, and R. J. Gutiérrez. 1986. The use of multivariate statistics for developing habitat suitability index models. Pp. 177–182 *in* J. Verner, M. L. Morrison, and C. J. Ralph, eds. Wildlife 2000: Modeling habitat relationships of terrestrial vertebrates. University of Wisconsin Press, Madison.

Brennan, L. A., A. N. Tri, and B. G. Marcot. 2019. Quantitative analyses in wildlife science. Johns Hopkins University Press, Baltimore.

Brown, C. F., and P. R. Krausman. 2003. Habitat characteristics of 3 leporid species in southeastern Arizona. Journal of Wildlife Management 67:83–89.

Burger, L. D., L. W. Burger Jr., and J. Faaborg. 1994. Effects of prairie fragmentation on predation on artificial nests. Journal of Wildlife Management 58:249–254.

Burnham, K. P., and D. R. Anderson. 2002. Model selection and multimodel inference: A practical information-theoretic approach, 2nd edition. Springer-Verlag, New York.

Capen, D. E., ed. 1981. The use of multivariate statistics in studies of wildlife habitat. General Technical Report RM-87. US Department of Agriculture, Forest Service, Rocky Mountain Forest and Range Experiment Station, Fort Collins, CO.

Carnes, B. A., and N. A. Slade. 1982. Some comments on niche analysis in canonical space. Ecology 63:888–893.

Chandler, S. K., J. D. Fraser, D. A. Buehler, and J. K. D. Seegar. 1995. Perch trees and shoreline development as predictors of bald eagle distribution on Chesapeake Bay. Journal of Wildlife Management 59:325–332.

Cook, C. W., and J. Stubbendieck, eds. 1986. Range research: Basic problems and techniques. Society for Range Management, Denver.

Cooley, W. W., and P. R. Lohnes. 1971. Multivariate data analysis. John Wiley & Sons, New York.

Cooper, C. F. 1957. The variable plot method for estimating shrub density. Journal of Range Management 10:111–115.

Crisol-Martínez, E., G. Ford, F. G. Horgan. P. H. Brown, and K. R. Wormington. 2017. Ecology and conservation of insectivorous bats in fragmented areas of macadamia production in eastern Australia. Austral Ecology 42:597–610.

Cumming, G. 2014. The new statistics: Why and how. Psychological Science 25(1):7–29.

Cutler, D. R., T. C. Edwards, K. H. Beard, A. Cutler, and K. T. Hess. 2007. Random forests for classification in ecology. Ecology 88:2783–2792.

Dillon, W. R., and M. Goldstein. 1984. Multivariate analysis: Methods and applications. John Wiley & Sons, New York.

Dodd, M. G., and T. M. Murphy. 1995. Accuracy and precision of techniques for counting great blue heron nests. Journal of Wildlife Management 59:667–673.

Draper, N. R., and H. Smith. 1981. Applied regression analysis, 2nd edition. John Wiley & Sons, New York.

Duchesne, T., D. Fortin, and N. Courbin. 2010. Mixed conditional logistic regression for habitat selection studies. Journal of Animal Ecology 79:548–555.

Efron, B. 1975. The efficiency of logistic regression compared to normal discriminant analysis. Journal of the American Statistical Association 70:892–898.

Ellison, A. M. 2004. Bayesian inference in ecology. Ecology Letters 7:509–520.

Endler, J. A. 1977. Geographic variations, speciation and clines. Princeton University Press, Princeton, NJ.

Fielding, A. H., and P. F. Haworth. 1995. Testing the generality of bird-habitat models. Conservation Biology 9:1466–1481.

Findley, J. S. 1993. Bats: A community perspective. Cambridge University Press, Cambridge.

Findley, J. S., and H. Black. 1983. Morphological and dietary structuring of a Zambian insectivorous bat community. Ecology 64:625–630.

Fretwell, S. D. 1972. Populations in a seasonal environment. Princeton University Press, Princeton, NJ.

Fretwell, S. D., and H. L. Lucas Jr. 1970. On territorial behavior and other factors influencing habitat distribution of birds: I. Theoretical Development. Acta Biotheoretica 19:16–36.

Ganey, J. L., and W. M. Block. 1994. A comparison of two techniques for measuring canopy closure. Western Journal of Applied Forestry 9:21–23.

Godvik, I. M. R., L. E. Loe, J. O. Vik, V. Veiberg, R. Langvatn, and A. Mysterud. 2009. Temporal scales, trade-offs, and functional responses in red deer habitat selection. Ecology 90:699–710.

Gotfryd, A., and R. I. C. Hansell. 1985. The impact of observer bias on multivariate analyses of vegetation structure. Oikos 45:223–234.

Graham, M. H. 2003. Confronting multicollinearity in ecological multiple regression. Ecology 84:2809–2815.

Green, R. H. 1971. A multivariate statistical approach to the Hutchinsonian niche: Bivalve molluscs of central Canada. Ecology 52:543–556.

Green, R. H., and R. C. Young. 1993. Sampling to detect rare species. Ecological Applications 3:351–356.

Grissino-Mayer, H. D., and T. W. Swetnam. 2000. Century-scale climate forcing of climate regimes in the American Southwest. Holocene 10:213–220.

Guthery, F., L. A. Brennan, M. Peterson, and J. Lusk. 2005. Information theory in wildlife science: Critique and viewpoint. Journal of Wildlife Management 69:457–465.

Harrell, F. E. 2015. Regression modeling strategies, 2nd edition. Springer Series in Statistics. Springer, New York.

Harris, R. J. 1985. A primer on multivariate statistics, 2nd edition. Academic Press, Orlando, FL.

Hatton, T. J., N. E. West, and P. S. Johnson. 1986. Relationships of error associated with ocular estimation and actual cover. Journal of Range Management 39:91–92.

Hazzah, L., S. Dolrenry, D. Kaplan, and L. Frank. 2013. The influence of park access during drought on attitudes toward wildlife and lion killing behaviour in Maasailand, Kenya. Environmental Conservation 40:266–276.

Hendry, A. P. 2019. A critique for eco-evolutionary dynamics. Functional Ecology 33:84–94.

Hijmans, R. J., S. E. Cameron, J. L. Parra, P. G. Jones, and A. Jarvis. 2005. Very high resolution interpolated climate surfaces for global land areas. International Journal of Climatology 25:1965–1978.

Hooten, M. B., and Hobbs, N. T. 2015. A guide to Bayesian model selection for ecologists. Ecological Monographs 85:3–28.

Hutchinson, G. E. 1957. Concluding remarks. Cold Spring Harbor Symposium on Quantitative Biology 22:415–427.

Hutchinson, G. E. 1978. An introduction to population ecology. Yale University Press, New Haven, CT.

James, F. C. 1971. Ordinations of habitat relationships among breeding birds. Wilson Bulletin 83:215–236.

James, F. C., and C. E. McCollough. 1990. Multivariate analysis in ecology and systematics: Panacea or Pandora's box? Annual Review of Ecology and Systematics 21:129–166.

James, F. C., and H. H. Shugart Jr. 1970. A quantitative method of habitat description. Audubon Field Notes 24:727–736.

Jennings, N., S. Parsons, and M. J. Pocock. 2008. Human vs. machine: Identification of bat species from their echolocation calls by humans and by artificial neural networks. Canadian Journal of Zoology 86:371–377.

Jochimsen, D. M., C. R. Peterson, and I. J. Harmon. 2014. Influence of ecology and landscape on snake road mortality in a sagebrush-steppe ecosystem. Animal Conservation 17:583–92.

Johnson, C. M., L. B. Johnson, C. Richards, and V. Beasley. 2002. Predicting the occurrence of amphibians: An assessment of multiple-scale models. Pp. 157–170 *in* J. M. Scott, P. J. Heglund, M. L. Morrison, J. B. Haufler, M. G. Raphael, W.A. Wall, and F. B. Samson, eds. Predicting species occurrences: Issues of accuracy and scale. Island Press, Washington, DC.

Johnson, D. H. 1981. The use and misuse of statistics in wildlife habitat studies. Pp. 11–19 *in* D. E. Capen, ed. The use of multivariate statistics in studies of wildlife habitat. General Technical Report RM-87. US Department of Agriculture, Forest Service, Rocky Mountain Forest and Range Experiment Station, Fort Collins, CO.

Johnson, R. A. 1992. Applied multivariate statistical analysis, 3rd edition. Prentice-Hall, Englewood Cliffs, NJ.

Jorde, D. G., G. L. Krapu, R. D. Crawford, and M. A. Hay. 1984. Effects of weather on habitat selection and behavior of mallards wintering in Nebraska. Condor 86:258–265.

Kasten, E. P., P. K. McKinley, and S. H. Gage. 2010. Ensemble extraction for classification and detection of bird species. Ecological Informatics 5:153–166.

Kaufman, D. W., and G. A. Kaufman. 1989. Population biology. Pp. 233–270 *in* G. L. Kirkland Jr. and J. N. Layneeds, eds. Advances in the study of *Peromyscus* (Rodentia). Texas Tech University Press, Lubbock.

Keating, K. A., and Cherry, S. 2004. Use and interpretation of logistic regression in habitat-selection studies. Journal of Wildlife Management 68:774–789.

Kelt, D. A., P. L. Meserve, and B. K. Lang. 1994. Quantitative habitat associations of small mammals in a temperate rainforest in southern Chile: Empirical patterns and the importance of ecological scale. Journal of Mammalogy 75:890–904.

Kent, M., and P. Coker. 1994. Vegetation description and analysis: A practical approach. John Wiley & Sons, New York.

Kepler, C. B., and J. M. Scott. 1981. Reducing count variability by training observers. Studies in Avian Biology 6:366–371.

Korkmaz, S., D. Goksuluk, and G. Zararsiz. 2014. MVN: An R package for assessing multivariate normality. R Journal 6:151–162.

Kuehl, A. K., and W. R. Clark. 2002. Predator activity related to landscape features in northern Iowa. Journal of Wildlife Management 66:1224–1234.

Lack, D. 1947. Darwin's finches. Cambridge University Press, Cambridge.

Lafontaine, A., P. Drapeau, D. Fortin, and M. St. Laurent. 2017. Many places called home: The adaptive value of seasonal adjustments in range fidelity. Journal of Animal Ecology 86:624–633.

Lee, J., I. Kwak, E. Lee, and K. Kim. 2007. Classification of breeding bird communities along an urbanization gradient using an unsupervised artificial neural network. Ecological Modelling 203:62–71.

Leigh, R. A., and A. E. Johnston, eds. 1994. Long-term experiments in agricultural and ecological sciences. CAB International, Oxford.

Lemoine, N. P., A. Hoffman, A. J. Felton, L. Bur, F. Chaves, J. Gray, Q. Yu, and M. D. Smith. 2016. Underappreciated problems of low replication in ecological field studies. Ecology 97:2554–2561.

Lindenmayer, D. B. 2018. Why is long-term ecological research and monitoring so hard to do? (And what can be done about it). Australian Zoologist 39:576–580.

Lindenmayer, D. B., G. E. Likens, A. Andersen, D. Bowman, C. M. Bull, E. Burns, C. R. Dickman, A. A. Hoffmann, D. A. Keith, M. J. Liddell, A. J. Lowe, D. J. Metcalfe, S. R. Phinn, J. Russell-Smith, N. Thurgate, and G. M. Wardle. 2012. Value of long-term ecological studies. Austral Ecology 37:745–757.

Likens, G. E. 1983. A priority for ecological research. Bulletin of the Ecological Society of America 64:234–243.

Lindsey, A. A., J. D. Barton, and S. R. Miles. 1958. Field efficiencies of forest sampling methods. Ecology 39:428–444.

Littell, R. C. 2002. SAS for linear models, 4th edition. SAS Institute, Cary, NC.

Lozano-Cavasos, E. A., L. A. Brennan, W. P. Kuvlesky Jr., F. Hernández, and W. C. Harrell. 2016. Land bird abundance and brush reduction in the Texas Coastal Prairie. Bulletin of the Texas Ornithological Society 49:2–32.

Ludwig, J. A., and J. F. Reynolds. 1988. Statistical ecology: A primer on methods and computing. John Wiley & Sons, New York.

Manly, B. F. J., and J. A. Navarro Alberto. 2016. Multivariate statistical methods: A primer, 4th edition. CRC Press / Taylor & Francis Group, Boca Raton, FL.

Marcot, B. G. 1985. Habitat relationships of birds and young-growth Douglas-fir in northwestern California. PhD diss., Oregon State University, Corvallis OR.

Marcot, B. G., M. G. Raphael, and K. H. Berry. 1983. Monitoring wildlife habitat and validation of wildlife-habitat relationships models. Transactions of the North American Wildlife and Natural Resources Conference 48:315–329.

Marra, P. P., E. B. Cohen, S. R. Loss, J. E. Rutter, and C. M. Tonra. 2015. A call for full annual cycle research in animal ecology. Biology Letters 11(8):2015.0552, doi:10.1098/rsbl.2015.0552.

Martin, J. A., and M. L. Morrison. 1999. Distribution, abundance, and habitat characteristics of the buff-breasted flycatcher in Arizona. Condor 101:272–281.

Martínez-Padilla, J., S. M. Redpath, M. Zeineddine, and F. Mougeot. 2013. Insights into population ecology from long-term studies of red grouse *Lagopus lagopus scoticus*. Journal of Animal Ecology 83:85–98.

Martinka, R. 1972. Structural characteristics of blue grouse territories in southwestern Montana. Journal of Wildlife Management 36:498–510.

Marzluff, J. M., J. J. Millspaugh, P. Hurvitz, and M. S. Handcock. 2004. Relating resources to a probabilistic measure of space use: Forest fragments and Steller's jays. Ecology 85(5):1411–1427.

McGarigal, K., S. A. Cushman, and S. Stafford. 2000. Multivariate statistics for wildlife and ecology research. Springer, New York.

Meents, J. K., J. Rice, B. W. Anderson, and R. D. Ohmart. 1983. Nonlinear relationships between birds and vegetation. Ecology 64:1022–1027.

Miles, D. B. 1990. A comparison of three multivariate statistical techniques for the analysis of avian foraging data. Studies in Avian Biology 13:295–308.

Morrison, D. F. 1967. Multivariate statistical methods, 2nd edition. McGraw-Hill, New York.

Morrison, L. W. 2016. Observer error in vegetation surveys: A review. Journal of Plant Ecology 9:367–379.

Morrison, M. L. 1981. The structure of western warbler assemblages: Analysis of foraging behavior and habitat selection in Oregon. Auk 98:578–588.

Morrison, M. L. 1984a. Influence of sample size and

sampling design on analysis of avian foraging behavior. Condor 86:146–150.

Morrison, M. L. 1984b. Influence of sample size on discriminant function analysis of habitat use by birds. Journal of Field Ornithology 55:330–335.

Morrison, M. L., and R. G. Anthony. 1989. Habitat use by small mammals on early-growth clear-cuttings in western Oregon. Canadian Journal of Zoology 67:805–811.

Morrison, M. L., W. M. Block, M. D. Strickland, B. A. Collier, and M. J. Peterson. 2008. Wildlife study design, 2nd edition. Springer-Verlag, New York.

Morrison, M. L., B. G. Marcot, and R. W. Mannan. 2006. Wildlife-habitat relationships: Concepts and applications, 3rd edition. Island Press, Washington, DC.

Morrison, M. L., C. J. Ralph, J. Verner, and J. R. Jehl Jr. 1990. Avian foraging: Theory, methodology, and applications. Studies in Avian Biology 13. Cooper Ornithological Society, Los Angeles.

Mueller-Dombois, D., and H. Ellenberg. 1974. Aims and methods of vegetation ecology. John Wiley & Sons, New York.

Nadeau, S., R. Décarie, D. Lambert, and M. St.-Georges. 1995. Nonlinear modeling of muskrat use of habitat. Journal of Wildlife Management 59:110–117.

Neff, N. A., and L. F. Marcus. 1980. A survey of multivariate methods for systematics. Privately published, printed by the American Museum of Natural History, New York.

Noon, B. R., and C. M. Biles. 1990. Mathematical demography of spotted owls in the Pacific Northwest. Journal of Wildlife Management 54:18–27.

Otis, D. L. 2002. Survival models for harvest management of mourning dove populations. Journal of Wildlife Management 66:1052–1063.

Peterson, A. T., and J. Soberón. 2012. Species distribution modeling and ecological niche modeling: Getting the concepts right. Natureza a Conservação: Brazilian Journal for Nature Conservation 10:102–107.

Pimentel, R. A. 1979. Morphometrics. Kendall/Hunt, Dubuque, IA.

Pitt, A. L., J. J. Tavano, R. F. Baldwin, and B. S. Stegenga. 2017. Movement ecology and habitat use of three sympatric anuran species. Herpetological Conservation and Biology 12:212–224.

Press, S. J., and S. Wilson. 1978. Choosing between logistic regression and discriminant analysis. Journal of the American Statistical Association 73:699–705.

Provencher, L., B. J. Herring, D. R. Gordon, H. L. Rodgers, G. W. Tanner, J. L. Hardesty, L. A. Brennan, and A. R. Litt. 2001. Longleaf pine and oak responses to hardwood reduction techniques in fire-suppressed sandhills in northwest Florida. Forest Ecology and Management 148:63–77.

Qiao, H., J. Soberón, and A. T. Peterson. 2015. No silver bullets in correlative ecological niche modelling: Insights from testing among many potential algorithms for niche estimation. Methods in Ecology and Evolution 6:1126–1136.

Queheillalt, D. M., J. W. Cain III, D. E. Taylor, M. L. Morrison, S. L. Hoover, N. Tuatoo-Bartley, L. Rugge, K. Christopherson, M. D. Hulst, M. R. Harris, and H. L. Keough. 2002. The exclusion of rare species from community-level analyses. Wildlife Society Bulletin 30:756–759.

Raphael, M. G. 1990. Use of Markov chains in analyses of foraging behavior. Studies in Avian Biology 13:288–294.

Raphael, M. G., and B. G. Marcot. 1986. Validation of a wildlife-habitat relationships model: Vertebrates in a Douglas-fir sere. Pp. 129–138 *in* J. Verner, M. L. Morrison, and C. J. Ralph, eds. Wildlife 2000: Modeling habitat relationships of terrestrial vertebrates. University of Wisconsin Press, Madison.

Ratkowsky, D. A. 1990. Handbook of nonlinear regression models. Marcel Dekker, New York.

Rexstad, E. A., D. D. Miller, C. H. Flather, E. M. Anderson, J. W. Hupp, and D. R. Anderson. 1988. Questionable multivariate statistical inferences in wildlife habitat and community studies. Journal of Wildlife Management 52:794–798.

Rexstad, E. A., D. D. Miller, C. H. Flather, E. M. Anderson, J. W. Hupp, and D. R. Anderson. 1990. Questionable multivariate statistical inferences in wildlife habitat and community studies: A reply. Journal of Wildlife Management 54:189–193.

Sanderlin, J. S., W. M. Block, and B. E. Strohmeyer. 2016. Long-term post-wildfire correlates with avian community dynamics in ponderosa pine forests. Pp. 89–101 *in* B. E. Ralston, ed. Proceedings of the 12th Biennial Conference of Research on the Colorado Plateau. Scientific Investigations Report 2015-5180. US Department of the Interior, US Geological Survey, Reston, VA.

Schmid, J. R., A. B. Bolten, K. A. Bjorndal, W. J. Lindberg, H. F. Percival, and P. D. Zwick. 2003. Home range and habitat use by Kemp's ridley turtles in west-central Florida. Journal of Wildlife Management 67:196–206.

Schooley, R. L. 1994. Annual variation in habitat selection: Patterns concealed by pooled data. Journal of Wildlife Management 58:367–374.

Schradin, C., and L. D. Hayes. 2017. A synopsis of long-term field studies in mammals: Accomplishments, future directions, and some advice. Journal of Mammalogy 98:670–677.

Schultz, A. M., R. P. Gibbens, and L. Debano. 1961. Artificial populations for teaching and testing range techniques. Journal of Range Management 14:236–242.

Scott, J. M., P. J. Heglund, M. L. Morrison, J. B. Haufler, M. G. Raphael, W. A. Wall, and F. B. Samson, eds. 2002. Predicting species occurrences: Issues of accuracy and scale. Island Press, Washington, DC.

Scott, J. M., F. L. Ramsey, and C. P. Kepler. 1981. Distance estimation as a variable in estimating bird numbers from vocalizations. Studies in Avian Biology 6:334–340.

Seber, G. A. F. 1989. Nonlinear regression. John Wiley & Sons, New York.

Shaw, P. J. A. 2003. Multivariate statistics for the environmental sciences. Oxford University Press, New York.

Shew, J., B. Greene, and F. Durbian. 2012. Spatial ecology and habitat use of the western foxsnake (*Pantherophis vulpinus*) on Squaw Creek National Wildlife Refuge (Missouri). Journal of Herpetology 46:539–548.

Shugart, H. H., Jr. 1981. An overview of multivariate methods and their application to studies of wildlife habitat. Pp. 4–10 *in* D. E. Capen, ed. The use of multivariate statistics in studies of wildlife habitat. General Technical Report RM-87. US Department of Agriculture, Forest Service, Rocky Mountain Forest and Range Experiment Station, Fort Collins, CO.

Sollmann, R., B. Gardner, J. L. Belant, C. M. Wilton, and J. Beringer. 2016. Habitat associations in a recolonizing, low-density black bear population. Ecosphere 7(8):e01406, doi:10.1002/ecs2.1406.

Stauffer, D. F. 2002. Linking populations and habitats: Where have we been? Where are we going? Pp. 53–61 *in* J. M. Scott, P. J. Heglund, M. L. Morrison, J. B. Haufler, M. G. Raphael, W. A. Wall, and F. B. Samson, eds. Predicting species occurrences: Issues of accuracy and scale. Island Press, Washington, DC.

Strayer, D., J. S. Glitzenstein, C. G. Jones, J. Kolasa, G. E. Likens, M. J. McDonnell, G. G. Parker, and S. T. A. Pickett. 1986. Long-term ecological studies: An illustrated account of their design, operation, and importance to ecology. Occasional Paper No. 2. Institute of Ecosystem Studies, New York Botanical Garden, Mary Flagler Cary Arboretum, Millbrook, NY.

Stuber, E. F., C. C. Chizinski, J. L. Lusk, and J. J. Fontaine. 2019. Multivariate models and analyses. Pp. 32–62 *in* L. A. Brennan, A. N. Tri, and B. G. Marcot, eds. Quantitative analyses in wildlife science. Johns Hopkins University Press, Baltimore.

Tabachnick, B. G., and L. S. Fidell. 1983. Using multivariate statistics. Harper & Row, New York.

Tacha, T. C., W. D. Warde, and K. P. Burnham. 1982. Use and interpretation of statistics in wildlife journals. Wildlife Society Bulletin 10:355–362.

Taylor, J. 1990. Questionable multivariate statistical inferences in wildlife habitat and community studies: A comment. Journal of Wildlife Management 54:186–189.

Thessen, A. 2016. Adoption of machine learning techniques in ecology and earth science. One Ecosystem 1:e8621. https://doi.org/10.3897/oneeco.1.e8621/.

Thompson, W. L., ed. 2004. Sampling rare or elusive species. Island Press, Washington, DC.

Thuiller, W. 2003. BIOMOD—optimizing predictions of species distributions and projecting potential future shifts under global change. Global Change Biology 9:1353–1362.

Timm, N. H. 2002. Applied multivariate analysis. Springer-Verlag, New York.

Touchon, J. C., and M. W. McCoy. 2016. The mismatch between current statistical practice and doctoral training in ecology. Ecosphere 7(8):1–11.

Towler, E., V. A. Saab, R. S. Sojda, K. Dickinson, C. L. Bruyère, and K. R. Newlon. 2012. A risk-based approach to evaluating wildlife demographics for management in a changing climate: A case study of the Lewis's woodpecker. Environmental Management 50:1152–1163.

Verner, J., M. L. Morrison, and C. J. Ralph, eds. 1986. Wildlife 2000: Modeling habitat relationships of terrestrial vertebrates. University of Wisconsin Press, Madison.

Vilizzi, L., G. H. Copp, and J. M. Roussel. 2004. Assessing variation in suitability curves and electivity profiles in temporal studies of fish habitat use. River Research and Applications 20:605–618.

Ward, R. L., and C. L. Marcum. 2005. Lichen litterfall consumption by wintering deer and elk in western Montana. Journal of Wildlife Management 69:1081–1089.

Wasserstein, R. L., and N. A. Lazar. 2016. The ASA's statement on *P*-values: Context, process, and purpose. American Statistician 70:129–133

Wesolowsky, G. O. 1976. Multiple regression and analysis of variance. John Wiley & Sons, New York.

Wester, D. B. 2019. Regression: Linear and nonlinear, parametric and nonparametric. Pp. 9–31 *in* L. A. Brennan, A.N. Tri, and B. G. Marcot, eds. Quantitative analyses in wildlife science. Johns Hopkins University Press, Baltimore.

Whitney, J. E., K. B. Gido, T. J. Pilger, D. L. Probst, and T. F. Turner. 2015. Metapopulation analysis indicates native and non-native fishes respond differently to effects of wildfire on desert streams. Ecology of Freshwater Fish 25:376–392.

Wiens, J. A. 1984. The place of long-term studies in ornithology. Auk 101:202–203.

Wiens, J. A. 1989. The ecology of bird communities. Vol. 1, Foundations and patterns. Cambridge University Press, Cambridge.

Weins, J. A. 2016. Ecological challenges and conservation conundrums. John Wiley & Sons, Hoboken, NJ.

Wiens, J. A., and J. T. Rotenberry. 1981. Habitat associations and community structure of birds in shrubsteppe environments. Ecological Monographs 51:21–41.

Williams, B. K. 1983. Some observations of the use of discriminant analysis in ecology. Ecology 64:1283–1291.

Wingfield, J. C. 2007. Organization of vertebrate annual cycles: Implications for control mechanisms. Philosophical Transactions of the Royal Society B, Biological Sciences 363(1490):425–441.

Woodward, A. A., A. D. Fink, and F. R. Thompson III. 2001. Edge effects and ecological traps: Effects on shrubland birds in Missouri. Journal of Wildlife Management 65:668–675.

Zar, J. H. 1999. Biostatistical analysis, 4th edition. Prentice-Hall, Upper Saddle River, N.J.

Zink, R. M. 2015. Genetics, morphology, and ecological niche modeling do not support the subspecies status of the endangered southwestern willow flycatcher (*Empidonax traillii extimus*). Condor: Ornithological Applications 117:76–86.

4 Measuring Behavior

In science, precise definitions are important.

Levitis et al. (2009:104)

Introduction

Many studies of animal-habitat use have focused on behavior. Understanding animal behavior is of obvious importance in understanding their distribution and abundance, as well as implementing management actions that meet their needs. Although this type of research is usually more informative than correlational investigations of species occurrence and environmental conditions, behavioral studies of animals, without considering other factors, are fraught with uncertainties and potential biases.

As noted by Morrison et al. (2020:Chapter 1), research in animal ecology begins with observations of individuals that belong to a recognized species. The authors emphasized that a project involving animal behavior should start with the development of a sound theoretical framework that clearly states how the researcher thinks that an individual animal perceives and then uses its environment and the associated resources. An investigator should determine the relevant biological population of interest and then sample characteristics of interest to that population. In animal ecology, 3 general areas of behavioral research have contributed to our understanding of how animals view their environment: the development of theoretical models, laboratory experiments, and field studies. Each of these 3 areas is incorporated into this chapter.

Regardless of the theoretical basis on which behavioral studies are constructed, animal ecologists must employ proper means of data collection and analysis. Such considerations include specific methods by which animals are located and watched, recognition of the biases associated with these observations, the application of statistical procedures, and an adequate sample size. Unfortunately, scant attention has historically been given to these concerns in too many behavioral investigations.

In this chapter we review the theoretical framework for research about animal behavior and then discuss the principal methods used to observe it and assess resource abundance and use. Because of the obvious importance that food gathering plays in the life of an individual animal, ways of studying foraging behavior and diet are also covered. Additionally, we look at material related to activity budgets, the influence of thermal constraints on behaviors, concepts related to foraging and diet, and time-series analyses of sequential data. We conclude with a description of an environmental-habitat model that can be used for testing behavioral hypotheses in animal ecology.

Theoretical Framework

How individual animals in a population select and use habitat reflects their evolutionary history, the current ecological conditions of the environment, and the influence of other animals, including humans. In the end, all behaviors—whether for foraging, dispersal, migration, reproduction, or predator avoidance—are modified and guided by the selective advantage of increased fitness for individuals and the vitality of populations. If either or both lack such advantages, they are less likely to persist over time. The ultimate challenge to behavioral researchers is to identify the linkages between the variability of behaviors by individuals (that presumably influence their fitness) and the vitality of populations (that presumably drive their persistence). This is a tall order, indeed.

The theoretical framework developed by Morrison et al. (2020) provides a background for observing and then analyzing and interpreting animal behavior. How an individual animal views its environment—that is, as a series of different spatial scales, ranging from a general vegetation type down to specific features of a microsite—are all reflected in its behavior. By studying their actions in a systematic manner, we can learn much about why animals persist, or fail to persist, in a particular environment.

Much like the word "habitat," behavioral biologists have used different and often conflicting definitions of the term "behavior" over the years (Levitis et al. 2009). For the purposes of both this chapter and this book, we define behavior, in the most fundamental sense, as the self-directed movements of an animal. These movements—or behaviors—can be instinctive or they can be learned. They can also be influenced by environmental conditions, by the activities of other individuals, and by habitual use patterns. The study of animal behavior is rich in both theory and application, which provides a foundation for employing interdisciplinary approaches—for example, integrating neuroscience, mathematical modeling, and field experiments on organisms—to advance animal ecology (Giuggioli and Bartumeus 2010).

The modern foundations of animal behavior, or ethology (Martin and Bateson 1993), were developed by Konrad Lorenz, Nikolaas Tinbergen, and Karl von Frisch, who shared the 1973 Nobel Prize in Physiology or Medicine (unfortunately, there is not a Nobel Prize for Biology or Ecology). Lorenz and Tinbergen focused their work on instinctive activities of vertebrates (primarily birds), while von Frisch devoted his attention to actions and communication in honeybees (*Apis* sp.). All 3 of these behavioral research pioneers lived and worked in Western Europe during the first two-thirds of the twentieth century: Lorenz was from Austria, Tinbergen from the Netherlands, and von Frisch from Germany.

The themes developed by Lorenz, Tinbergen, and von Frisch provide the conceptual background for nearly all animal behavior research conducted today (with the exception of genetics). At the time, the idea of a newly born (or hatched) animal becoming "imprinted" on an animate or inanimate object (Lorenz 1937) was a stunning scientific breakthrough. Tinbergen built on Lorenz's body of work and developed his hierarchical stimulus model of animal behavior (Tinbergen 1951). Using a series of elegant experiments, von Frisch (1927) demonstrated that honeybees not only recognized colors and could smell, but they also communicated directions and distances to food sources to other bees by behaviors that he described as various dance movements.

Much like the organizational theme of this book, Lorenz (1937) and Tinbergen (1951) focused their work on an individual organism and then scaled up to develop inferences for larger groups (e.g., populations) and, ultimately, species. While von Frisch (1927) studied groups of animals with a highly complex social structure, it was also his astute observations of individual bees that allowed him to make and integrate inferences about how behaviors were tied to the survival and persistence of populations, as well as species. All of this foundational work is linked to an evolutionary construct that can only be understood in the context of the fitness of individuals, which allows them to place their genes into the next generation of their descendants.

Tinbergen's (1951) hierarchical model identified the parts of a major instinctual behavior (feeding) by identifying an internal stimulus (an empty stomach) needed to trigger the removal of an internal releasing mechanism, or block. This allowed an individual to use visual and olfactory means to locate food and then procure it, via mechanical methods and morphological adaptations (Fig. 4.1). His work formed the basis not only of contemporary foraging theory, but also the broader field of behavioral ecology, which remains extremely active today (see Birkehead et al. 2014:Chapter 7).

Ploger and Yasukawa (2003) used many of the classic behavioral concepts developed by Tinbergen (1951), Lorenz (1937), and von Frisch (1927) and synthesized a contemporary perspective of animal behavior based on 4 components. What causes an animal to perform a certain behavior? How does the behavior change as the animal develops? What is the evolutionary history of the behavior? And how does the behavior help the animal survive and reproduce successfully? Below we briefly discuss each of these 4 essential components of animal behavior as we set the stage for describing fundamental principles for recording behavioral data and then using the results to manage wildlife.

1. *Causation of behavior*. Behavioral impetuses have multiple sources, including environmental stimuli, mental processes, neural pathways, and hormones. Thus the study of causation involves ecology, psychology, neurobiology, endocrinology, ethology, and other areas. Both individually and combined, these

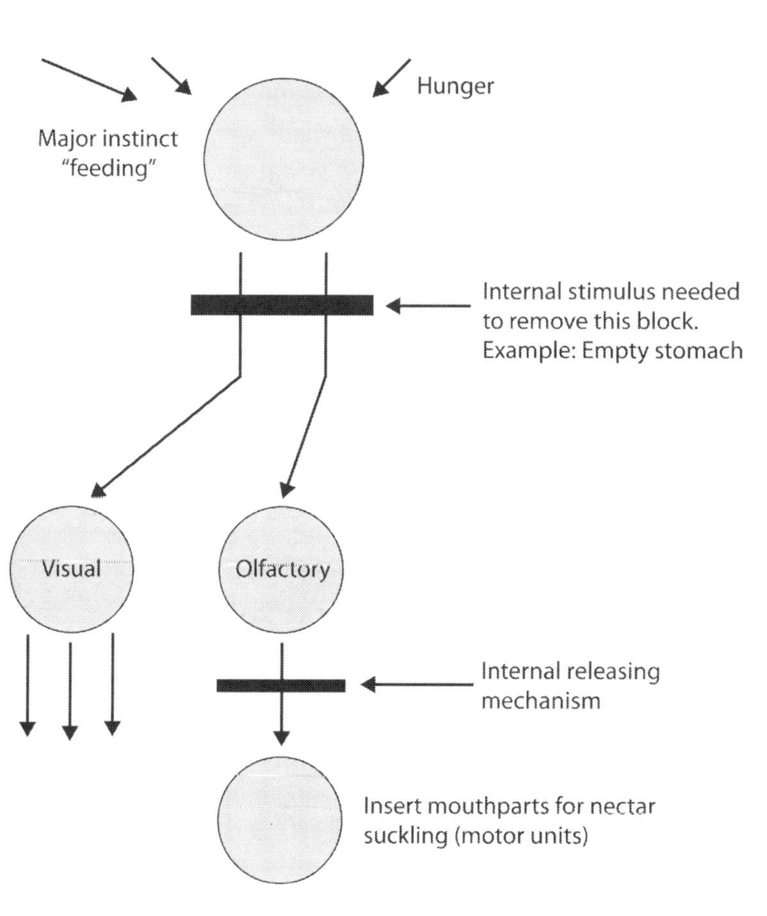

Figure 4.1. A diagram of Tinbergen's (1951) hierarchical stimulus model of animal behavior. (Courtesy of the Leonard A. Brennan Collection)

disciplines examine the way in which a behavior is triggered—the proximate mechanisms. Why an action is elicited usually has a direct application to wildlife study, such as the reason for an individual to abandon a nest, flee at certain noises, choose to settle in a particular location, and so on.

2. *Development of behavior*. Behaviors emerge as an individual goes through its life cycle. Initial activities are primarily innate and focus principally on growth and survival (e.g., the begging response of a young animal). Depending on the species, animals then learn additional modes of conduct, based on events they experience and survive. In reality, most actions have a fundamental root that can be modified by experience. As such, there is usually not a clear dichotomy between innate and learned behaviors. The specific patterns of activity an individual learns, when it discovers them, and the intensity of stimuli under which those behaviors were acquired, all provide biologists with valuable information when developing management guidelines.

3. *Evolution of behavior*. Behaviors are based on evolutionary history, and studies on the basis for them are concerned with the ultimate mechanism or reason for the activity (versus the proximate mechanism, noted above). Understanding the particular evolutionary history of an organism, or at least the setting in which a species evolved, is useful in animal ecology investigations, because it places behavioral observations in a broad causal context, as well as helping us understand why the abundance and distribution of a species is changing.

4. *Function of behavior*: Understanding the function of the behavioral repertoire of animals tells us how those actions promote survival and reproduction (plus a host of related activities). Determining function is a core component of the ultimate reason for a particular behavior. Once we know the function of different activities, we can focus our attention on those that promote survival and reproductive success.

In addition, 3 key terms that often used in behavioral studies are "use," "selection," and "requirement." In this chapter we define "use" as the demonstrated presence of a particular item (e.g., a den, perch, or foraging site) in an animal's behavioral repertoire or a prey item in its diet. "Selection," in contrast, has typically been defined as "use" coupled with evidence that the frequency with which the behavior occurs is significantly greater (statistically and biologically) than the amount of time some kind of resource is available in the animal's environment. More recently, Hall et al. (1997) suggested that "selection" should be used in reference to the process by which innate and learned behavioral decisions are made. They further defined "preference" as the consequence of this selection process, which is measured as the disproportional use of some resources over others. The converse is "avoidance," but this may be problematic to differentiate from preference as the motivation for a behavior. Finally, "requirement" is the presence (or a particular minimum amount), of a resource an animal must obtain so it can live and reproduce. Unfortunately, few research efforts have identified such requirements, which may best be determined through controlled ex situ experiments. Observational studies in the wild usually cannot be designed so the researcher can conclude whether an animal selects a resource because of a behavioral or a physiological requirement. As such, most investigations of wildlife behavior report the use of various resources, although the data may inappropriately be interpreted as showing a partiality for those items (Hall et al. 1997). If many studies, conducted across different time periods and locations, consistently show preferences for a particular resource or behavior, however, then one can probably infer that the species is exhibiting an action that has adaptive significance. This line of evidence therefore implicates that activity as a requirement for a particular organism (Ruggiero et al. 1988).

Measuring Behavior

Martin and Bateson (1993:19–23) and Ploger (2003) outlined steps to follow in designing and implementing a behavioral study. Here, we are specifically interested in the choice of variables, selection of recording methods, accumulation of adequate sample sizes, and form of data analysis. Additional reviews of research methods in behavioral studies are available in many publications (see especially Altmann 1974; Hazlett 1977; Colgan 1978; Kamil and Sargent 1981; Kamil et al. 1987; Gottman and Roy 1990; Morrison et al. 1990; Sommer and Sommer 1991; Lehner 1996; Ploger and Yasukawa 2003). Every issue of major behavioral journals (e.g., *Journal of Animal Behaviour*, *Behaviour*, *Behavioral Ecology*)—including those devoted to the study of specific taxa, such as primates (e.g., *International Journal of Primatology*, *Primates*)—present numerous papers that contain examples of behavioral methods and analyses.

Most of these observations can be broken down into 3 categories: structure, consequence, and spatial relation (Martin and Bateson 1993:57–58). Researchers must recognize these categories before defining specific variables to record. Otherwise, problems are likely to arise in the analysis and interpretation of data. "Structure" refers to the appearance, physical form, or temporal pattern of an animal's behavior, described in terms of posture and movements. For example, saying that a deer "reaches down and removes a leaf from a bush" indicates the structure of that behavior. "Consequence," in contrast to structure, describes the effects of an animal's behavior. Here, the action is identified without reference to how its results are achieved (see also Dewsbury 1992). Stating that a deer is "browsing" notes the consequences of the activity but says nothing about how it occurred (i.e., the structure). "Spatial relation" describes behavior in terms of an animal's proximity to features in the environment, including other animals. For example, "approaching a bush" depicts how close an animal is to a potential foraging substrate. Recording the relationship between the 2 objects or the orientation of that animal adds substantial information to the behavioral data being collected.

A common mistake is for activities to be described in terms of presumed consequences, rather than in neutral terms that do not apply a meaning that is probably tainted by observer biases. For instance, labeling an action as "stalking" implies a consequence that may be unwarranted after further, more objective study (see Martin and Bateson 1993:57–58). Such anthropocentrisim is a common mistake in behavioral studies (Lehner 1996:85; Ploger 2003). Technological advances can help overcome these obstacles. In a study of foraging by northern elephant seals (*Mirounga angustirostris*), Kuhn et al. (2009) used implanted sensors having satellite uplinks to quantify changes in behavioral states, such as transit, searching, and feeding.

Martin and Bateson (1993:62–66) noted several commonly used terms in behavioral studies that must be defined before observations can be recorded. "Latency" is the length of time from some specified event (e.g., the beginning of the recording session) to the onset of the first occurrence of the activity of interest. "Frequency" is the number of times that pattern happens per unit of time. The total number of occurrences, however, should not be used as a measure of frequency unless accompanied by a time unit. "Duration" is the length of time for which a single instance of the behavior pattern lasts. "Intensity" is widely used but has no specific definition. It is best viewed as a measure of the amplitude or magnitude of an action (Martin and Bateson 1993:65), such as how loud a call was, or how high was a jump.

What to Measure: Selection of Variables for a Behavioral Study

In Chapter 2, we addressed the selection of variables used in habitat analyses. Much of that discussion also generally applies to the choice and analysis of these elements in studies of animal behavior. Whereas habitat analysis focuses on a depiction of the environment surrounding an animal, the behavioral descriptions of habitat selection and use concentrate on specific actions of that animal within its habitat. The context of how these activities—such as courtship, nesting, denning, brooding, roosting, foraging, and so on—happen are also crucial to the inferences that can be drawn from such observations. Operational definitions, or operationalization, allow researchers to translate behavioral concepts into specific terms that can be measured objectively over time by different workers. Without clear delineations, comparisons between studies are difficult, and entire datasets can be rendered useless. Also, once a concept is operationalized, it must be used consistently by all observers (Lehner 1996:115; Glover 2003).

Animals perform myriad activities during a 24-hour period. They sleep, groom, engage in intra- and interspecific interactions, feed, and so on. To quantify these behaviors in any practical way obliges us to devise some form of record keeping. A useful starting point is to create an "ethogram" (*etho-*, as in "ethology"), which is a complete catalog of all activities exhibited by a species. This is a time-intensive process—normally using captive animals in a laboratory setting or, occasionally, individually marked animals in the wild—that requires meticulous attention to detail. A simple GoogleScholar search provides access to scores of ethograms, but most are of domesticated animals or, in some cases, captive wild animals in zoos. The role and value of ethograms in the behavioral sciences has been the subject of some debate (Gordon 1985), and Schleidt et al. (1984) called for standardization in how ethograms are compiled and developed. First, this requires intensive observation and cataloging of the behaviors of individual animals (often in captivity). Second, further evaluation is needed, using additional individuals (of a population), with the assumption that the ethogram represents a particular species. Such a hierarchical organization, with a focus on individual animals at its base, is consistent with the themes covered by Morrison et al. (2020).

The literature on animal behavior makes it clear that there are relatively few ethograms that have been developed and cataloged for wild vertebrates, which is unfortunate. Having ethograms, especially for species of conservation concern, has merit in adding to our understanding of their basic biology. Even species that have been subjected to intense research scrutiny for more than half a century, such as northern bobwhites (*Colinus virginianus*) or wild turkeys (*Meleagris gallopavo*), still do not have complete ethograms (Brennan et al. 2014; Kuvlesky et al. 2020). Given that millions of dollars are spent annually on conservation and management activities for these 2 species of galliforms, it would seem wise to invest a modest sum for research support to develop complete ethograms for them, as well as for other species of high conservation concern.

Ethograms are a valuable research tool, because they allow a behavioral scientist to organize the activities of a species into functional groups (locomotion, maintenance, foraging, roosting, mating, etc.) that provides instant and objective operational contexts for analysis (Gordon 1985). Ethograms become even more useful when these behaviors are organized into sequences of activities that allow them to be analyzed in relation to other actions, rather than simple analyses based on the individual movements themselves. Leonard (1984) described this strategy as a top-down analytical approach. The idea of context with respect to behavioral organization—and responses in specific situations—is absolutely critical to any study involving the behavioral ecology of wild animals.

Classifying Behaviors

Slater (1978) provided 5 basic points for classifying behaviors. First, behavioral categories must be discrete. This means that acts within a classification must have clear points of similarity that are not shared with those outside that category. Second, behavioral acts must be homogeneous. All the acts within a bracket should be similar in form, so there is little danger of conflating 2 different behaviors. Third, it is better to split than to combine. Within reason, 2 similar activities with possibly different consequences should be placed into separate categories. They can always be combined later. Fourth, the names of these classifications should not possess a causal or functional implication (as discussed above). Rather, names should clinically describe a behavior in terms of actions, not outcomes or motives. Fifth, the number of categories must be manageable, as including too many will reduce an observer's reaction time and lower the accuracy of the information that is collected. Automated recording devices, including computer-aided data entry programs, can help overcome this problem to a degree (see Raphael 1988; Martin and Bateson 1993:Chapter 7; Paterson et al. 1994; Samuel and Fuller 1994).

Lehner (1996) recommended that the selection of behavioral variables should begin with a list of all of those known or suspected to affect the activities under investigation:

1. Environmental variables
   a. Biotic
      i. Members of a social group
      ii. Predator-prey relationships
      iii. Interspecific interactions
      iv. Vegetation
   b. Abiotic
      i. Temperature
      ii. Wind speed
      iii. Moisture
      iv. Cloud cover
      v. Topography
      vi. Substrate
2. Organismal variables
   a. Genotype
      i. Sex
      ii. Parental stock
   b. Phenotype
      i. Behavioral characteristics
      ii. Description of the behavior
      iii. Frequency
      iv. Rate
      v. Duration

By listing all of the potentially important variables for a behavioral study, researchers can work toward determining the potential cause for an activity, as well as identifying possible sources of shifts in a behavioral parameter that can be eliminated or measured. Because the number of variables that could affect a particular action can be large, you should expect to spend a good deal of time developing this list and narrowing it down with pilot studies. It is always useful to first look at previous work that involved the same or similar species and have several colleagues review your list prior to continuing your investigation.

Recording Methods

Once researchers have developed the behavioral categories for a particular study, they must rigorously define the methods for recording these activities. Martin and Bateson (1993:Chapter 6) and Ploger (2003) have provided excellent frameworks for noting down behavioral observations by defining sets of sampling and recording rules (Fig. 4.2). We summarize these rules below and provide brief examples of their applications to animal behavior research.

Sampling Rules

Before collecting data, you must objectively decide which individuals to observe (i.e., sample) and when to view them. In Chapter 2 we have described reasons for defining and selecting species and their habitats for study, as well as the potential use of model species or systems for experimentation. In Chapter 3 (and elsewhere) we have discussed the importance of temporal and spatial stratification for data collection. Here we concentrate on observations of individual animals within the species chosen for a particular investigation and within the context of a proper spatial and temporal research design.

Ad Libitum Sampling

When rare events are the activity of interest, a generalized behavior sampling, or *ad libitum* methodology, is usually preferred. This involves watching the entire group for long periods of time and recording each

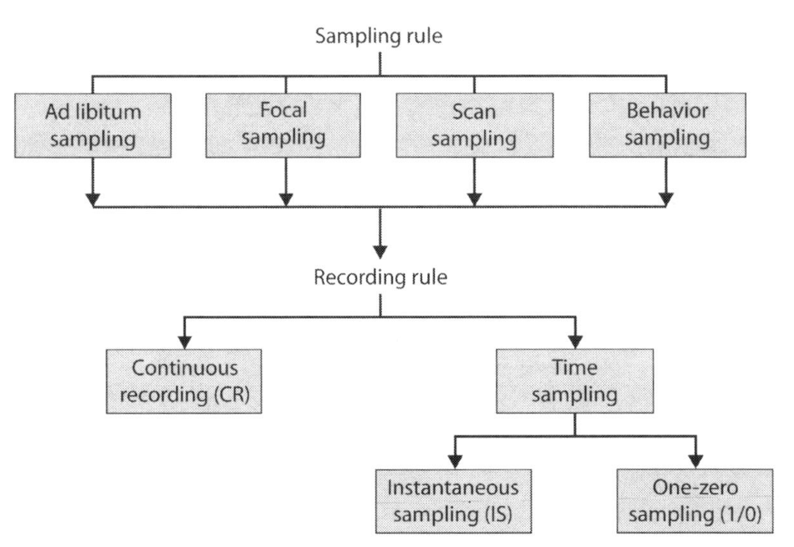

Figure 4.2. A diagram showing the hierarchy of sampling rules (i.e., determining what is watched and when) and recording rules (i.e., determining what is recorded and how often). (Courtesy of the Leonard A. Brennan Collection)

instance of the relevant behaviors (Ploger 2003). Masataka and Thierry (1993), for example, used *ad libitum* sampling in their research on Tonkean macaques (*Macaca tonkeana*). This form of sampling is seldom the primary method used in a behavioral study, because other methods provide a more structured means of gathering data from the population being investigated. Nonetheless, *ad libitum* sampling can be incorporated as a subpart of either focal or scan sampling (Martin and Bateson 1993:87).

Focal-Animal Sampling

Focal-animal sampling, or just focal sampling, involves observing 1 individual—even if it is located within a group of animals—for a specified period of time. This method has the clear advantage of allowing adequate records to be collected on various classes of the species of interest, such as data on the sex and age of the focal animal. Also, recording sequences of behaviors by an individual over time allows an investigator to compute transitional probabilities, in order to gain further insight on predicting subsequent behaviors (Kabak 1970). Further, if animals are individually tagged or show unique markings (e.g., coat colors and patterns for wild horses, or the size and shape of antlers or horns), you can accumulate information on each individual's behavioral variations, rather than simply combining the data across several animals in a group. For example, Carranza (1995) used the shape and branching pattern of antlers to identify individual red deer (*Cervus elaphus*). Stamps and Krishnan (1994) focused on an individual bronze anole lizard (*Anolis aeneus*) to determine the outcome of social interactions within a group of these lizards.

Focal-animal approaches are widely used to quantify behavior, competitive interactions, macrohabitat and microhabitat use, transitional probabilities, and other variables. The focal target for observation can be an individual animal or a group of individuals. For example, Hale et al. (2003) used assemblages of brown jays (*Cyanocorax* [*Psilorhinus*] *morio*) for focal observations, and Jones and Bock (2003) used mixed-species foraging flocks; other examples abound (e.g., Trainer et al. 2002; Salewski et al. 2003). A potential bias with focal sampling involves selecting the more conspicuous animals. This is especially the case when dominant individuals are making displays that are more noticeable or vocalizations that are louder than those by subdominant individuals.

Losito et al. (1989) developed a modification of focal sampling to address the problem of individuals disappearing from view, which is especially prevalent in areas with limited visibility (e.g., dense brush).

Termed "focal-switch sampling," their method allows shifting to a new focal individual (the nearest neighbor) if the original animal went out of sight. They found that focal switching was more efficient than standard focal sampling, because it saved 24 percent of their samples from premature termination. When using focal-switch sampling, the recorded data should denote when the switch to another individual occurred, to avoid errors in analyzing sequences of behaviors.

Scan Sampling

Although used less than focal-animal sampling, scan sampling entails observers "viewing" or "scanning" across a large group and collecting data at regular intervals, recording the behavior of either each individual or a subset of the organisms—in the latter case, usually without identifying 1 animal in particular. The behavior of each scanned individual is intended to be an instantaneous sample, so this methodology usually restricts the observer to recording only a small number of simple categories of behavioral variables. It takes only a few seconds to several minutes to collect an observation by scan sampling, depending on the amount of information being recorded, the group's size, and the activity level of the group (e.g., a grazing versus a stampeding herd). For example, Davoren et al. (2003) used scan sampling to study the foraging strategies of common murres (*Uria aalge*), as observed from boats. They established survey transects and scanned for birds from fixed points for a set period of time (Fig. 4.3). Through this method, they were able to relate the foraging behavior of the murres to the presence of their primary prey species.

Bias in scan sampling is influenced largely by the period of time that elapses between scans, as rare or inconspicuous behaviors will probably be observed at a lower frequency than the instances when they actually occur. For example, Harcourt and Stewart (1984) showed that studies using scan sampling of foraging gorillas resulted in underestimates of their actual foraging time. Focal-animal sampling potentially resulted in a more accurate estimate, because the researchers could continuously follow individual animals. Other biases when using scan sampling include (1) differential visibility between activity categories and seasons, (2) poor correlation between group composition and the representation of age and sex classes in the scans, (3) diurnal variation in sample sizes (i.e., the number of scans), and (4) animals' increased habituation to the observers over the duration of a study (Newton 1992).

Scan sampling can also be combined with focal sampling during the same session. In this situation, researchers record the behavior of focal individuals in detail, but at specific intervals they also scan sample the group for simpler behavioral categories. Workers have shown that for flocking birds and herding mammals, the time an individual spends foraging is related to the number and proximity of group members. For example, an individual must watch more vigilantly for predators as the group's size decreases (Hamilton 1971; Pulliam 1973; Caraco et al. 1980; Kildaw 1995). Using both sampling methods in this manner allows an investigator to place focal-animal observations into the overall context of group activities.

L'Heureux et al. (1995) used scan and focal sampling to study mother-yearling associations in bighorn sheep (*Ovis canadensis*) in Alberta, Canada. Scan sampling, at 15-minute intervals, was employed to record the distance between a mother and her yearling. Focal-animal sampling was applied by observing individual sheep for at least 40 minutes before they were out of sight, or for a maximum of 2 hours. Carranza (1995) also used both forms of sampling to study male-female interactions in red deer, with the focal method used to record interactions among territorial males, and the scan procedure (sampling at 1-minute intervals) to record the number of females in a territory.

Behavior Sampling

By focusing on individuals, researchers may record actions that vary considerably in length, as well as

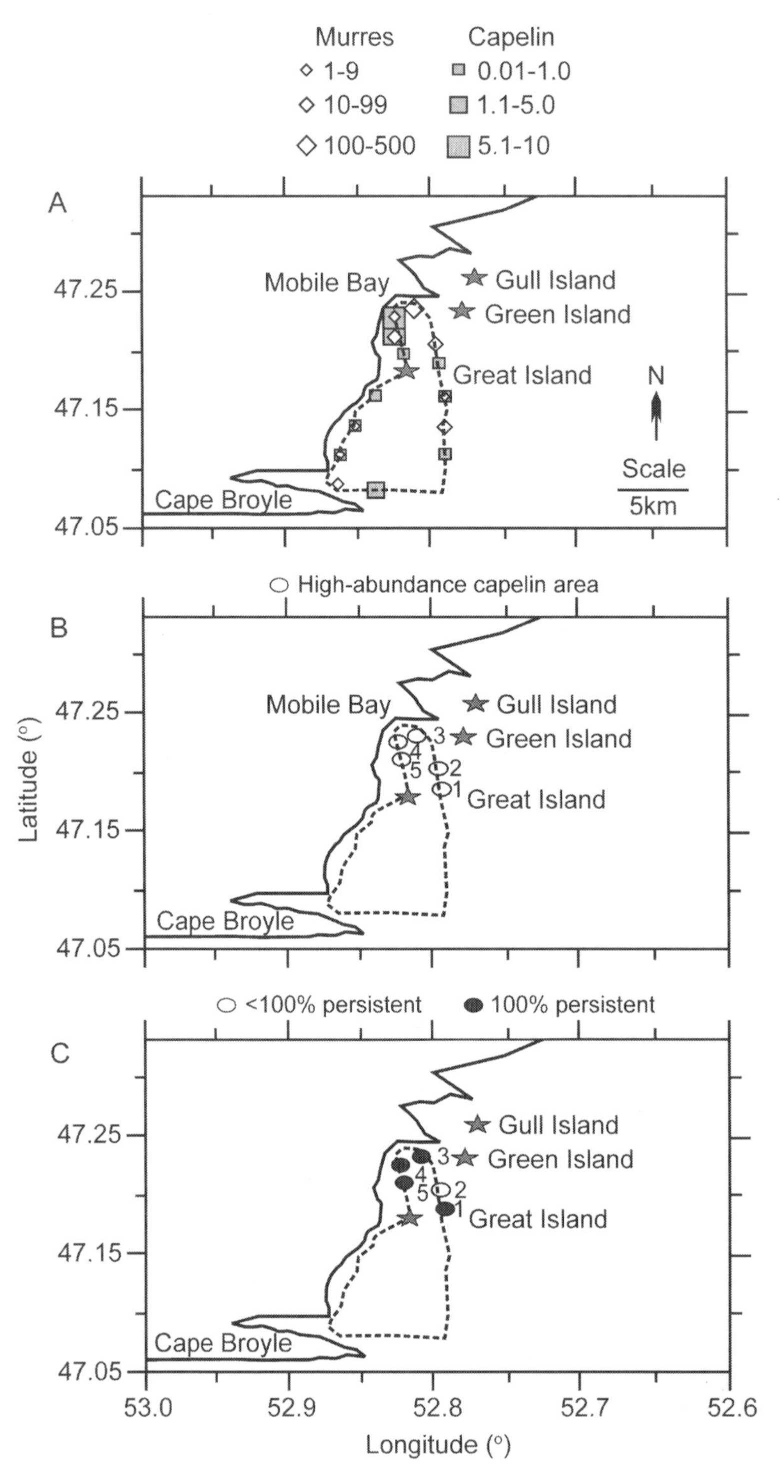

Figure 4.3. An example of data from scan sampling to study foraging by common murres (*Uria aalge*). (*A*) The distribution and abundance of common murres and capelin (*Mallotus villosus*) around Great Island in Witless Bay, Newfoundland, Canada. (*B*) The location of capelin hot spots (i.e., capelin abundance higher than the mean). (*C*) The persistence of capelin presence at each hot spot on all visits. The *dashed lines* represent transect routes. (Reproduced from Davoren et al. (2003:Fig. 3), with the permission of the Ecological Society of America)

those that might differ from the average behaviors of the rest of the group or the population. For example, foraging birds are obvious when they are near branch tips and at lower places on a tree, but they usually disappear from view when they move upward and inward into the tree's foliage. Likewise, the response of ungulates to disturbance (e.g., number of steps or bounds, head movements) is easy to observe and quantify when they are at the edge of an opening, but this becomes difficult when they move into vegetation cover. Further, sometimes simply following animals and looking at their activities can influence their behavior in ways that are largely unknown to us. Visibility bias (i.e., actions influenced by the relatively close presence of an observer) can be introduced into the data when trying to track animals in dense vegetation.

Because of the challenges described above, behavioral sampling periods must be short, and thus are biased toward the most conspicuous individuals. Unfortunately, no sampling method employing only the observer's eyes can overcome this problem, although many researchers have made attempts to do so. For example, students interested in bird foraging behavior often look at an individual for a short period of time (usually a few seconds) before beginning to record data. The rationale here is that they are allowing the bird to settle after possibly being disturbed by the worker, or to move to a less conspicuous position from the place where it was initially seen. Birds that are singing are usually ignored, because such individuals normally are in a high, conspicuous spot and thus may not represent the behavior of the rest of a group (Heil et al. 1990).

Small radio transmitters have the potential to at least partially solve this problem. Although investigators still cannot see exactly what the out-of-sight animal is doing, at least its general location and movement can be approximated without causing undue disturbance. Depending on the size of the transmitter, global positioning system (GPS) units can be incorporated, allowing fairly accurate location data to be collected. Many radios can also be fitted with activity switches or vital rate sensors to detect body position, heart rate, body temperature, and other variables and activities. Event recorders can automate much of the data recording (Berdoy and Evans 1990). Further, miniaturized video cameras are being developed that will allow behavioral observations from the animal's point of view. Clearly, technology is rapidly enhancing our ability to follow the activities of organisms in increasingly unbiased ways. Williams (1990) provided a review of the use of telemetry in studies of bird foraging behavior, and many publications and books are available on radio-tagging wildlife (e.g., Mech 1983; Kenward 1987; White and Garrott 1990; Samuel and Fuller 1994; Millspaugh and Marzluff 2001).

Recording Rules

How you sample is defined by sampling rules, and what you record is delineated by recording rules. Figure 4.2 shows the relationship between these sets of rules. We can divide recording rules into 2 basic categories: continuous recording and time sampling. Although we would prefer to be consistent and use the term "time recording" to clearly identify its classification as a recording rule, we will follow the terminology used by Martin and Bateson (1993), as changing the wording would only add further confusion. Note that focal sampling (a sampling rule) is not synonymous with continuous recording (a recording rule), and scan sampling (a sampling rule) is not synonymous with instantaneous sampling (a recording rule).

1. *Continuous recording.* In continuous recording, researchers note each occurrence of a behavioral act, the time at the start and end of the activity, and information on specific environmental conditions. This provides data on true frequencies, latencies, and durations of behavior *if* an animal can, indeed, be watched continuously for a sufficient period of time ("sufficient" must be determined through sample-size analyses or perhaps sampling pilot studies). Premature termination of a recording session will usu-

ally result in an unreliable estimation of the time span for certain activities. Continuous recording is necessary when their sequence and length is of interest. This is a vital part of determining the energy budgets of individuals (discussed further below). The continuous recording method is tedious, however, and in practice only a few categorical variables can be reliably measured.

In gathering this type of data on foraging behavior, workers record the activities of an animal for a specific period. Then, after some time has elapsed, the same or a different individual is observed for the same timespan. Within continuous recording periods, the duration of each behavior is written down. For example, Block (1990) watched foraging white-breasted nuthatches (*Sitta carolinensis*) for 10–15 seconds at a stretch, noting how long a bird spent searching for and procuring prey. Because a foraging bird can seldom be continuously observed for more than several seconds to a few minutes, accurate estimates of latency and, especially, duration are difficult to obtain. In another example, Loughry (1993) used 5-minute continuous focal-animal sampling to study the behavior of black-tailed prairie dogs (*Cynomys ludovicianus*). In their investigation of foraging elk (aka wapiti) (*Cervus elaphus* [*C. e. canadensis*]), Wilmshurst et al. (1995) delineated a "cropping sequence" as beginning when an animal put its head down to graze and ending when it lifted its head. The cropping rate (but not its duration) was calculated as the number of bites taken during the cropping sequence, divided by time. Weckerly (1994) watched foraging black-tailed deer (*Odocoileus hemionus columbianus*) for periods of 7–10 minutes.

2. *Time sampling*. Time sampling is a general category that involves recording behavior periodically. Each observation session is divided into successive short periods of time, called "sample intervals." Time sampling has been criticized because there is no standardized timespan, so the data on frequencies of behavioral acts lack any universal interpretation (Quera 1990). Therefore, time sampling is reported as the relative duration or the prevalence of the behavioral acts that take place.

There are 2 basic types of time sampling. The first, "instantaneous sampling," sometimes referred to as "point sampling," records the instant, or point, at which the activity of an individual organism is recorded. The structure of the data that are obtained largely depends on the length of the sample interval. If it is long, relative to the average duration of the behavior, then one can obtain a measure of the proportion (but not the frequency) of all sample points at which that action occurred. If the sample interval is short, again relative to the average length of the activity, then instantaneous sampling can approximate the results of continuous recording. Poysa (1991) concluded that the averages for samples from a vast number of individuals (hundreds or even more) gave reliable estimates of true time budgets, even for behavioral sequences of short duration. He also showed that if the length of a particular activity is very short, compared with other acts under investigation, then rate measures should be used instead of instantaneous sampling. It is critical that the time samples from each individual being observed are long and roughly equal.

Instantaneous sampling is commonly used to record wildlife behavioral data. Typically, a researcher follows an animal while it remains in view (or a limit is placed on the observation period) and records specific behavioral acts at set intervals. In a study of territoriality, Stamps and Krishnan (1994) recorded the location of each marked bronze anole lizard every 20 minutes. Students observing bird foraging commonly have used instantaneous sampling (Heil et al. 1990). Paterson et al. (1994), however, recommended focal-animal sampling, instantaneous scan sampling, and focal-time sampling (see Baulu and Redmond 1978) as more flexible and meaningful techniques.

The second type of time sampling, "one-zero sampling," is similar to instantaneous sampling in that the recording session is divided into short intervals when data are gathered. Here, however, an observer merely notes whether a particular behavior occurred during the sample period, recording no information on the frequency or duration of the act. Unlike instantaneous sampling, one-zero sampling consistently overestimates duration, because a behavior is recorded as though it occurred throughout the entire time interval, when it needed to have occurred only once during that period to be registered. Because of this problem, Altmann (1974) argued against the use of one-zero sampling in all applications. In contrast, Martin and Bateson (1993:97–98) provided a strong rationale for using this method in certain situations (e.g., male birds in lek that are or are not engaged in a set of reproductive displays) and included examples of how it should be applied. Bernstein (1991) concluded that one-zero sampling may be preferred when activities are clustered and the goal is to predict the probability of at least 1 such act occurring within a given time interval. Clearly, circumspection should be employed when using any method, especially if it involves controversy among researchers.

Concerns

As noted above, there are numerous elements of caution that must be considered with respect to biases associated with specific methods of sampling and recording behavioral data. In the following sections we discuss issues that are common to all behavioral observations, including independence, observer bias, and sample size requirements. Duijns et al. (2015) noted that understanding the fundamental mechanisms that underlie a particular set of behaviors (e.g., prey detection and prey processing) is only possible under controlled experimental conditions.

Independence

Most statistical analyses carry the assumption that the data points represent random samples from populations and, thus, that each item is statistically independent from the other records (Zar 1999). Thus the number of *individuals* from which information is collected characterizes or defines the entire sample, and the *total number of animals* in the location, group, or biological population from which these observations are made defines the samples' statistical population. Machlis et al. (1985) noted that the objective of research is to obtain measurements from an adequate number of individuals, not to obtain a large set of samples per se. They called this the "pooling fallacy." Although pooling might not be a fatal error in certain situations (see below), studies of behavior are replete with examples of such errors.

Another consideration in behavioral sampling is the independence of observed activities. Researchers commonly collect a series of instantaneous records on a single individual and then statistically treat each point sample (or each interval) as an independent sample, or pseudoreplicate (see Hurlbert 1984). Recording 6 instantaneous samples for a specified time period from the same animal, however, correctly constitutes only a sample size of 1. A procedure termed "aggregation" entails averaging samples of activities from the same individual, but this can lead to false conclusions if the behavior has a bimodal (or, more generally, non-normal) distribution. In such instances, the mean score for that organism would be for an activity that rarely occurs (Leger and Didrichsons 1994). Likewise, a second sample involving the same animal, taken a short time later, would not represent a sample taken from a different individual. This would also be the case if, a posteriori, a researcher randomly selected 1 sample from the repeated samples originally taken from a single animal.

The problem, then, lies in determining when samples become independent. Martin and Bateson (1993:86) noted that samples "must be adequately spaced out over time." Defining "adequate" is difficult, however. It is related to both the goal of the investigation and the particular species of animal being studied. That is, if your objective is to describe behavior at the population level (i.e., individual variations within a population), then sampling new organisms across the population is crucial for obtaining unbiased, independent observations. Resampling an individual should be avoided, so the results are not biased toward a certain segment of the population. Independence is thus assured. Nonetheless, this statement must be tempered in cases where the "population" is localized and rare. Likewise, if the goal is to examine a small segment of a larger population, independence *and* an adequate sample size will be difficult to achieve. Time-series analyses and repeated-measures designs (described below) are clearly indicated in such situations. Defining objectives and selecting appropriate methodologies do not create a source of bias or indicate a problem in a study design, and these practices are separate from "independence," as discussed here. They constitute a fundamental and logical process of doing science, including ethological studies.

Leger and Didrichsons (1994) noted that pooling observations is especially common in field studies of endangered species or other small populations, because multiple samples are taken from each individual, in order to accumulate large sample sizes. They stated that the central question about pooling is whether a population can be represented in an unbiased manner by repeated sampling of the same organisms. If, for instance, all animals in a theoretical population had the same mean and variance for the behavior of interest, then it would not matter if a researcher obtained 100 data points by sampling 100 individuals once each, 50 of them twice, or even 1 individual 100 times. Because organisms vary, the degree to which the selections represent an entire group or population can become problematic and remain uncertain.

Studies using radio telemetry have wrestled with the issue of independence between successive fixes. Swihart and Slade (1985) were perhaps the first to address this. They used the ratio of the mean squared distance between successive observations to the mean squared distance from the activity center in determining when autocorrelation among the sightings became negligible. This 1 paper influenced numerous subsequent investigations using radio telemetry to estimate home-range size and habitat use by terrestrial and aquatic vertebrates.

Leger and Didrichsons (1994) hypothesized that pooling was a reliable procedure *if* intrasubject variance exceeded between-subject variance. They evaluated several datasets and concluded that pooling could provide estimates of population means and variances that were at least as dependable as those supplied by single sampling (of individuals) and aggregation, as long as the samples sizes were about equal among individuals or the observed variance for 1 animal was higher than the variations seen for several individuals. The lesson here is that when intrasubject variance exceeds between-subject variance, unequal sample sizes among the organisms being studied become problematic.

McNay et al. (1994), using telemetry data, examined the independence of movements by black-tailed deer. They found that, even with 6-week intervals between samples (8 samples/year), observations were still not independent for over 50 percent of the tested deer. Because most animal location datasets are likely to have a skewed distribution of data points, the authors recommended that researchers emphasize sampling animal placements systematically through time, rather than trying to determine a time interval that would provide independent location samples. In other words, do not try to achieve independence in your samples at the expense of gathering an adequate understanding of animal behavior. Swihart and Slade (1985) and White and Garrott (1990:148) provided additional discussions on this topic.

Within-individual variation is an important, although seldom analyzed, aspect of behavioral research. Investigators have tended to treat populations as homogeneous units, thus largely ignoring differences in actions among their multiple members. Such an attitude implies that individual variance about a mean (i.e., behavior) is irrelevant and not of biological interest. In a monograph on this subject, Lomnicki (1988) contended that further advances in population ecology would require a consideration of differences between individuals, such as unequal access to resources. Thus repeatedly collecting samples from specifically identified animals supplies important ecological data. Knowing whether they are from the same or separate populations can provide further inferences about behaviors among them (Dingemanse et al. 2012). Additionally, changes in foraging strategies are common during ontogenetic size development, a phenomenon especially evident in reptiles (see Webb and Shine 1993 on blindsnakes [Typhlopidae]; Wikelski and Trillmich 1994 on iguanas [*Amblyrhynchus*]). Likewise, Bachman (1993) noted that changes in energy stores can influence the tradeoff between foraging time and vigilance in Belding's ground squirrels (*Citellus* [*Spermophilus; Urocitellus*] *beldingi*).

In most cases, advance planning will avoid the majority of problems involving data independence. Below, we define some recording rules that seek to maximize the independence of observations, while realizing that complete independence will not be achieved in all cases:

- Only 1 individual within a group (e.g., flock, herd) should be recorded per sampling session.
- Observers should systematically cover the study area (which ought to be as large as possible, especially if site-specific data are not needed), seeking out new individuals or groups—that is, avoiding a resampling of the same ones.
- Sampling sessions should be stratified by time period, both within and between days.
- Sessions should be distributed throughout the identified biological period (e.g., breeding) to avoid grouping samples within short spans of time.
- In small populations, an attempt should be made to individually identify animals.
- The need for spatial and temporal independence in telemetry fixes had become immutable dogma, ignoring the intrinsic value of dependence in addressing biologically relevant questions. Hooten et al. (2017) provided a comprehensive review of the statistical treatment of radio telemetry, including using autocorrelated data, to address a number of questions related to a species' use of space through time. They suggested that temporal sequences of radio fixes, although not independent, provided valuable information on organisms' use of space and environmental conditions.

Observer Bias

Avoiding bias in studies of animal behavior is virtually impossible. Bias takes various forms: the influence of observers on an animal; intra- and interobserver consistency in recording data; preconceived notions regarding how an animal "should" behave; differences in workers' abilities; and so forth. Each individual bias is serious, and, when combined, they can make reliable and repeatable interpretation of results difficult.

Researchers conducting behavioral studies should be aware that their presence and their activities probably will influence an animal's activities. Investigators often state that their presence "did not markedly alter an individual's behavior," or that "we waited until the animal returned to normal activities before recording data." This is clearly an exercise in self-delusion. Wild animals are constantly on the lookout for predators and competitors. The presence of an observer is likely to heighten the awareness of the animal being observed. Such heightened awareness or responsiveness is termed "sensitization." Further, that individual probably knew you were there long before you ever saw it. The waning of responsiveness is called "habituation" and is considered by most to be a form of learning (Immelmann and Beer 1989). Animal species vary widely in their ability to learn. Research has shown that birds and mammals can perform both temporal and numerical operations in parallel (Roberts and Mitchell 1994). For example, many corvids can remember the location of food caches for months, and they also can recall which caches they visited previously. Shettleworth (1993) reviewed the general topic of learning in animals. Rosenthal (1976) presented a detailed analysis of effects by observers in behavioral studies. Sutherland (2007) discussed population-level responses to disturbances by humans. He suggested that areas of needed research should include discovering what the patterns of these disruptions are, how to best determine population-level responses, what the importance of disturbance-caused ecological traps is, what the interaction of predation and disturbance is, when habituation occurs, how individual physiological responses to disturbance affect population sizes, and other factors.

Thus animals that appear to have become habituated to the presence of observers have adopted a modified pattern of behavior that allows them to keep them under surveillance. Animals adjust their behaviors according to the costs and benefits associated with different courses of action available to them (e.g., hiding, fleeing). Further, there may be a significant period of time between the detection of a potential predator (or human observer) and a visible response. For example, Roberts and Evans (1993) showed that sanderlings (*Calidris alba*), when encountering a human, were acting to minimize both the number of flights they made and the distance of each flight, although they did not tolerate any close approaches. We reviewed some of the responses of wildlife to human activities in Chapter 3, and numerous publications have documented such actions, including to people's recreational activities (e.g., Knight and Gutzwiller 1995). Delaney et al. (1999) used remote cameras to record the behavioral response of Mexican spotted owls (*Strix occidentalis lucida*), a federally listed threatened species, to military helicopter overflights and chainsaws. Their study was initiated over concerns that military helicopters would alter the owls' nesting activities. The authors found that helicopters flying over the forest had no effect on these owls, but they did note a reaction to chainsaws operating within 40 m of their roosts. The authors concluded that these owls had evolved to respond to perceived threats from within the forest, but not from above. Such factors should be considered when designing behavioral studies, as the reactions of animals to the presence of humans may affect the conclusions of an investigation that is not concerned with such impacts per se.

*Intra*observer reliability is a measure of the ability of a specific person to obtain the same data when measuring a repeated behavior on different occasions (Martin and Bateson 1993:32–34). It indicates the ability of workers to be consistent in their measurements, and greater congruency could indicate a higher degree of precision in those data. Because we seldom know what the actual behavioral pattern of an organism is—that is, the "true" pattern—we cannot directly measure an observer's accuracy. Moreover, assessing intraobserver reliability in field-based studies is difficult, because animals seldom repeat their behavior in exactly the same fashion, although the early ethological studies by Konrad Lorenz used the concept of "fixed action patterns," which are stereotyped behavioral responses to specific stimuli. A test of intraobserver reliability is to videotape animals and then repeatedly present (in some random fashion) individual sequences to a worker. Researchers can use the results of such trials to estimate the degree of observer reliability in their investigation.

*Inter*observer reliability measures the ability of 2 or more workers to obtain the same results on the same occasion (Martin and Bateson 1993:117; Glover 2003). This, however, may present difficulties in field studies. Ford et al. (1990) found that comparisons of foraging behaviors among individuals of the same species in disparate areas or years, recorded by different people, needed to be treated cautiously. Problems were particularly evident when the observers had not previously agreed on standard methods to use or a similar classification of terms. Dissimilar amounts of experience among workers apparently accounted for much of the interobserver variability that was noted. Further concerns pertain to comparing or combining the results of behavioral observations on a given species from different studies.

Intra- and interobserver reliabilities have been substantially improved by careful, rigorous, repeated training. Each new member is taught how data should be recorded, initially working with others in the field, comparing data and discussing reasons for decisions. Videos (as noted above) or captive animals can also be used. In our experience, observer reliability increases when workers are informed about and comfortable with why a particular behavior is categorized in a certain manner. Training should continue throughout the study, with frequent sessions to recalibrate the data-gathering techniques of the observers. Each behavior to be investigated should be carefully defined in writing. It usually helps to delineate an action by its structure (e.g., *probe* indicates "to insert the bill beneath the surface of the substrate," as opposed to *glean*, meaning "to pluck something from the surface of the substrate"). In protracted studies, definitions and criteria commonly tend to "drift," or change with the passage of time, as observers become more familiar with behaviors and possibly lazier in their evaluations. Careful and repeated training and recalibration exercises will reduce this problem. Further, efforts have been made to standardize terminology in various disciplines. For example, Remsen and Robinson (1990) developed a precise set of terms and definitions for avian foraging studies. The methodology for a standard ethogram, developed by Schleidt et al. (1984), is a useful starting point for behavioral studies, as noted earlier in this chapter.

Sample Size Requirements

Researchers *must* incorporate an evaluation of sample size requirements into the design phase of each behavioral study. Such planning guides the collection of data and provides observers with an objective way of avoiding both under- and overkill in their sampling efforts. Further, determining an appropriate sample size lets you know if you are trying to accomplish too much with the amount the time and resources available. In our experience with graduate students, we have found that when reviewing research proposals, our first job is to prevent a well-meaning student from attempting too much. It is far better to do 1 thing very well than to do many things poorly. Many workers have been forced to combine data, sometimes inappropriately, across seasons, years, ages, and sexes, because they did not plan properly in advance.

We have previously discussed how temporal and sexual differences affect unequal habitat use by animals. Studies of behavior show that, even within what we consider to be a "season," combining data over just a few months can obscure important patterns of resource use. In designing investigations, researchers must carefully determine the number and types of variables for which they will have enough time to collect data.

For example, Brennan and Morrison (1990) found that biologically and statistically significant variations in the use of tree species by foraging chestnut-backed chickadees (*Parus* [*Poecile*] *rufescens*) occurred throughout the year (Fig. 4.4). Using some of the same data, Morrison (1988) showed how lumping can result in inappropriate interpretations, as it tends to "average-out" many potentially important ecological relationships. Likewise, Sakai and Noon (1990) discovered that Pacific-slope flycatchers (*Empidonax difficilis*) significantly altered their foraging behavior within their breeding period.

These results are not surprising. Animals must respond to changes in resource availability and abundance, as well as to the demands placed on them by both abiotic and biotic factors. Many equivocal results obtained in behavioral studies probably resulted from lumping the data. Thus, not only must researchers determine the appropriate sample sizes needed for reliable interpretations, but they must also carefully evaluate these samples over relevant timespans—periods that are usually shorter than those presented in most papers.

Activity Budgets and Energetics

Naturalists have long been interested in quantifying how animals divide their time among requirements for foraging, sleeping, moving, breeding, and so on. Formulating these activity (or time) budgets, using data recorded from observations of individual animals, is the first step in developing an understanding

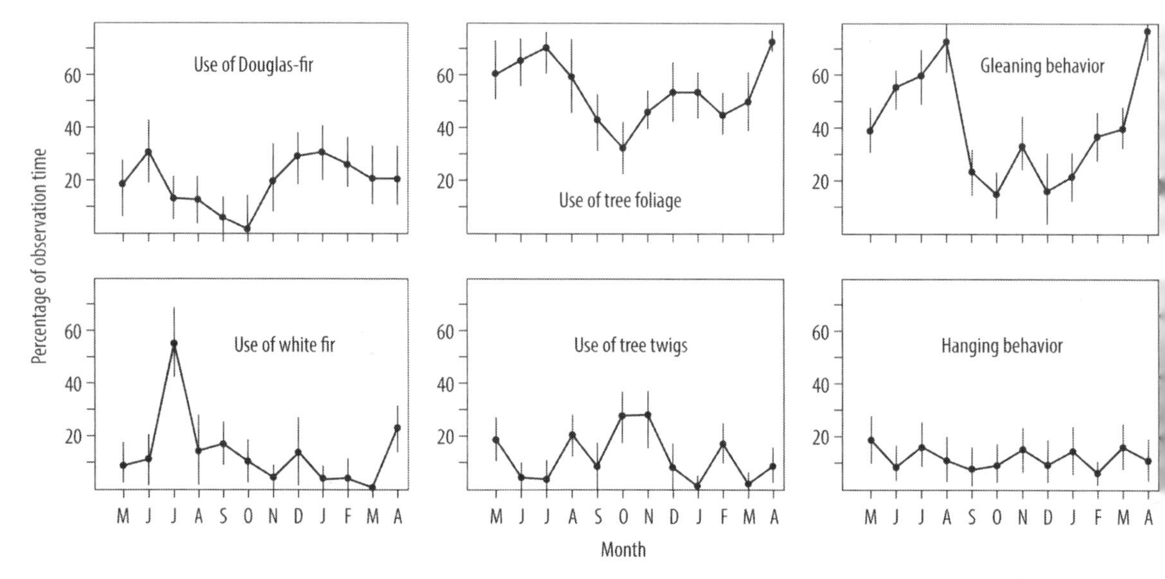

Figure 4.4. A set of graphs indicating the relative percentages of chestnut-backed chickadee (*Poecile rufescens*) foraging times on different substrates across a 12-month period. *Dots* represent mean values, and *vertical bars* represent 1 standard deviation. (Reproduced from Brennan and Morrison (1990:Fig. 4), with the permission of the American Ornithological Society)

of the relationship between time allocation and survivorship, reproductive success, body condition, and other aspects of the natural history of a population. Defler (1995) presented a good example of time budgets in woolly monkeys (*Lagothrix lagotricha*) (see also Newton 1992). As noted by Weathers et al. (1984), however, quantifying allotments of time is easy, relative to assessing the energy expended by the organisms while conducting their various activities. In this section we present a brief review of time-energy budgets as they relate to advancing our understanding of animal behavior, and, ultimately, survival and fitness.

Assessments of animal energetics often commonly assume that the goal of an organism is to maximize its net energy balance between the amount expended for various activities and the quantity stored from food consumption. Many theoretical and empirical studies have examined how animals achieve this equilibration. Thus, the neurological and physiological capabilities of an individual animal link it to its environment. From this perspective, one should think of the environment in its functional relation to an organism, rather than merely as the geographic area and physical structure of the habitat in which that animal lives (Moen 1973:21).

Thermal energy is exchanged between an animal and its environment by radiation, conduction, convection, and evaporation, as well as by excretion. These methods shift, relative to each other, as that individual's environment alters. Rain, wind, and other abiotic factors affect energy exchange in the short term, and alterations in factors such as plant cover influence it over longer periods of time.

Porter and Gates (1969) were arguably the first animal ecologists to formally quantify and model how organisms are required to be in some sort of averaged thermodynamic equilibrium over a certain length of time in order to survive. The interplay among environmental factors and how they influence energy exchange within an animal is complex, but this also offers a window into potential factors that account for the proximate behavior and ultimate survival of individuals in a population through time. Sunlight, reflected light, and thermal radiation, as well as air temperature and movement, all operate on the environmental side of the equilibrium equation. From an animal's perspective, its metabolic rate when standing still versus when it is active; the rate of moisture loss; the amount of fat, fur or feathers it has; and its body size and shape are all factors influencing its thermal dynamics (Fig. 4.5).

A wild animal has an ecological metabolic rate that is an expression of the "cost of living" (in terms of energy use) for daily activities and other physiological processes. Ecological metabolisms vary from 1 activity to another and from 1 species to another. There are multiple direct and indirect methods for determining the metabolic rates of organisms under diverse environmental conditions, both in the laboratory and in the field.

Total basic metabolic rates (BMRs) generally increase with body size. Thus overall energy and food requirements should be higher in large animals, relative to small ones (everything else being equal). Mass-specific metabolic rates, however, decrease with increasing body size. Therefore, energy requirements and food consumption *per unit of body mass* should be greater in small animals than in large ones. Such relationships influence the ways in which organisms behave, including the amount of time spent foraging, their degree of territoriality, their movement patterns (including seasonal shifts and migrations), and a host of related factors. A knowledge of energy requirements can thus be fundamental in explaining why animals use specific habitat patches, as well as their choice of activities within those patches. For example, Rychlik and Jancewicz (2002) related body size, food habits, and activity patterns in 4 species of shrews with differing body sizes. Organisms also adjust their energy reserves according to food availability, which, in turn, is often based on environmental conditions, such as season and weather (Kelly and Weathers 2002). Here again, we see that knowledge about energy needs is a factor in understanding activity patterns. Animals' mobility gives them options for selecting—and then using—different elements of environmental space across time. Because environments can change hourly, daily, and monthly across an annual cycle, adaptability, as a means of coping with these changes, is essential for the survival and reproduction of an individual organism and, ultimately, the persistence of a population.

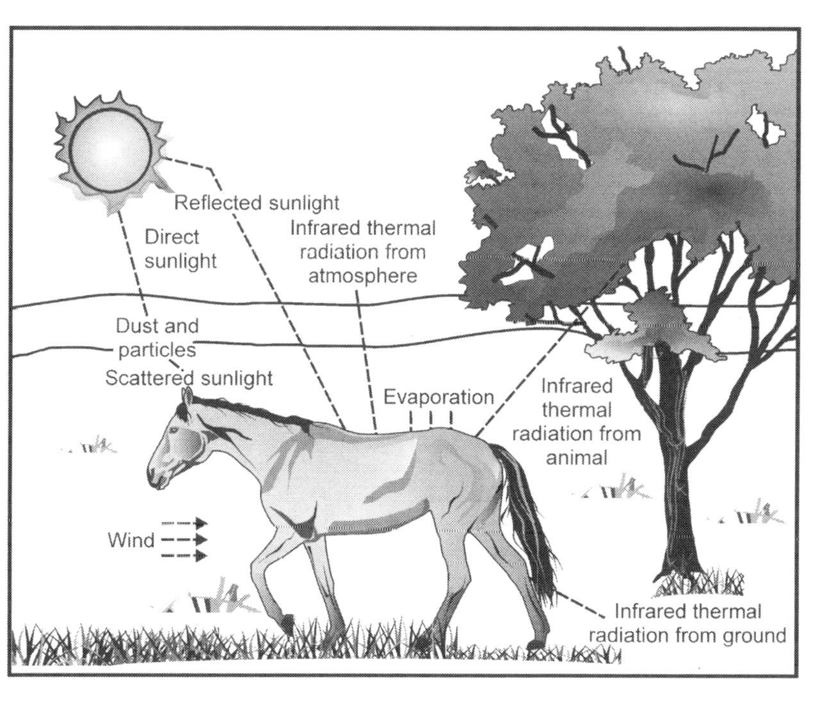

Figure 4.5. A schematic representation of environmental factors that influence animal behaviors. (Reproduced from Porter and Gates (1969:Fig.1), with the permission of the Ecological Society of America)

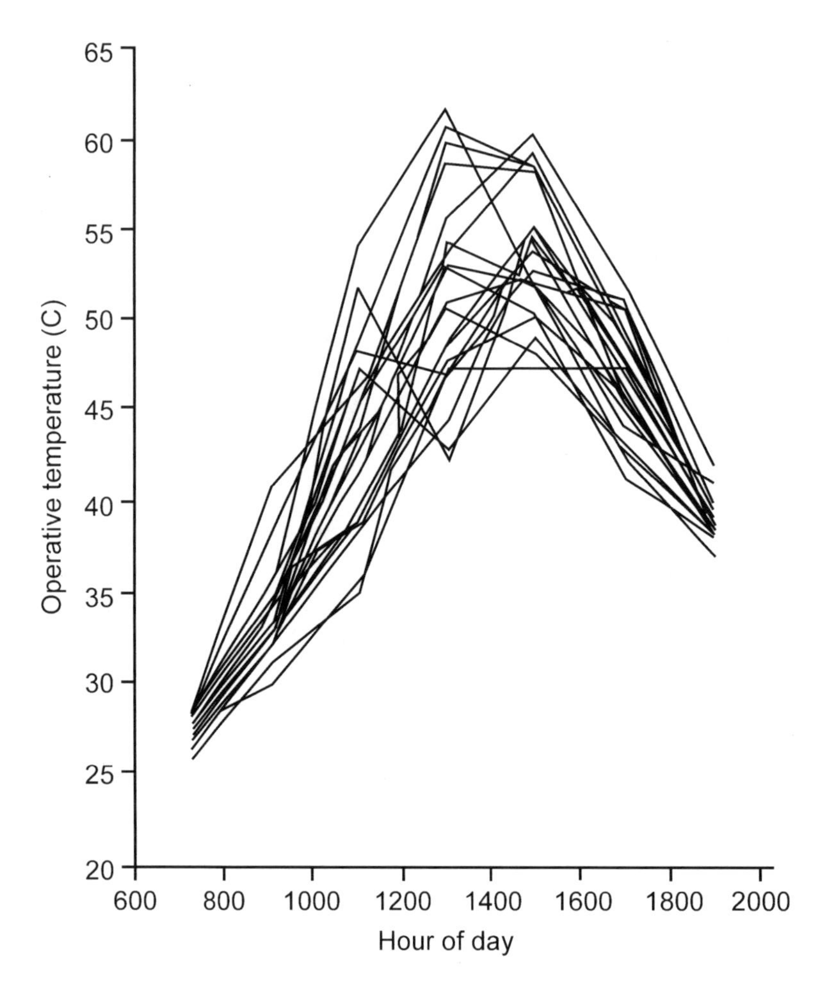

Figure 4.6. A graph of hourly trends in operative temperatures at ground level at 23 points in a thicket composed of honey mesquite (*Prosopis glandulosa*) in Webb County, Texas, on 11 June 1996. (Reproduced from Guthery et al. (2000:Fig 2), courtesy of Tall Timbers Research Station)

In some instances, the amount of usable space available to an animal varies drastically, hour by hour, within a given day. For example, shorebirds that winter on an estuary encounter ebbing and flowing tide patterns that constantly change the amount of mudflat area available for foraging (Brennan et al. 1985). Terrestrial environments also can be dynamic with respect to the amount of usable space available on the landscape within a single day. Guthery et al. (2000) documented that in South Texas during the summer, for 23 points at ground level under honey mesquite (*Prosopsis glandulosa*) less than 3 meters in height, operative temperatures can exceed the lethal temperature tolerance (>38.7°C) of northern bobwhites for as much as 2 to 4 hours in the middle of the day (Fig. 4.6). To survive in such an environment, northern bobwhites seek refuge in dense woody cover to mitigate heat stress (Guthery et al. 2005). Such habitat components are essential for ensuring the population's persistence. Similar relationships exist for other species of quail and ground-dwelling birds. For example, Goldstein (1984) documented that Gambel's quail (*Callipepla gambeli*) can perish from hyperthermia in as little as 1 minute after leaving protective shady coverts on a hot summer day.

An animal's orientation, via behaviors or movements that allow it to select a more favorable thermal environment, is 1 of the few options it has to adjust its actions because of an environmental stressor. Many species of birds will exhibit gular flutter—a rapid, panting-type behavior similar to that in canids—to eliminate excess heat and potentially avoid hyperthermia. Case and Robel (1974) documented that during cool nighttime conditions, roosting northern bobwhites huddled tightly in a ring, with their tails pointing inward and beaks facing outward. During warmer conditions, these birds, while still roosting in a circle, were less organized, with little or no lateral contact. Habitat management strategies that mitigate the effects of excess cold or heat—where feasible—can potentially be effective, such as the recommendations by Guthery et al. (2005) for northern bobwhites.

A time-energy budget (TEB) method is a commonly used technique for estimating total daily energy expenditure in animals. TEB has 2 parts. First, you develop a time activity budget, as described above. Second, you convert the activity data to energy equivalents, using estimates of the energy costs for each activity, as determined from controlled studies or the literature (Haufler and Servello 1994). Numerous workers have conducted laboratory investigations to determine the relationship between activity and energy expenditure, and approximations can usually be made for the species you are studying, based on general allometric equations. These values are only approximations, but they do provide a general understanding of activity-energy relationships (described below). Weathers et al. (1984), however, showed that time-budget estimates that assign energy equivalents derived from the literature to behaviors, rather than calculating them empirically, are subject to errors of up to 40 percent.

Finch (1984) determined the daily activity budget of Abert's towhees (*Pipilo* [*Melozone*] *aberti*) by quantifying the duration of 4 activities: perching, ground foraging, flying, and attending to a nest. These data were transformed into percentages of the observation periods and the day's activities. The daily energy expenditure for this species was then determined, using published estimates of their basal and thermostatic requirements, as well as assessments of the energy requirements for these 4 activities. Hobbs (1989) developed a model of energy balances in mule deer (*Odocoileus hemionus*) that predicted both alterations in the body conditions of does and fawns and the relationship of those changes to rates of mortality. He used values in the literature to craft his detailed model, based on an animal's activity level. Another good example of the creation of an activity budget and a subsequent estimation of energy expenditure was provided by Dasilva (1992) for western black-and-white colobus monkeys (*Colobus polykomos*).

A more direct method of determining energy expenditures involves doubly labeled water, in which both hydrogen and oxygen have been replaced (i.e., "labeled") with uncommon isotopes of these elements. This technique involves injecting oxygen-18 (^{18}O) and tritium or deuterium isotopes into an animal prior to its release, taking an initial sample after the isotopes have reached equilibrium in its body, later recapturing and resampling that animal, and then calculating the rate of CO_2 production. This latter information can be equated to the animal's metabolic rate, determined from the relative turnover rates of the isotopes, measured from the resampling. This turnover rate is termed the "field metabolic rate," or FMR (Haufler and Servello 1994).

There is a large body of reports in the literature on FMR, as determined by using doubly labeled water (e.g., Nagy 1987; Nagy and Obst 1989; Speakman and Racey 1987). In birds, for example, FMR estimates (in Kj/day) were 118 in 22 g tree swallows (*Tachycineta bicolor*), 343 in 220 g Eurasian kestrels (*Falco tinnunculus*), 997 in 1,089 g little penguins (*Eudyptula minor*), and 2,401 in 3,706 g grey-headed albatross (*Diomedea* [*Thalassarche*] *chrysostoma*) (Nagy and Obst 1989).

Mock (1991) used the daily allocation of time and energy in western bluebirds (*Sialia mexicana*) to examine the tradeoff between parental survival and that of their young. She concluded that individuals regulate their overall daily energy expenditures through a differential use of thermal environments and activity budgets. Speakman and Racey (1987; see also Entwistle et al. 1995) used doubly labeled water to study the energetics of brown long-eared bats (*Ple-*

cotus auritus). Other methods of determining energy costs include estimating the loss of mass after activity periods (e.g., long flights), oxygen consumption, and heart-rate telemetry (see Goldstein 1990 a for review).

Foraging and Diet

Optimal Foraging Theory

Research on foraging behavior has advanced our knowledge of animals and animal management in many important ways. Analyses of these behaviors have improved our understanding of habitat assessments and the design of nature reserves, as well as livestock husbandry and grazing management. These analyses have also resulted in better ways to match dietary preferences with resource availability, as well as in a way to comprehend and predict a forager's impact on its resources and environment (Ash et al. 1996). For example, we now know that herbivore diets are closely linked to habitat and patch use, and these are objects that can be manipulated through management.

Organisms are generally adapted to efficiently exploit specific types of foraging substrates. A "substrate" is the specific location and surface at which an animal directs its feeding. It must compare the risks associated with one manner of approaching this activity against those connected with other processes and locations. The choice of both method and site is based not only on the number or density of food items present, but also on the quality of that food and the predation risks involved with consuming it at a particular location. That is, do numerous low-quality items that are easy to acquire have a net energy benefit over scarce but high-quality ones that are riskier and more difficult to obtain?

Schoener (1969) hypothesized that animals can achieve a net balance by 2 extreme strategies: energy maximization or time minimization. Energy maximizers try to obtain the greatest amount of energy possible within a given period of time. Time minimizers seek to substantially reduce how long it takes to procure a given amount of food and, thus, have more time available for other activities, such as grooming and parental care (see Morse 1980:53–54). Stephens and Krebs (1986) recognized, however, that a maximization of net energy was not necessarily a desirable goal. The disparity they observed lay in equating the net rate of energy consumption with its cost-benefit ratio (i.e., foraging efficiency). They noted that maximizing efficiency ignored the amount of time required to harvest resources, and it failed to distinguish between tiny gains made at a small energy cost and larger ones made with a greater expenditure. For example, 0.01 calories acquired at a loss of 0.001 calories gives the same benefit-cost ratio as the accumulation of 10 calories using up 1 calorie. The 10-calorie alternative, however, yields 1,000 times the net gain of the 0.01-calorie option (Stephens and Krebs 1986:8–9).

These and related considerations of cost-benefit ratios form the basis of a large area of scientific investigation, collectively known as "optimal foraging theory." Some researchers consider the term "optimal foraging" to be misleading, because it implies that there is a preeminent strategy for an animal to follow, but foraging theory per se makes no such claim (Stephens 1990). The idea that the search for and consumption of foods was a way to maximize the fitness of individual animals emerged in the 1960s from work by Emlen (1966) and MacArthur and Pianka (1966). A decade later, Charnov (1976) used a series of mathematical models to show how foraging behavior could be explained by an "optimal" context (based on how an individual accumulated food resources) and an assumption (i.e., increased breeding success and, hence, greater fitness).

Although some have questioned its usefulness (e.g., Pierce and Ollason 1987), we contend that all studies must be firmly based in theory, and that the optimal forging concept provides a guide for the development of a research strategy. In addition, comparing actual field data with a model that represents some kind of optimal condition or circumstance allows one to assess the divergence between theoretically optimal and empirically realized strategies. Stephens and Krebs's (1986) book provides excellent coverage of foraging theory. Here, we examine the ways in which these theories and their associated models relate to our descriptions of how animals might view their environment. Ideas about perception lead, in turn, to how we should observe and record organisms as they exploit their food resources. Thus we will briefly explore how models of foraging activities can help us design studies of animal behavior.

There are 2 rather distinct alternative views when considering how organisms choose whether to forage in a particular area. Is the individual selecting among various food sources distributed throughout a generally suitable area? Or is it distinguishing between various patches of food sources? Foraging theorists thus differentiate food-choice models and patch-choice models, although an animal can "optimally" acquire nutrients within either of these 2 basic constructs. These models have clear implications for the design of foraging studies. That is, if food is distributed in a patchy manner, then an assessment of its abundance or availability must have the proper spatial and temporal scales to recognize these patches. An overall average of these characteristics across a large area would mask their sporadic nature. Likewise, investigations conducted at too small a scale could fail to identify any patches, or identify ones that an individual might not recognize.

Both food-choice and patch-choice models assume that a foraging animal sequentially searches for food, encounters it, and then decides whether to try to consume it. But the form of the decision taken by that individual when encountering a food item differs between these 2 models. In the food-choice model, an organism must decide whether it should take the item or continue searching. One can then state various rules on which it should base its decision (e.g., food abundance). In the patch-choice model, an animal decides how long it should forage in the patch where it encountered the food. Both models consider how an individual can best make these decisions, with a goal of maximizing its average rate of energy intake over the long term.

These models thus develop a general theoretical constraint on which we can base foraging studies. Clearly, the rate and method of searching, the frequency of encounters with a food item, and the method that was used to eat it all tell us a great deal about the ecology of a species. Further, variations in an organism's search, encounter, and consumption behaviors may lend insight into its current physiological condition. Thus measuring aspects of the foraging rate, the frequency and types of food encounters, and related activities is important in understanding an animal's distribution, survival, and reproductive success in occupied locations.

Foraging theory predicts when an organism will consider food items, but its decision to take a specific item will depend not only on that item's abundance, but also on the plenitude of other foods. These can be ranked by the ratio of their nutritional value to handling time (Morse 1980:54). "Handling time"—which applies to predator-prey relationships, rather than to herbivory—is the amount of time necessary to pursue, capture, and consume prey (Stephens and Krebs 1986:14). Foraging theory suggests that we should be concerned not only with how an animal goes about the process of foraging, but also with the types and relative rankings of the plenitude and quality of prey encountered by that individual. Sampling only the abundance of items consumed by an organism tells us little about the reasons for that decision.

The foraging patterns of animals has also been evaluated using the concept of "giving-up densities" (GUD), which is the level at which an organism stops using a particular resource-patch location. In other words, most of that patch had contained some type of food that was spatially mixed within a matrix of other materials (e.g., sand), resulting in a diminishing rate of return (to that individual) as the food was consumed (Yunger et al. 2002). Without data on the availability of food resources, investigators are limited to making inferences based on direct observations of factors (e.g., foraging substrate); spe-

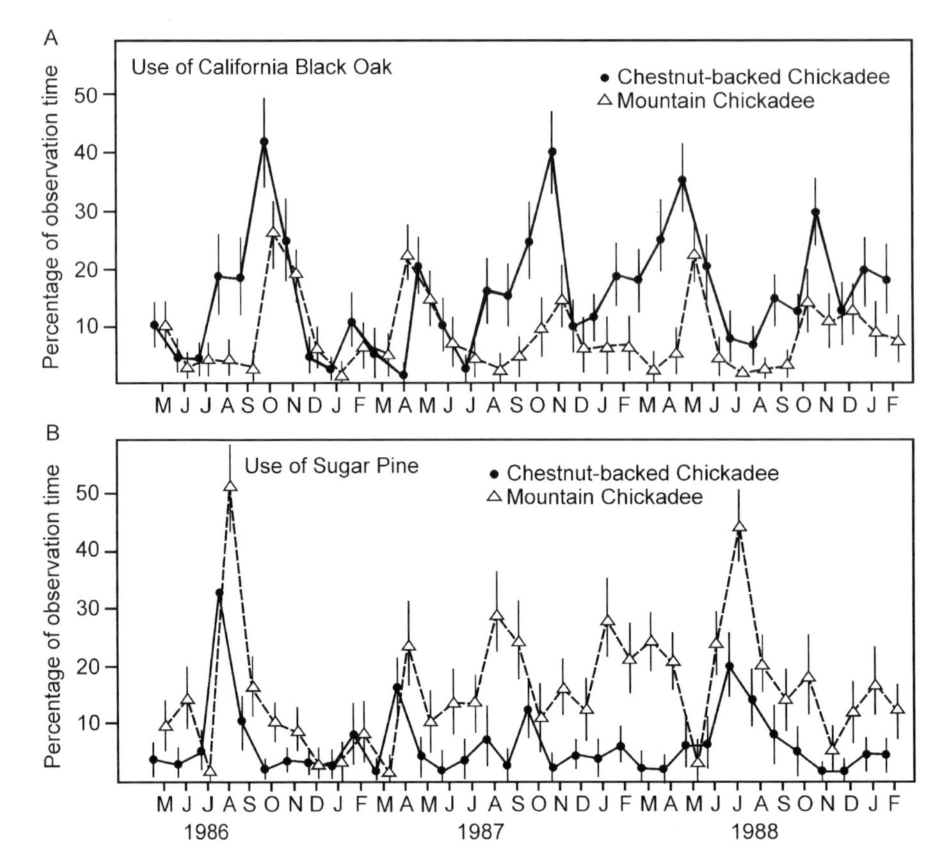

Figure 4.7. Graphs of comparative foraging dynamics of chestnut-backed chickadees (*Poecile rufescens*) and mountain chickadees (*Poecile gambeli*) on 2 species of trees—(*A*) California black oaks (*Quercus kelloggii*), and (*B*) sugar pines (*Pinus lambertiana*)—at the Blodgett Forest Research Station in El Dorado County, California, over 34 consecutive months. *Dots* represent mean values, and *vertical bars* represent 1 standard deviation. (Reproduced from Brennan et al. (2000), with the permission of the Pacific Northwest Bird and Mammal Society)

cific location, selection, and consumption behaviors; or how such activities change over time. Whereas such information is potentially noteworthy from the standpoint of classical natural history, it tells us little about the implications it has regarding the fitness of an individual organism. Although foraging studies that use such comparisons are useful in quantifying how animals shift their use of space and substrate over time (Brennan et al. 2000; Fig. 4.7), such investigations fail to reveal the underlying aspects of why such changes take place.

To get around this problem, Brown (1988) developed the GUD concept, which describes patches of food as depletable resources that could be exploited differentially to maximize fitness. In its simplest form, a GUD model is related to the harvest rate (H, assumed to decrease over time); predation risk (P); the costs (C) associated with searching, handling, processing, and digesting the food, as well as with thermoregulation; and the implications of missed opportunities elsewhere (MOC). These factors can be expressed in a harvest rate equation:

$$H = C + P + MOC$$

GUD has been frequently applied in foraging studies since Brown (1988) proposed this theory. In a review of publications ranging from 1988 to 2012, Bedoya-Perez et al. (2013) documented 192 papers in the peer-reviewed literature that directly made use of

GUD theory. In foraging research, GUD does not replace the need for data on the availability of different food resources or patches, nor is it a substitute for information on their profitability for an individual foraging animal. What GUD does provide, however, is a clear theoretical construct, where an investigator can quantify the tradeoffs in how animals forage in different patches by using observational data in a laboratory or other experimental environment (e.g., an aviary), where patches of different nutritional profitability have been constructed by that researcher.

Kohlmann and Risenhoover (1996) performed 2 controlled experiments to examine how northern bobwhites responded to different GUDs. They documented that the birds abandoned patches sooner when food supplementation offered a better alternative foraging resource, or when a lack of protective cover apparently enhanced the risk of predation. In other words, northern bobwhites adjusted their foraging behavior in predictable ways, which allowed them to procure more food faster when greater quantities of such items became available, while also minimizing the risk of being exposed to predation.

Abu Baker and Brown (2009) used a series of experiments to test how mourning doves (*Zenaida macroura*) and eastern cottontail rabbits (*Sylvilagus floridanus*) responded to GUDs of food resources. Using patches of low, medium, and high food abundance, the doves and rabbits reacted similarly. The proportion of foods harvested from the low, high, and medium patches by the mourning doves ranged from 40.7 percent, to 43.8 percent, to 48.3 percent, respectively, while the proportions harvested by the eastern cottontails ranged from 34.9 percent, to 35.8 percent, to 38.4 percent, respectively. Both species were negatively influenced by the depths of the foraging substrates, with the doves being more strongly affected than the rabbits. This was expected, because mourning doves typically are weak scratchers.

Using European starlings (*Sturnus vulgaris*) and American robins (*Turdus migratorius*) as subjects for their investigation, Oyugi and Brown (2003) documented that GUDs of both species increased in relation to the distance to the edge from a protective shrub or tree canopy, thus indicating that predation risk was a factor that strongly influenced the foraging behavior of both of these species.

In an experimental study that directly addressed predation hazards in relation to GUD, Holtcamp et al. (1997) examined how the presence of red imported fire ants (*Solensopsis invicta*) influenced foraging by deer mice (*Peromyscus maniculatus*). They found that these mice foraged more efficiently (i.e., at a faster rate) on the patches with fire ants than they did on patches without the ants. Evidently the deer mice were able to compensate for the presence of fire ants—which envenomate small rodents—by increasing their use of food-rich patches and stepping up their in-patch harvest rates.

In something of a companion study to Holtcamp et al. (1997), Darracq et al. (2016) used hispid cotton rats (*Sigmodon hispidus*) and red imported fire ants in an experimental setting and documented that GUDs of foods on the fire ant patches were 2.2 times greater than GUDs on the non–fire ant patches. Unlike the deer mice–fire ant study by Holtcamp et al. (1997), cotton rats were apparently unable to compensate for the presence of fire ants. Darracq et al. (2016) inferred that indirect risks, such as envenomation and an opportunity for encounters with other predators, could reduce the fitness of individual cotton rats.

Foraging theory has also been applied as a framework for the design of studies in herbivore ecology. As outlined by Wilmshurst et al. (1995), optimal foraging theory assumed that such decisions by herbivores should be strongly influenced by physiological and environmental constraints on the rates of nutrient uptake. There are 2 such constraints that are commonly invoked for vertebrate grazers: the effect of plant density on the short-term rate of consumption (i.e., an availability constraint), and the effect of digestive capacity on the long-term rate of energy assimilation (i.e., a processing constraint). The short-term rate of nutrient intake should be positively related to plant size, bite size, and plant density. Using this model as a guide, investigations could be designed to measure plant size and density, and then relate these environmental variables to bite size and actual food intake.

Working with elk, Wilmshurst et al. (1995) hypothesized that at low forage biomass, the processing rate should be high but the short-term rate of intake low. At high forage biomass, the opposite should occur. Their "foraging maturation hypothesis" stated that the net rate of energy intake from food should accordingly be maximized on patches of intermediate plant biomass. They concluded that a preference for grass patches with intermediate biomass and fiber content could help explain patterns of animal aggregations and seasonal migration.

Schmitz (1992) tested whether white-tailed deer (*Odocoileus virginianus*) selected their diet in accordance with a foraging model predicting that in order to maximize fitness, they had to balance maximizing their growth and offspring production with minimizing their risk of starvation. He found that white-tailed deer apparently show plasticity in their diet in response to temporal and spatial changes in perceived risks and gains. The animals seemed to balance gains in fitness (due to reproduction) with losses (due to energetic shortfalls). When a starvation risk was eliminated, the deer tended to select a broad range of foods that simply maximized their mean rate of energy intake.

Foraging theory directly relates to wildlife management. The type of foraging model developed by Nudds (1980) for ungulates most closely followed by an animal (energy maximization, equal food value, nutrient optimization, unequal food value) can guide management decisions concerning the amounts and types of food to emphasize. He concluded that white-tailed deer, as well as other temperate latitude ungulates, are primarily habitat specialists but become food generalists in the winter. The foraging behavior of these deer in that season adhered most closely to the predictions of the energy maximizing models, where it seemed to be energetically less costly to remain in sheltered areas and not eat than to forage in exposed areas. Translating these conclusions into a management scenario, the author suggested that manipulating these animals' winter habitats by increasing only the abundance of "preferred" food would not be warranted. Management efforts would be more beneficial if they were directed toward the physical structure of winter habitat for the deer, in order to provide more shelter. Although the logic of some of his suggestions have been criticized (Jenkins 1982; but see Nudds 1982), Nudd's 1980 paper was (and is) important in that it directly linked theory with management. Likewise, Kotler et al. (1994) showed how research on patch use in Nubian (or Alpine) ibex (*Capra ibex*) could be used by managers to modify their habitat to reduce predation risk, thus allowing the animals to more efficiently use the available food. As noted by Jenkins (1982), foraging theory, combined with good empirical work on food preferences, may lead to valuable new insights about problems in wildlife management.

Foraging models used within this context should address the spatial and temporal scales at which management activities can be applied. These management scales are usually larger than those of the models, as the former are often determined by socioeconomic factors not under the control of a researcher. Although management may not specifically address the details of grazing behavior, an understanding of the intricacies of the foraging process and the requirements of the species of interest are essential to provide a contextual framework for decision making (Ash et al. 1996).

Diet

A dichotomy exists in the literature regarding the emphasis placed on the quantification of animal diets. Wildlife biologists and economic entomologists have expended much effort to determine the actual food items consumed by animals. Korschgen (1980) observed that, in the late 1800s and early 1900s, studies of animal diets examined the economic importance of birds' feeding habits, seeing wild animals as pests in terms of agricultural crops, poultry, and livestock. The greatest activity in such investigations took place in the 1930s and 1940s, emphasizing waterfowl and upland game birds. Papers dealing with ungulate diets dominated the literature prior to the 1950s. The proportion of research reporting on food availability, food digestibility, and food requirements has grown steadily since that time.

In contrast, scientists studying the ecological relationships of animals have seldom attempted to quantify the occurrence of specific food items in their diets. Rather, this research has concentrated on indirect measures of food use, such as foraging location (Hutto 1990; Rosenberg and Cooper 1990). Although morphological differences between species undoubtedly reflect some degree of evolutionary response to resources, such divergences may not necessarily be good predictors of the species' diets, especially under local environmental conditions (Rosenberg and Cooper 1990).

Although studies of animal diets abound, most are sole-species investigations from single locations that were conducted over a short period of time. Thus little generalization is possible from such short-term research at relatively isolated locations. Further, as noted by Rosenberg and Cooper (1990), 1 of the reasons that studies of avian diets have been neglected by modern ornithologists is that they fear the detail, tedium, and technological expertise thought to be necessary for such investigations. Regardless of the rationales, only a limited amount of literature exists on the diets of most species of animals in the world, at least publications that are useful in describing and, especially, predicting their patterns of habitat use. The preponderance of such studies has been largely relegated to game species, and they have provided helpful linkages between behaviors and habitat use of certain species of galliforms, waterfowl, and, in a few cases, raptors. Ungulates, such as white-tailed deer and mule deer, have also been the foci of many such investigations.

Sampling Techniques

Methods used to study the diets of vertebrates can be divided into 3 basic categories: those involving the collection of individual animals, those involving capture of or other temporary disturbances to an animal, and those requiring little or no disturbance (Rosenberg and Cooper 1990).

Several reviews of dietary assessments are available. Although many of these studies directly concern specific groups of organisms (e.g., seabirds), many of the methods also apply to other groups. Rosenberg and Cooper (1990) provided a thorough overview of methods used to sample bird diets. Ratti et al. (1982:765–913) reprinted papers on the food habits and feeding ecology of waterfowl, and they included a bibliography of other important references on diet. Each new edition of The Wildlife Society's *Wildlife Management Techniques Manual* includes reviews of assessment methods for birds and mammals (e.g., Haufler and Servello 1994; Litvaitis et al. 1994). Riney (1982:124–137) summarized studies of mammalian food habits, and many of the authors in Cooperrider et al. (1986) covered dietary research in wildlife.

The most frequently used method of sampling organisms' diets is a direct examination of stomach and esophageal contents. Its primary advantage is that statistically adequate numbers of digestive tracts are usually relatively easy to obtain through trapping or shooting. If shot, an animal can be collected after the researcher has observed its specific foraging behavior. The investigator can then attempt to relate the particular food items in its stomach to those sampled from the foraging substrate and connect that to the behavior that individual used to gather food. For game animals, researchers often take stomach samples from hunter check stations. Another advantage of this method is that all of the stomach contents can be obtained. Kill sampling, however, has numerous disadvantages. The organism obviously cannot be resampled at a later date, thus eliminating a quantification of any temporal (and possibly spatial) changes in its food habits. This method can also have a potentially substantial impact on the population under investigation, negating studies of other aspects of the its ecology (i.e., abundance, reproductive performance, behavior). Lastly, researchers are often subjected to severe criticism from certain segments of the public when killing animals is part of their research.

Nondestructive means of sampling food habits are also available for wildlife. Live-caught organisms can be forced to regurgitate, using a variety of chemical emetics. Although some mortality can occur, methods are available that will minimize losses. For many animals, the most easily obtained samples of their diets come from their feces, collected either from the environment or during live trapping. In live-trapping studies, droppings can be obtained year-round from animals of any age or reproductive state, and repeated sampling from known individuals is possible. Ralph et al. (1985) described a technique for collecting and analyzing bird droppings. This and related methods have been used successfully in many species of birds (Davies 1976, 1977a, 1977b; Waugh 1979; Tatner 1983; Waugh and Hails 1983; Ormerod 1985) and small mammals (Meserve 1976; Dickman and Huang 1988). Other researchers have detailed methods of fecal collection and analysis in ungulates (Riney 1982:129–131; Haufler and Servello 1994).

Such nondestructive methods also involve viewing the animals' foraging behaviors and analyzing their food removal rates. Direct observation of the items that are eaten is possible with some species in certain vegetation types. For example, many studies have been designed to quantify the amount of plant material removed by foraging ungulates. Investigators assess the height, weight, and condition of the plants over periods of time and then relate the results to the types and amounts of vegetation consumed (Dasmann 1949; Severson and May 1967; Willms et al. 1980). Large ungulates can sometimes be studied when grazing or browsing, with "bite counts"—calculated as the number of bites per plant species—being recorded (Willms et al. 1980, Thill 1985). Observations of food removal and bite counts can then be combined to develop a picture of dietary habits. Diurnal birds of prey, such as eagles, vultures, and large hawks, can be observed when they forage in open areas. Other methods of assessing food ingestion include inspecting castings or pellets from birds of prey (Williams et al. 2012), examining prey remains in raptor nests, or using stable isotopes and molecular genetics techniques (Mumma et al. 2015). Unfortunately, non-destructive methods of assessing diets do not present a panacea for researchers. In most cases, foraging animals cannot be observed closely without adversely influencing their behavior, thus negating a direct observation of their food habits. Because trapping most species is difficult, the food in their stomachs is often highly digested, so an investigator does not know where or when an animal was feeding. Moreover, researchers have little control over which animals are captured.

As with most aspects of wildlife research, a combination of carefully selected methods is usually necessary. A useful strategy is to first determine if the species under investigation shows any feeding activities that depart markedly from those of other closely related species. For example, studies of foraging behavior in the Sierra Nevada revealed that many small insectivorous birds, such as chickadees (*Parus* [*Poecile*]) significantly increased their use of bark foraging during winter, which warranted a closer examination of food availability and the birds' dietary habits (Morrison et al. 1989). Taking a few preliminary stomach samples (using direct or indirect methods) will indicate the sample sizes necessary and the time required to analyze them, as well as the likely level of resolution possible after a full investigation is conducted. There is no excuse for simply collecting large numbers of stomachs that will either sit on a laboratory shelf or yield no useful information if the purpose of the research is management of a species.

Differential digestion rates of dietary items impose a large potential bias in any study of gut contents. Various kinds of foods, consumed at about the same time, are often digested at dissimilar rates. Thus further steps must be taken to prevent excessive postmortem food digestion. For example, small-bodied insects may be gone from a bird's gizzard within 5 minutes, whereas hard seeds may persist for many days (Swanson and Bartonek 1970). Therefore, several authors have developed correction factors for the differential rates of digestion shown by organisms (e.g., Mook and Marshall 1965). Nor are these divergent rates confined to the intertaxonomic level. Rosenberg and Cooper (1990) discussed data showing that second and third instar moth larvae were digested in less than half the time it took to process fourth and fifth instars.

Level of Identification

As noted throughout this book, the goals of a specific study should determine the level of identification required to reach useful conclusions, but the topic of a proper taxonomic identification of food items has received little attention in the literature. The taxonomic level selected for a diet analysis can have substantial impacts on the ecological interpretation of the results. This problem is analogous to the selection of variables for use in habitat models—that is, how finely should we divide categories? If identification to the species level was a simple matter, then this issue would be of minor concern. Not only are many foods difficult to identify when in excellent condition, but their mastication and digestion further complicate the task. The level at which comestibles are identified is likely to affect measurements of their presence, frequency, abundance, and similarity, as well as any conclusions drawn from these data.

Greene and Jaksic (1983) examined the influence of prey identification levels on measures of niche breadth and niche overlap in raptors, carnivores, and snakes. They calculated the niche metrics as narrowly as possible (usually at the species or genus level), using what was reported in diet studies, and then recalculated these numbers after combining the prey lists to the ordinal level (i.e., the taxonomic ranking 2 steps higher). For all vertebrate groups that were examined, they found that niche breadth was consistently larger at the 2 finer prey-identification resolutions than at the broader distinction by taxonomic order. Computations at the ordinal level underestimated niche breadth at the species and genus levels by more than an order of magnitude, and they did so even more spectacularly when just a single species was involved. On the other hand, their calculations overestimated food-niche overlaps, which rose to infinity when 2 predator species did not actually coincide in their use of any prey species but appeared to do so because of identification at the level of taxonomic order. The authors clearly showed that using an ordinal level of prey classification can result in serious misinterpretations of some of the potentially most important food-niche and community parameters in assemblages of many organisms. Further, even evaluations at the species level could mask finer differences in the selection of individual prey or food items, but they can still serve as a useful standard for comparison. Cooper et al. (1990) offered several guidelines regarding the level of taxonomic identification to choose in a study. Be sure, however, that it contains enough observations to make your analysis meaningful. There are several practical considerations here. First, variables for food and prey categories with high numbers of zero counts will not be normally distributed and usually cannot be transformed to normality. Thus parametric multivariate statistical procedures lose validity, and non-parametric ones, including ordinal-scale comparisons, may be more appropriate (see Chapter 3). Second, you should decide if there would be any benefit to dividing a particular order into finer levels. That is, if the ecological and behavioral characteristics of 2 groups within an order do not differ substantially, then it is unlikely that subdivision will provide much additional information. Food items do not necessarily need to be identified using Linnaean nomenclature, as morphometric characteristics of phenotypically distinguishable taxa (such as grasshopper A, B, etc.) are an adequate substitute (see also Greene and Jaksic 1983; Wolda 1990). The chosen level of food identification should be based on the goals of the study, not decided a posteriori because of constraints on funds and time, or difficulties in identifying the items. Simple preplanning prevents later disappointments, and, more importantly, the potential waste of animal life when kill sampling is involved.

Analyses

Choosing which statistical analyses to use for behavioral data should be an integral part of the initial study design. A large and varied number of methods have been employed, based largely on the goals of the researcher and the form of the data. Clearly, sufficient and appropriate planning should guide your decision. Virtually any general statistical text can offer direction on analyzing behavioral data. Especially useful techniques are those of Siegel (1956), Conover (1980), Snedecor and Cochran (1980), Sokal and Rohlf (1995), and Zar (1999). Texts dealing more specifically with quantitative methods in ethology include Hazlett (1977), Colgan (1978), Siegel and Castellan (1988), Weinberg and Goldberg (1990), Sommer and Sommer (1991), and Lehner (1996). Martin and Bateson (1993:Chapter 9) presented a concise but thorough review of fundamental univariate techniques for the study of animal behavior. (See Chapter 3 for multivariate methods.) Although not dealing specifically with statistical analyses, Kamil (1988) discussed the application of experimental methods from the perspective of a behavioralist. Many of the papers included in Morrison et al. (1990), which concentrated on statistical analyses in foraging studies, can be applied to any type of behavioral data. Roa (1992) and Manly (1993) discussed potentially helpful analyses for use in experiments on food preference.

Behavioral data are usually recorded as categorical variables, such as types of activity (e.g., walking, running). Later, continuous data are often further subdivided for analysis (e.g., speed of movement). When data are classified this way, the result is a contingency table, in which cells contain frequencies of the various combinations (e.g., activity type by sex). The null hypothesis of homogeneity of categories can then be tested using a contingency analysis, such as a chi-square test or *G* log–likelihood statistic, or contingency tables can be compared using a Mantel test (or others). When 3 or more categories are involved, a multidimensional contingency table analysis should be used (Colgan and Smith 1978).

Researchers usually record the behaviors of animals in a sequential manner, such as walk-pause-walk-bite-swallow-walk. Using contingency tables, chi-square tests, and related analyses may not be strictly valid for such data, however. This is because such observations are likely to be connected and thus violate the important assumption of independence in most statistical tests (Raphael 1990). The often-sequential nature of data collection can, however, have a benefit with respect to the inferences that can be made from recorded behaviors. Examining a series of actions by a single animal in the context of transition probabilities (see below) potentially provides much more information on how that individual exploits its environment than would an overall lumping or averaging of its behaviors.

A sequence in which the behavioral pattern always occurs in the same order is termed "deterministic." As mentioned above (in the section on "Observer Bias"), classical ethologists often refer to such sequences as "fixed action patterns." Vertebrates seldom, if ever, repeat behaviors in the same order, but they can exhibit some types of actions that may be predictable. These sequences are called "stochastic" (or "probabilistic," to identify the statistical probability that can be assigned to each behavior). Groups of activities that show no temporal pattern—that is, the component behavior patterns are sequentially independent—are considered to be "random sequences." In a random sequence, 1 behavior (or 1 set of behaviors) can be followed by any other behavior with equal probability. The conditional probability that 1 behavior follows another—that is, the probability that behavior B will follow behavior A, given that A has occurred—is denoted as *P*(B/A) and is called a "transition probability" (Martin and Bateson 1993:152–154).

There are several methods for analyzing sequential data, such as time-series analyses. Of particular interest to us are those involving Markov chains. Markov analysis is a method for distinguishing whether a sequence is random or contains some degree of temporal order. In a first-order Markov process, the probability of occurrence for the next event depends on the immediately preceding event. If the probability hinges on the 2 preceding events, then the process is considered to be second order. Higher-order processes are involved as additional events are considered. Sequences are analyzed by comparing the observed frequency of each transition with the frequency of transitions that would be expected if the sequences were random (Martin and Bateson 1993:152). A simple example of the Markov analysis was given by Martin and Bateson (1993:153) and is shown in Figure 4.8. Raphael (1990) presented a review of this method and a detailed example of applying it to foraging data. Colwell (1973), Riley (1986), Diggle (1990), and Gottman and Roy (1990) used it for sequential data. More sophisticated Markov chain Monte Carlo (MCMC) methods (Table 2.1) comprise a group of statistical techniques used for analyzing sequences using a simulated probability distribution.

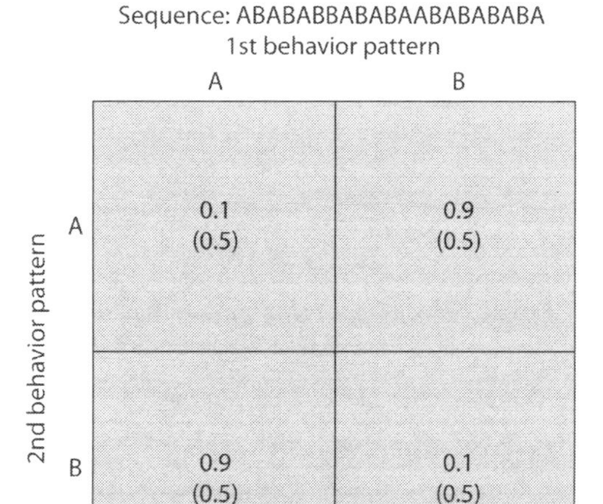

Figure 4.8. A simplified transition matrix, analyzing the sequence shown above it, which consists of 2 behavior patterns (A and B). The matrix shows the empirical transition probabilities for the 4 different types of transition (A|A, B|A, A|B and B|B). It also indicates that transitions from A to B or B to A tend to alternate. The probability of these transitions is 0.9, compared with a random sequence (0.5), while repeat probabilities of A to A or B to B are rare (the probability of these transitions is 0.1). (Courtesy of the Leonard A. Brennan Collection)

Analyzing the sequential nature of data can identify changes in the behavior of individuals that might initially be obscured by only examining overall averages. Further, such analyses can identify which aspect of an organism's behavior is being impacted by a natural or human-induced change in its environment. Each step in a sequence of behaviors can also be related to measurements of the environment encountered during that step. For example, the behavioral sequence of a foraging bird (e.g., hop-hop-glean-probe-hop-hop) can be related to the foraging substrate being encountered (e.g., ground-ground-leaf-bark-ground-ground). Behavioral sequences might lend insight into an animal's response to treatment effects that might not be evident by using averages. Such changes in behavior might also explain alterations in activity times and perhaps even changes in fecundity. Here again, the goal of a particular study should determine the amount of detail required.

Indices

Scientists have developed a methodology to quantify the use of resources in relation to their availability. Widely known as indices of "preference," "selectivity," or "electivity," these measures seek to compare the frequency of use of a resource with the occurrence (or "availability") of those items in an animal's environment by representing the data as a single index value. Manly et al. (2002) reviewed the developmental history of these indices. Many of them have received a good deal of attention in the wildlife literature, especially with regard to analyses of the food habits of ungulates. They have also been applied to vegetation and other measurements of the environment (Morrison 1982). Because of this attention, as well as the potential application of indices to a wide variety of situation, we will discuss some important considerations in their use.

The general approach to employing electivity indices is to establish a ratio between the frequency of an item consumed by an animal and the amount available for use. For example, Ivlev's (1961) electivity index, E, compared the relative availability of an item in the environment (p) with the relative use of that item (r):

$$(r): E_i = (r_i - p_i) / (r_i + p_i)$$

Other indices have a similar form, as reviewed by Lechowicz (1982) and Manly et al. (2002). If r and p are equal for all items, then the organism is choosing the items randomly. If r and p differ, then you can usually conclude that the animal is either avoiding (a negative index value) or selecting (a positive value) a particular element.

The most straightforward indices consist of the estimated percentage of use for an item, divided by the total amount of all of the ones available for use. Values usually range from −1 to 0 (denoting avoidance), and 0 to infinity (denoting preference). Such indices have been termed the "forage ratio" (Jacobs 1974) and have been widely used (e.g., Van Dyne and Heady 1965; Chamrad and Box 1968; Petrides 1975; Hobbs and Bowden 1982). The major drawback to these simple indices, however, is their intrinsic asymmetry—that is, their unbounded positive values. A log transformation is used for the forage ratio (Jacobs 1974; Strauss 1979; Lechowicz 1982). Unfortunately, however, the forage ratio changes with the relative abundance (p) of items in the environment. Thus this index cannot be used if one wishes to examine the relationship between the relative abundance of substances and an organism's preference for them (Jacobs 1974; Lechowicz 1982).

Works by Ivlev (1961), Jacobs (1974), Chesson (1978, 1983), Strauss (1979), Vanderploeg and Scavia (1979a, 1979b), and Johnson (1980) described the development of the more widely used electivity indices. Lechowicz (1982) compared the characteristics of 7 of the electivity indices proposed by these researchers, most of which were permutations of Ivlev's original index. He found that Ivlev's (1971), Jacob's (1974), and Strauss's (1979) indices could not potentially obtain the full range for all values of r and p, because those for their intermediate values depended on the relative abundance of either other items in the environment or those used by the animal. A crucial problem with most of the indices is that direct comparisons between those derived from samples that differ in their relative abundances are inappropriate (but see Chesson 1983 for exceptions).

Aebischer et al. (1993) presented a compositional analysis that has been widely used in wildlife home-range studies, although it can also be employed in diet or time-budget analyses. It is based on the proportional use of a resource (e.g., a food item, vegetation type) by individual animals. Given that the data are multivariate, an analysis is done using multivariate linear models. Applications of compositional analysis can examine relationships with food abundance or compare use versus the occurrence of vegetation classes to test for preferences. A potential pitfall of compositional analysis, however, is that it is mathematically impossible to divide by zero. The substitution of small, non-zero numbers in zero use cells, as recommended by Aebischer et al. (1993), results in large Type I error rates (Bingham and Brennan 2004). The only solution to this problem is to categorize the initial composition matrix so that no non-zero use cells exist, which may be hard or impossible to accomplish under certain circumstances. To avoid the difficulties associated with the inclusion or exclusion of specific elements in the calculation and evaluation of indices, Johnson (1980) developed a procedure based on ranks. He proposed employing the difference between the rank of an item's use and the rank of its availability, as statistical methods based on ranks are nearly as efficient as those based on the original data. This is especially true if the assumptions necessary to treat the original data (e.g., an assumption of normality) are not met. He also provided methods for determining the statistical significance among components in the data. The compositional analysis method discussed in Chapter 2 can also be used to analyze diet and activity data.

The Concept of Wildness and Animal Behavior

The concept of wildness seems to be either underappreciated or overlooked by many people in the worlds of wildlife science and animal ecology. Perhaps this issue flows from a tautological perspective for the term "wildlife." We too often consider the obvious—wildlife is wild—and leave it at that. It may be useful, however, to take things a step further and consider where the "wild" in wildlife comes from, vis-à-vis the concept of wildness, which dictionary.com defines as "[animals] living in a state of nature; not tamed or domesticated."

We are addressing the concept of wildness here because animals that are not tamed, acclimated to human presence, or domesticated have vastly different behaviors compared with those that are. Wild animals exhibit elements of this attribute primarily for 1 reason: to survive and get their genes into the population's next generation. Everything else is secondary. What we witness when we observe the behaviors of wild animals is the result of millennia of evolution. Lose the *wild*, and you lose the *wild*life. For example, many species—such as corvids, raccoons, deer, bears, and others—that have become acclimated to towns as human commensals have adopted foraging and other behaviors unlike their truly wild counterparts.

The saga of the decline, recovery, and restoration of wild turkey populations in the United States is a textbook case history of wildlife conservation success. A combination of overharvests and excessive habitat loss (from overharvesting forests) wiped out wild turkey populations in most of their original US geographic range. Remnant populations persisted only in places that were largely inaccessible to humans, such as the vast southeastern bottomland forests. This phenomenon led many people to believe that the only places suitable for wild turkeys were those with large bottomland roost trees in standing water. As it turned out, this was not the actual issue. Rather, it was a simple artifact of their refugia.

In the early part of the twentieth century, efforts at wild turkey population restoration focused on using pen-raised birds from game farm facilities. These efforts were completely unsuccessful, as well as a vast waste of time and money. During the late 1930s, the Missouri Department of Conservation decided to investigate whether there were biological differences between truly wild turkeys (i.e., those that could persist in the wild) and pen-raised ones. Leopold (1944) documented that wild turkeys, when compared with those from game farms, had larger pituitary glands, along with other, subtler biological differences that functioned to maintain what he called the "wildness" in these birds. This attribute was quickly lost, even in the first generation of pen-raised birds. Keeping turkeys in captivity rapidly weeded out those individuals with the greatest amount of wildness and quickly selected for traits favoring docile behaviors. Furthermore, the lack of opportunity for newly hatched turkey poults to imprint on an adult hen further compromised their ability to exhibit the wary behaviors needed to detect and elude predators, as well as find food efficiently. The outcome of his findings—that the behavioral dissimilarities between pen-raised and wild turkeys were rooted in biological differences—provided a basis for state wildlife agencies to abandon the use of game farm birds for population restoration and focus instead on trapping and translocating wild birds. The result was a resounding success, which illustrates the importance of understanding the implications of animal behavior or physiology—or both—in conservation efforts.

Page and Whitacre (1975) were the first contemporary ornithologists to document that raptors ate a considerable proportion of shorebirds on their wintering grounds. Over the course of their study, during 2 winters at Bolinas Lagoon in California, the authors discovered that raptors consumed the following percentages of wintering populations of their prey: dunlins (*Calidris alpina*), 21 percent; least sandpipers (*C. minutilla*), 12 percent; western sandpipers (*C. mauri*), 8 percent; sanderlings (*C. alba*), 13 percent; and dowitchers (*Limnodromus* spp.), 16 percent. The implications of these findings were that numerous species of shorebirds in the family Scolopacidae endured near constant predation, primarily by merlins (*Falco columbarius*), but also by as many as 8 other species of raptors on their wintering grounds.

In a study focused on the predator-prey behaviors of merlins and dunlins in western Washington, Buchanan et al. (1988) quantified 7 behaviors the merlins used to attack and capture dunlins, as well as 3 flocking behaviors used by dunlins to evade merlins. As predators, merlins were successful in capturing dunlins about 5 percent of the time, typically by either making a direct stoop, or dive, from a high altitude into a flock of dunlins, or by separating a single bird from the group and chasing it down. Dunlins typically aggregated into flocks, ranging from 50 to more than 10,000 individuals, and used what are known as "distraction-evasion behaviors" to avoid predation. These actions involved a chorus-line effect of individual birds coordinating the rotation of their bodies in flight, with what was a visible shifting between their dark dorsal plumage and light ventral feathers, to produce a flashing or rippling pattern that presumably would make it difficult for an individual merlin to isolate a single dunlin for attack. The success of these behaviors allows dunlins to evade about 70–90 percent of the attacks by merlins.

The predator-prey dynamics between raptors and shorebirds can be explained from an evolutionary perspective, in that raptors have evolved specific behaviors to capture prey, and shorebirds have developed ones to avoid capture. The primary goal of an individual shorebird, such as a dunlin, is to survive the winter and return to the High Arctic to breed. The tendency of this species to coalesce into huge (>10,000 individuals) flocks most likely is a mechanism for finding food and minimizing the risk of being eaten by a raptor. This is "wildness," operating in a coevolutionary context.

Muskoxen (*Ovibos moschatus*) also inhabit the High Arctic and have evolved to persist in extreme winter conditions. They do not defend territories, probably because they need to roam over large areas to procure food. Nevertheless, muskoxen have both male and female social hierarchies, based on age, with older animals being dominant (Tener 1965). Wolves (*Canis lupus*) have been their most common predator. When muskoxen sense danger from predation, they coalesce into a tight ring, surrounding the calves and juveniles, with the heads and horns of the adults facing outward. During the rut, or mating season, bulls determine the structure of the defensive circle, while cows shape it for the other months of the year (Freeman 1971).

The evolved, highly complex social structure of African wild dogs (*Lycaon pictus*) has allowed them to number among the most efficient predators in the world (Courchamp et al. 2002). This species is 1 of the few diurnal mammalian predators in Africa, primarily going after midsized animals, first by stealth and then by a rapid chase. Pups, which are often cared for by subdominant individuals, are given first access to a kill, even before dominant pack members. Although lions (*Panthera leo*) are considered by many to be the ultimate predator in Africa, their hunting success rate is only about 10 percent. In contrast, around 80 percent of all African wild dog hunts are successful.

This concept of wildness is clearly linked to the persistence of individuals and, therefore, to populations of such organisms over time. Either directly or indirectly, it is clearly at the center of any behavioral investigation of a wild animal.

An Environmental-Habitat Model for Testing Behavioral Hypotheses

Berger-Tal et al. (2011) have developed what many consider to be a useful model for animal behavior and conservation science (Fig. 4.9). The behavioral domains at the center of their model—movement, the use of space, foraging, vigilance, social organization, and reproduction—mirror many of the themes addressed in this chapter. There are 3 extrinsic factors—anthropogenic impacts on animal behavior, behavioral indicators, and behavior-based management—that are identified as drivers. They influence the pathways that underlie how this conceptual model works. Its overriding goal is "aimed at investigating how proximate and ultimate aspects of animal behavior can be of value in preventing the loss of biodiversity" (Berger-Tal et al. 2011:236). While some writers have dismissed the implications of behavior as an important component of conservation science (e.g., Caro 2007), their role as a factor in wildlife conservation has been appreciated for at least the past half century (Geist and Walter 1974), and maybe even longer.

One can also argue that there are behavioral factors embedded in the philosophy of "fair chase," which is a fundamental component of the North American Model of Wildlife Conservation (Krausman and Mahoney 2015). Although Geist and Walter (1974) provided a clear conceptual bridge between this and the behavioral conservation model offered by Berger-Tal et al. (2011), as well as the North American Model (Mahoney and Jackson 2015), there are philosophical differences between these 2 camps that are worth noting. For example, there is often a bifurcation between the perspectives of a conservation

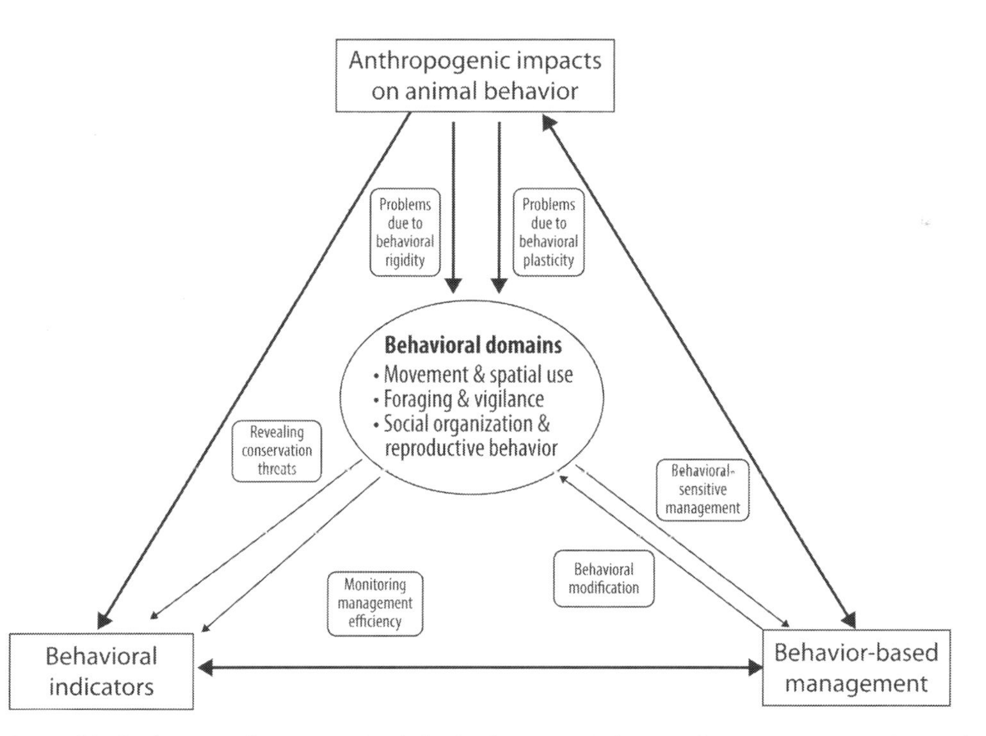

Figure 4.9. A schematic of a conservation behavior framework, composed of 3 interrelated themes: (1) anthropogenic (i.e., human-influenced) impacts on animal behavior, (2) behavior-based management, and (3) behavioral indicators. Arrows represent pathways that connect each theme to the behavioral domains or interactions between the conservation themes. (Courtesy of the Leonard A. Brennan Collection)

biologist (whose goal may focus more on protecting and restoring diversity) and those of a wildlife biologist (whose goal is often directed toward protecting and restoring individual species), even though both camps aspire to do essentially the same thing: provide conditions for the natural functioning and persistence of native animals.

We contend, therefore, that the conservation behavior framework model offered by Berger-Tal et al. (2011) implicitly represents an extension of the North American Model of Wildlife Conservation, since the former includes species that are not exploited via harvest. Additionally, environmental-habitat components can be incorporated into this model by way of the impetus at the apex of their model: anthropogenic (i.e., human-influenced) impacts on animal behavior. In our prodigal society, these are legion and can obviously have major repercussions that affect both individual organisms and populations of wild animals.

Management Actions and Animal Behavior

Some animal ecologists tend to overlook or underappreciate the study of animal behavior, because they find it difficult to make direct linkages between it and management actions. The roots of this disconnect most likely lie in a perception that behaviors are "fixed" and cannot, in most cases, be influenced by management activities that manipulate vegetation structure or otherwise affect environmental resources. The animal ecology literature, however, contains numerous instances of how such actions can be linked to specific behaviors of individual organisms which result in benefits for populations and, ultimately, species, if conducted on a large enough scale. Some examples are presented in the next few paragraphs.

Grassland bird populations have been declining during the past 50–100 years at a greater rate than any other group of birds in North America. Reasons for the declines in their populations are myriad and complex (Brennan and Kuvlesky 2005). In the northeastern United States, farmers have been harvesting hay from pastures earlier in the summer than they traditionally did, which has resulted in negative effects on breeding grassland birds. In response to this problem, Perlut et al. (2006) designed a study to examine how the timing of hay harvests impacted these birds in the Champlain Valley of Vermont and New York. Focusing on savannah sparrows (*Passerculus sandwichensis*) and bobolinks (*Dolichonyx oryzivorus*), they found that early haying (before 11 June) caused 99 percent of the active savannah sparrow nests and 100 percent of the bobolink nests to fail. Savannah sparrows immediately re-nested after haying, whereas bobolinks left the area for about 2 weeks and then returned. On fields where haying was delayed until the period between 21 June and 10 July, savannah sparrows fledged 3.5 (±0.42) young, and bobolinks fledged 2.2 (±0.26). Savannah sparrow and bobolink reproduction was similar on late-hayed (after 1 August) and rotationally grazed pastures. In a companion investigation that examined the survival of these 2 species, Perlut et al. (2008) found that the overall rate was about 25 percent higher for birds that nested in late-hayed fields, compared with those that used early-hayed fields and rotationally grazed pastures. The implication of these studies was that farmers should make their first hay cutting as early as possible—perhaps even during May, before most of the birds arrive—and then allow a 65-day window until the next cutting, in order to give savannah sparrows, bobolinks, and other grassland birds time to breed. Making a simple adjustment in the timing of these hay harvests to accommodate the breeding behaviors of these avian species can have a major payoff with respect to the conservation of their populations. In this way, understanding animal behavior can directly translate into management recommendations.

In another example, wetland managers often make huge efforts to manipulate marshland vegetation for waterfowl. Kaminski and Prince (1981) were among the earliest wildlife scientists to examine how waterfowl and their macroinvertebrate foods responded to the manipulation of wetlands. Using a study site in the Delta Marsh of southcentral Manitoba in Canada, they created experimental wetland areas with 3 different percentage ratios of hydrophytes to open water (30:70, 50:50, and 70:30), along with 2 basin treatments (mowing and rototilling). Overall, they found that the greatest density and diversity of dabbling duck (*Anas*) pairs was on the 50:50 areas of the wetlands. The also observed the highest incidence of pursuit flights—males hustling after females during courtship displays—on the 50:50 and the mowed areas. A companion study by Murkin et al. (1982) found that the relatively high numbers of dabbling ducks on the 50:50 plots was strongly correlated with the abundance of macroinvertebrates in these areas. Evidently, what waterfowl and wetland managers consider a "hemi-marsh" configuration of freshwater wetlands and emergent vegetation supports the greatest biomass of invertebrate foods that, in turn, sustains the highest density and diversity of breeding waterfowl.

Harvest Regulations and Behavior

We often take for granted the contemporary harvest regulations for game animals by state and federal agencies. For example, most hunting seasons for large mammals take place during the fall and winter, outside the peak breeding time for most of these species. This is also the case for the majority of resident game birds, with wild turkeys being a notable exception. Spring wild turkey hunts are typically limited to bagging adult males, or gobblers, because of both their polygamous reproductive behavior and the ease of identifying gobblers versus hens (Hughes and Lee 2015). While it is legal to kill immature males, or jakes, this is often discouraged, because many hunters want males in this age category to be recruited into the breeding population.

During the early part of the twentieth century, it was once popular to hunt ducks during their northward migration to their breeding grounds (Leopold 1933). The negative biological implications of this policy—such as inflicting excessive mortality on birds that were about to breed—were quite obvious and have been modified accordingly, primarily as a function of the North American Model of Wildlife Conservation (Anderson and Padding 2015).

Mourning doves are the most widely hunted game birds in North America, with more than 14 million of them harvested from an estimated population greater than 270 million (Seamans and Sanders 2014; Brennan et al. 2017), or about a 5 percent annual harvest rate. Because this species is widely distributed across the conterminous United States, the US Fish and Wildlife Service has divided the country into 3 mourning dove management units—eastern, central, and western—primarily for setting hunting seasons. A curious aspect of these seasons is that they are often split into 2 time periods, with a break in between. For example, during recent years in Texas, mourning dove hunting season runs from around the third week in September to the third week in October. It is then closed until about the third week in December, when it reopens for around a month, and closes again circa the third week in January. This policy of split hunting seasons is typical of many other states within this species' range. Yet there is usually little or no biological basis for the division into 2 different periods. Often, agency biologists and administrators will posit the behavioral assertion that after a month or so of hunting, the birds need a "rest" from being shot at. What is probably more likely is that hunter satisfaction is thought to be elevated by having 2 separate seasons in which to shoot mourning doves. In this case, it is the behavior of the hunters that forms the basis for designing and administering a hunting season.

Summary

While animal behavior, in the most fundamental sense, can be defined as the movement(s) of an organism, it is the context in which these movements take place that points us toward understanding how such actions are related to the fitness of an individual, the trajectory of a population, and, ultimately, the persistence and successful management of a species. Describing and quantifying—in other words, measuring—the activities of wild animals is a far from trivial task. Behavioral studies are among the most daunting aspects of animal ecology research. In the wild, it often takes hundreds (if not thousands) of hours of meticulous data collection. In the laboratory, there are often massive logistical challenges to maintaining humane and effective animal care and experimentation procedures, as well as to relating ex situ investigations in the laboratory to in situ conditions in the field. Furthermore, no behavioral study, whether in the wild or in a laboratory, has a chance of being successful in the absence of a rigorous comparative or experimental design. As much as we may romanticize the image of a naturalist with binoculars or a spotting scope jotting down random observations in a field notebook, that era has long since been passed as a basis for the modern science of animal behavior.

Analyses of behavioral data from wild animals can provide a window into how they are able to subsist. As such, incorporating and measuring behavior can be an important component of a habitat-relationships study, because it provides direct information on how an animal uses its available resources. The problem, however, lies in quantifying the extent to which these resources are available and, to complicate things further, the degree to which others can also be obtained. This is why the concept of "giving up density" has been something of a breakthrough in foraging theory. Getting a handle on GUD can be disconcerting, however, when it comes to assigning levels of profitability to various patches used for finding, choosing, and consuming food, especially outside of controlled laboratory conditions.

The concept of wildness—and the ways in which behaviors that are linked to wildness may be related to survival and, ultimately, fitness—has often been overlooked (or at least underappreciated) by wildlife scientists. Vigilance is an important component of wildness, and it is experienced by both predators and prey. This type of awareness—and, by extension, wildness itself—is at the core of the behavioral conservation model developed by Berger-Tal et al. (2011). Examining the extent to which their behavioral conservation model works in nature would be a fruitful avenue to pursue in future behavioral ecology studies of animals.

LITERATURE CITED

Abu Baker, M. A., and J. S. Brown. 2009. Patch area, substrate depth, and richness affect giving-up densities: A test with mourning doves and cottontail rabbits. Oikos 118:1721–1731.

Aebischer, N. J., P. A. Robertson, and R. E. Kenward. 1993. Analysis of habitat use from animal radio-tracking data. Ecology 74:1313–1325.

Altmann, J. 1974. Observational study of behavior: Sampling methods. Behaviour 49:227–267.

Anderson, M. G., and P. I. Padding. 2015. The North American approach to waterfowl management: Synergy of hunting and habitat conservation. International Journal of Environmental Studies 72:810–829.

Ash, A., M. Coughenour, J. Fryxell, W. Getz, J. Hearne, N. Owen-Smith, D. Ward, and E. A. Laca. 1996. Second International Foraging Behavior Workshop. Bulletin of the Ecological Society of America 77:36–38.

Bachman, G. C. 1993. The effect of body condition on the trade-off between vigilance and foraging in Belding's ground squirrels. Animal Behaviour 46:233–244.

Baulu, J., and D. E. Redmond Jr. 1978. Some sampling considerations in the quantification of monkey behavior under field and captive conditions. Primates 19:391–400.

Bedoya-Perez, M. A., A. J. R. Carthey, V. S. A. Mella, C. McArthur, and P. B. Banks. 2013. A practical guide to avoid giving up on giving-up densities. Behavioral Ecology and Sociobiology 67:1541–1553.

Berdoy, M., and S. E. Evans. 1990. An automatic recording system for identifying individual small animals. Animal Behaviour 39:998–1000.

Berger-Tal, O., T. Polak, A. Oron, Y. Lubin, B. P. Kotler, and D. Salz. 2011. Integrating animal behavior and conservation biology: A conceptual framework. Behavioral Ecology 22:236–239.

Bernstein, I. S. 1991. An empirical comparison of focal and *ad libitum* scoring with commentary on instantaneous scans, all occurrence and one-zero techniques. Animal Behaviour 42:721–728.

Bingham, R. L., and L. A. Brennan. 2004. Comparisons of Type I error rates for statistical analyses of resource selection. Journal of Wildlife Management 68:206–212.

Birkhead, T., J. Wimpenny, and B. Montgomerie. 2014. Ten

thousand birds: Ornithology since Darwin. Princeton University Press, Princeton, NJ.

Block, W. M. 1990. Geographic variation in foraging ecologies of breeding and nonbreeding birds in oak woodlands. Studies in Avian Biology 13:264–269.

Brennan, L. A., J. B. Buchanan, T. M. Johnson, and S. G. Herman. 1985. Interhabitat movements of wintering dunlins in western Washington. Murrelet 66:11–16.

Brennan, L. A., F. Hernández, and D. Williford. 2014. Northern bobwhite, *Colinus virginianus*, revision. Birds of North America Online. American Ornithologists' Union and Cornell Lab of Ornithology.

Brennan, L. A., and W. P. Kuvlesky Jr. 2005. North American grassland birds: An unfolding conservation crisis? Journal of Wildlife Management 69:1–13.

Brennan, L. A., and M. L. Morrison. 1990. Influence of sample size on interpretations of foraging patterns by chestnut-backed chickadees. Studies in Avian Biology 13:187–192.

Brennan, L. A., M. L. Morrison, and D. L. Dahlsten. 2000. Comparative foraging dynamics of chestnut-backed and mountain chickadees in the western Sierra Nevada. Northwestern Naturalist 81:129–147.

Brennan, L. A., D. L. Williford, B. M. Ballard, W. P. Kuvlesky Jr., E. D. Grahmann, and S. J. DeMaso. 2017. The upland and webless migratory game birds of Texas. Texas A&M University Press, College Station.

Brown, J. S. 1988. Patch use as an indicator of habitat preference, predation risk, and competition. Behavioral Ecology and Sociobiology 22:37–47.

Buchanan, J. B., C. T. Schick, L. A. Brennan, and S. G. Herman. 1988. Merlin predation on wintering dunlins: Hunting success and dunlin escape tactics. Wilson Bulletin 100:108–118.

Caraco, T., S. Martindale, and T. S. Whittham. 1980. An empirical demonstration of risk-sensitive foraging preferences. Animal Behaviour 28:820–830.

Caro, T. 2007. Behavior and conservation: A bridge too far? Trends in Ecology & Evolution. 22:394–400.

Carranza, J. 1995. Female attraction by males versus sites in territorial rutting red deer. Animal Behaviour 50:445–453.

Case, R. M., and R. J. Robel. 1974. Bioenergetics of the bobwhite. Journal of Wildlife Management 38:638–652.

Chamrad, A. D., and T. W. Box. 1968. Food habits of white-tailed deer in south Texas. Journal of Range Management 21:158–164.

Charnov, E. L. 1976. Optimal foraging: The marginal value theorem. Theoretical Population Biology 9:129–136.

Chesson, J. 1978. Measuring preference in selective predation. Ecology 59:211–215.

Chesson, J. 1983. The estimation and analysis of preference and its relationship to foraging models. Ecology 64:1297–1304.

Colgan, P. W., ed. 1978. Quantitative ethology. John Wiley & Sons, New York.

Colgan, P. W., and J. T. Smith. 1978. Multidimensional contingency table analysis. Pp. 145–174 *in* P. W. Colgan, ed. Quantitative ethology John Wiley & Sons, New York.

Colwell, R. K. 1973. Competition and coexistence in a simple tropical community. American Naturalist 107:737–760.

Conover, W. J. 1980. Practical nonparametric statistics, 2nd edition. John Wiley & Sons, New York.

Cooper, R. J., P. J. Martinat, and R. C. Whitmore. 1990. Dietary similarity among insectivorous birds: Influence of taxonomic versus ecological categorization of prey. Studies in Avian Biology 13:104–109.

Cooperrider, A. Y., R. J. Boyd, and H. R. Stuart, eds. 1986. Inventory and monitoring of wildlife habitat. US Department of the Interior, Bureau of Land Management Service Center, Denver.

Courchamp, F., G. S. A. Rasmussen, and D. W. Macdonald. 2002. Small pack size imposes a trade-off between hunting and pup-guarding in the painted hunting dog *Lycaon pictus*. Behavioral Ecology 13:20–27.

Darracq, A. K., L. M. Conner, J. S. Brown, and R. A. McCleery. 2016. Cotton rats alter foraging in response to an invasive ant. PLoS ONE 11(9): e0163220, doi:10.1371/journal.pone.0163220.

Dasilva, G. L. 1992. The western black-and-white colobus as a low-energy strategist: Activity budgets, energy expenditure and energy intake. Journal of Animal Ecology 61:79–91.

Dasmann, W. P. 1949. Deer-livestock forage studies in the interstate winter deer range in California, Journal of Range Management 2:206–212.

Davies, N. B. 1976. Food, flicking, and territorial behavior of the pied wagtail (*Motacilla alba yarelli* Gould). Journal of Animal Ecology 45:235–252.

Davies, N. B. 1977a. Prey selection and the search strategy of the spotted flycatcher (*Muscicapa striata*): A field study on optimal foraging. Animal Behaviour 25:1016–1033.

Davies, N. B. 1977b. Prey selection and social behavior in wagtails (Aves: Motacillidae). Journal of Animal Ecology 46:37–57.

Davoren, G. K., W. A. Montevecchi, and J. T. Anderson. 2003. Search strategies of a pursuit-diving marine bird and the persistence of prey patches. Ecological Monographs 73:463–481.

Defler, T. R. 1995. The time budget of a group of wild woolly monkeys (*Lagothrix lagotricha*). International Journal of Primatology 16:107–120.

Delaney, D. K., T. G. Grubb, P. Beier, L. A. Pater, and M. H. Reiser. 1999. Effects of helicopter noise on Mexican spotted owls. Journal of Wildlife Management 63:60–76.

Dewsbury, D. A. 1992. On the problems studied in ethology, comparative psychology, and animal behavior. Ethology 92:89–107.

Dickman, C. R., and C. Huang. 1988. The reliability of fecal analysis as a method for determining the diet of insectivorous mammals. Journal of Mammalogy 69:108–113.

Diggle, P. J. 1990. Time series: A biostatistical introduction. Oxford University Press, Oxford.

Dingemanse, N. J., K. M. Bouwman, M. van de Pol, T. van Overveld, S. C. Patrick, E. Maththysen, and J. L. Quinn. 2012. Variation in personality and behavioural plasticity across four populations of the great tit *Parus major*. Journal of Animal Ecology 81:116–126.

Duijns, S., I. E. Knot, T. Piersma, and J. A. Gils. 2015. Field measurements give biased estimates of functional response parameters, but help explain foraging distributions. Journal of Animal Ecology 84:565–575.

Emlen, J. M. 1966. The role of time and energy in food preference. American Naturalist 100:611–617.

Entwistle, A. C., J. R. Speakman, and P. A. Racey. 1995. Effect of using the doubly labelled water technique on long-term recapture in the brown long-eared bat (*Plecotus auritus*). Canadian Journal of Zoology 72:783–785.

Finch, D. M. 1984. Parental expenditure of time and energy in the Abert's towhee (*Pipilo aberti*). Auk 101:473–486.

Ford, H. A., L. Bridges, and S. Noske. 1990. Interobserver differences in recording foraging behavior of fuscous honeyeaters. Studies in Avian Biology 13:199–201.

Freeman, M. M. R. 1971. Population characteristics of musk-oxen in the Jones Sound region of the Northwest Territories. Journal of Wildlife Management 35:103–108.

Geist, V. and F. Walther, eds. 1974. The behavior of ungulates and its relation to management: The papers of an international symposium held at the University of Calgary, Alberta, Canada, 2–5 November 1971. International Union for Conservation of Nature and Natural Resources, Morges, Switzerland.

Giuggioli, L., and F. Bartumeus. 2010. Animal movement, search strategies and behavioural ecology: A cross-disciplinary way forward. Journal of Animal Ecology 79:906–909.

Glover, T. 2003. Developing operational definitions and measuring interobserver reliability using house crickets (*Acheta domesticus*). Pp. 31–40 *in* B. J. Ploger and K. Yasukawa, eds. Exploring animal behavior in laboratory and field. Academic Press, San Diego.

Goldstein, D. L. 1984. The thermal environment and its constraint on activity of desert quail in summer. Auk 101:542–550.

Goldstein, D. L. 1990. Energetics of activity and free living in birds. Studies in Avian Biology 13:423–426.

Gordon, D. M. 1985. Do we need more ethograms? Zeitschrift für Tierpsychologie 68:340–342.

Gottman, J. M., and A. K. Roy. 1990. Sequential analysis: A guide for behavioral researchers. Cambridge University Press, Cambridge.

Greene, H. W., and F. M. Jaksic. 1983. Food-niche relationships among sympatric predators: Effects of level of prey identification. Oikos 40:151–154.

Guthery, F. S., N. D. Forrester, K. R. Nolte, W. E. Cohen, and W. P. Kuvlesky Jr. 2000. Potential effects of global warming on quail populations. Proceedings of the National Quail Symposium 4:198–204.

Guthery, F. S., A. R. Rybak, S. D. Fuhlendorf, T. L. Hiller, S. G. Smith, W. H. Puckett Jr., and R. A. Baker. 2005. Aspects of the thermal ecology of bobwhites in North Texas. Wildlife Monographs 159:1–36.

Hale, A. M., D. A. Williams, and K. N. Rabenold. 2003. Territoriality and neighbor assessment in brown jays (*Cyanocorax morio*) in Costa Rica. Auk 120:446–456.

Hall, L. S., P. R. Krausman, and M. L. Morrison. 1997. The habitat concept and a plea for standard terminology. Wildlife Society Bulletin 25:173–182.

Hamilton, W. D. 1971. Geometry for the selfish herd. Journal of Theoretical Biology 31:293–311.

Harcourt, A. H., and K. J. Stewart. 1984. Gorillas' time feeding: Aspects of methodology, body size, competition and diet. African Journal of Ecology 22:207–215.

Haufler, J. B., and F. A. Servello. 1994. Techniques for wildlife nutritional analysis. Pp. 307–323 *in* T. A. Bookhout, ed. Research and management techniques for wildlife and habitats, 5th edition. The Wildlife Society, Bethesda, MD.

Hazlett, B. A., ed. 1977. Quantitative methods in the studies of animal behavior. Academic Press, New York.

Heil, S. J., J. Verner, and G. W. Bell. 1990. Sequential versus initial observations in studies of avian foraging. Studies in Avian Biology 13:166–173.

Hobbs, N. T. 1989. Linking energy balance to survival in mule deer: Development and test of a simulation model. Wildlife Monographs 101:1–39.

Hobbs, N. T., and D. C. Bowden. 1982. Confidence intervals on food preference indices. Journal of Wildlife Management 46:505–507.

Holtcamp, W. N., W. E. Grant, and S. B. Vinson. 1997. Patch use under predation hazard: Effect of the red imported fire ant on deer mouse foraging behavior. Ecology 78:308–317.

Hooten, M. B., D. S. Johnson, B. T. McClintock, and J. M. Morales. 2017. Animal movement: Statistical models for telemetry data. CRC Press / Taylor & Francis Group, Boca Raton, FL.

Hughes, T. W., and K. Lee. 2015. The role of recreational hunting in the recovery and conservation of the wild turkey (*Meleagris gallopavo*). International Journal of Environmental Studies 72:797–709.

Hurlbert, S. H. 1984. Pseudoreplication and the design of ecological field experiments. Ecology 54:187–211.

Hutto, R. L. 1990. On measuring the availability of food resources. Studies in Avian Biology 13:20–28.

Immelmann, K., and C. Beer. 1989. A dictionary of ethology. Harvard University Press, Cambridge, MA.

Ivlev, V. S. 1961. Experimental ecology of the feeding of fishes. Yale University Press, New Haven, CT.

Jacobs, J. 1974. Quantitative measurement of food selection. Oecologia 14:413–417.

Jenkins, S. H. 1982. Management implications of optimal foraging theory: A critique. Journal of Wildlife Management 46:255–257.

Johnson, D. H. 1980. The comparison of usage and availability measurements for evaluating resource preference. Ecology 61:65–71.

Jones, Z. F., and C. E. Bock. 2003. Relationships between Mexican jays (*Aphelocoma ultramarina*) and northern flickers (*Colaptes auratus*) in an Arizona oak savanna. Auk 120:429–432.

Kabak, I. W. 1970. Wildlife management: An application of a finite Markov chain. American Statistician 24:27–29.

Kamil, A. C. 1988. Experimental design in ornithology. Pp. 312–346 *in* R. F. Johnston, ed. Current Ornithology, vol. 5. Plenum Press, New York.

Kamil, A. C., J. R. Krebs, and H. R. Pulliam, eds. 1987. Foraging behavior. Plenum Press, New York.

Kamil, A. C., and T. D. Sargent, eds. 1981. Foraging behavior: Ecological, ethological, and psychological approaches. Garland STPM, New York.

Kaminski, R. M., and H. H. Prince. 1981. Dabbling duck and aquatic macroinvertebrate responses to manipulated wetland habitat. Journal of Wildlife Management 45:1–15.

Kelly, J. P., and W. W. Weathers. 2002. Effects of feeding time constraints on body mass regulation and energy expenditure in wintering dunlin (*Calidris alpina*). Behavioral Ecology 13:766–775.

Kenward, R. 1987. Wildlife radio tagging. Academic Press, New York.

Kildaw, S. D. 1995. The effect of group size manipulation on the foraging behavior of black-tailed prairie dogs. Behavioral Ecology 6:353–358.

Knight, R. L., and K. J. Gutzwiller, eds. 1995. Wildlife and recreationists: Coexistence through management and research. Island Press, Washington, DC.

Kohlman, S. G., and K. L. Risenhoover. 1996. Using artificial food patches to evaluate habitat quality for granivorous birds: An application of foraging theory. Condor 98:854–857.

Korschgen, L. J. 1980. Procedures for food-habitat analyses. Pp. 113–127 *in* S. D. Schemnitz, ed. Wildlife management techniques manual, 4th edition. The Wildlife Society, Bethesda, MD.

Kotler, B. P., J. E. Gross, and W. A. Mitchell. 1994. Applying patch use to assess aspects of foraging behavior in Nubian ibex. Journal of Wildlife Management 58:299–307.

Krausman, P., and S. P. Mahoney. 2015. How the Boone and Crockett Club (B&C) shaped North American conservation. International Journal of Environmental Studies 72:746–755.

Kuhn, C. E., D. E. Crocker, Y. Tremblay, and D. P. Costa. 2009. Time to eat: Measurements of feeding behavior in a large marine predator, the northern elephant seal *Mirounga augustirostris*. Journal of Animal Ecology 78:513–523.

Kuvlesky, W. P., L. A. Brennan, J. A. Ortega-S., D. L. Williford, J. B. Hardin, H. L. Perotto-Baldivieso, L. C. Fritz, C. D. Hilton, F. C. Bryant, S. A. Nelle, B. M. Mitchell, and N. J. Silvy. 2020. The wild turkey in Texas: Ecology and management. Texas A&M University Press, College Station.

Lechowicz, M. J. 1982. The sampling characteristics of electivity indices. Oecologia 52:22–30.

Leger, D. W., and I. A. Didrichsons. 1994. An assessment of data pooling and some alternatives. Animal Behaviour 48:823–832.

Lehner, P. N. 1996. Handbook of ethological methods, 2nd edition. Cambridge University Press, Cambridge, MA.

Leonard, J. L. 1984. A top-down theoretical approach to the analysis of behavioral organization. Journal of Theoretical Biology 107:457–470.

Leopold, A. 1933. Game management. Charles Scribner's Sons, New York.

Leopold, A. S. 1944. The nature of heritable wildness in turkeys. Condor 46:133–197.

Levitis, D. A., W. Z. Lidicker, and G. Freund. 2009. Behavioral biologists don't agree on what constitutes behavior. Animal Behaviour 78:103–110.

L'Heureux, N., M. Lucherini, M. Festa-Bianchet, and J. T. Jorgenson. 1995. Density-dependent mother-yearling association in bighorn sheep. Animal Behaviour 49:901–910.

Litvaitis, J. A., K. Titus, and E. M. Anderson. 1994. Measuring vertebrate use of terrestrial habitats and foods. Pp. 254–274 *in* T. A. Bookhout, ed. Research and management techniques for wildlife and habitats, 5th edition. The Wildlife Society, Bethesda, MD.

Lomnicki, A. 1988. Population ecology of individuals. Princeton University Press, Princeton, NJ.

Lorenz, K. Z. 1937. The companion in the bird's world. Auk 54:245–273.

Losito, M. P., R. E. Mirarchi, and G. A. Baldassarre. 1989. New techniques for time-activity studies of avian flocks in view-restricted habitats. Journal of Field Ornithology 60:388–396.

Loughry, W. J. 1993. Determinants of time allocation by adult and yearling black-tailed prairie dogs. Behaviour 124:23–43.

MacArthur, R. H., and E. R. Pianka. 1966. On optimal use of a patchy environment. American Naturalist 100: 603–609.

Machlis, L., P. W. D. Dodd, and J. C. Fentress. 1985. The pooling fallacy: Problems arising when individuals contribute more than one observation to the data set. Zeitschrift für Tierpsychologie 68:201–214.

Mahoney, S. P., and J. J. Jackson III. 2015. Enshrining hunting as a foundation for conservation—the North American Model. International Journal of Environmental Studies 70:448–459.

Manly, B. F. J. 1993. Comments on design and analysis of multiple-choice feeding-preference experiments. Oecologia 93:149–152.

Manly, B. F. J., L. L. McDonald, D. L. Thomas, T. L. McDonald, and W. P. Erickson. 2002. Resource selection by animals: Statistical design and analysis for field studies, 2nd edition. Kluwer Academic, Boston.

Martin, P., and P. Bateson. 1993. Measuring behaviour, 2nd edition. Cambridge University Press, Cambridge.

Masataka, N., and B. Thierry. 1993. Vocal communication of Tonkean macaques in confined environments. Primates 34:169–180.

McNay, R. S., J. A. Morgan, and F. L. Bunnell. 1994. Characterizing independence of observations in movements of Columbian black-tailed deer. Journal of Wildlife Management 58:422–429.

Mech, L. D. 1983. Handbook of animal radio-tracking. University of Minnesota Press, Minneapolis.

Meserve, P. L. 1976. Food relationships of a rodent fauna in a California coastal sage community. Journal of Mammalogy 57:300–319.

Millspaugh, J. J., and J. M. Marzluff, eds. 2001. Radio tracking and animal populations. Academic Press, San Diego.

Mock, P. J. 1991. Daily allocation of time and energy of western bluebirds feeding nestlings. Condor 93:598–611.

Moen, A. N. 1973. Wildlife ecology. W. H. Freeman, San Francisco.

Mook, L. J., and H. W. Marshall. 1965. Digestion of spruce budworm larvae and pupae in the olive-backed thrush, *Hylocichla ustulata swainsoni* (Tschudi). Canadian Entomologist 97:1144–1149.

Morrison, M. L. 1982. The structure of western warbler assemblages: Ecomorphological analysis of the black-throated gray and hermit warblers. Auk 99:503–513.

Morrison, M. L. 1988. On sample sizes and reliable information. Condor 90:275–278.

Morrison, M. L., L. A. Brennan, B. G. Marcot, W. M. Block, and K. S. McKelvey. 2020. Foundations for advancing animal ecology. Johns Hopkins University Press, Baltimore.

Morrison, M. L., D. L. Dahlsten, S. M. Tait, R. C. Heald, K. A. Milne, and D. L. Rowney. 1989. Bird foraging on incense-cedar and incense-cedar scale during winter in California. Research Paper PSW-195. US Department of Agriculture, Forest Service, Pacific Southwest Forest and Range Experiment Station, Berkeley, CA.

Morrison, M. L., C. J. Ralph, J. Verner, and J. R. Jehl Jr., eds. 1990. Avian foraging: Theory, methodology, and applications. Studies in Avian Biology no. 13. Cooper Ornithological Society, Los Angeles.

Morse, D. H. 1980. Behavioral mechanisms in ecology. Harvard University Press, Cambridge, MA.

Mumma, M. A., C. Zieminski, T. K. Fuller, S. P. Mahoney, and L. P. Waits. 2015. Evaluating noninvasive genetic sampling techniques to estimate large carnivore abundance. Molecular Ecology Resources 15:1133–1144.

Murkin, H. R., R. M. Kaminski, and R. D. Titman. 1982. Responses by dabbling ducks and aquatic invertebrates to an experimentally manipulated cattail marsh. Canadian Journal of Zoology 60:2324–2332.

Nagy, K. A. 1987. Field metabolic rate and food requirement scaling in mammals and birds. Ecological Monographs 57:111–128.

Nagy, K. A., and B. S. Obst. 1989. Body size effects on field energy requirements of birds: What determines their field metabolic rates? International Ornithological Congress 20:793–799.

Newton, P. 1992. Feeding and ranging patterns of forest Hanuman langurs (*Presbytis entellus*). International Journal of Primatology 13:245–285.

Nudds, T. D. 1980. Foraging "preference": Theoretical considerations of diet selection by deer. Journal of Wildlife Management 44:735–740.

Nudds, T. D. 1982. Theoretical considerations of diet selection by deer: A reply. Journal of Wildlife Management 46:257–258.

Ormerod, S. J. 1985. The diet of dippers *Cinclus cinclus*

and their nestlings in the catchment of the River Wye, mid-Wales: A preliminary study of faecal analysis. Ibis 127:316–331.

Oyugi, J. O. and J. S. Brown. 2003. Giving-up densities and habitat preferences of European starlings and American robins. Condor 105:130–135.

Page, G., and D. F. Whitacre. 1975. Raptor predation of wintering shorebirds. Condor 77:73–83.

Paterson, J. D., P. Kubicek, and S. Tillekeratne. 1994. Computer data recording and DATAC6, a BASIC program for continuous and interval sampling studies. International Journal of Primatology 15:303–315.

Perlut, N. G., A. M. Strong, T. M. Donovan, and N. J. Buckley. 2006. Grassland birds in a dynamic management landscape: Behavioral responses and management strategies. Ecological Applications 16:2235–2247.

Perlut, N. G., A. M. Strong, T. M. Donovan, and N. J. Buckley. 2008. Grassland songbird survival and recruitment in agricultural landscapes: Implications for source-sink demography. Ecology 89:1941–1952.

Petrides, G. A. 1975. Principal foods versus preferred foods and their relation to stocking rate and range condition. Biological Conservation 7:161–169.

Pierce, G. J., and J. G. Ollason. 1987. Eight reasons why optimal foraging theory is a complete waste of time. Oikos 49:111–118.

Ploger, B. J. 2003. Learning to describe and quantify animal behavior. Pp. 11–30 *in* B. J. Ploger and K. Yasukawa, eds. Exploring animal behavior in laboratory and field. Academic Press, San Diego.

Ploger, B. J., and K. Yasukawa, eds. 2003. Exploring animal behavior in laboratory and field. Academic Press, San Diego.

Porter, W. P., and D. M. Gates. 1969. Thermodynamic equilibria of animals with environment. Ecological Monographs 39:227–244.

Poysa, H. 1991. Measuring time budgets with instantaneous sampling: A cautionary note. Animal Behaviour 42:317–318.

Pulliam, H. R. 1973. On the advantages of flocking. Journal of Theoretical Biology 38:419–422.

Quera, V. 1990. A generalized technique to estimate frequency and duration in time sampling. Behavioral Assessment 12:409–424.

Ralph, C. P., S. E. Nagata, and C. J. Ralph. 1985. Analysis of droppings to describe diets of small birds. Journal of Field Ornithology 56:165–174.

Raphael, M. G. 1988. A portable computer-compatible system for collecting bird count data. Journal of Field Ornithology 59:280–285.

Raphael, M. G. 1990. Use of Markov chains in analysis of foraging behavior. Studies in Avian Biology 13:288–294.

Ratti, J. T., L. D. Flake, and W. A. Wentz. 1982. Waterfowl ecology and management: Selected readings. The Wildlife Society, Bethesda, MD.

Remsen, J. V., and S. K. Robinson. 1990. A classification scheme for foraging behavior of birds in terrestrial habitats. Studies in Avian Biology 13:144–160.

Riley, C. M. 1986. Foraging behavior and sexual dimorphism in emerald toucanets (*Aulacorhynchus prasinus*) in Costa Rica. Master's thesis, University of Arkansas, Fayetteville.

Riney, T. 1982. Study and management of large mammals. John Wiley & Sons, New York.

Roa, R. 1992. Design and analysis of multiple-choice feeding-preference experiments. Oecologia 89:509–515.

Roberts, G., and P. R. Evans. 1993. Responses of foraging sanderlings to human approaches. Behaviour 126:29–43.

Roberts, W. A., and S. Mitchell. 1994. Can a pigeon simultaneously process temporal and numerical information? Journal of Experimental Psychology: Animal Behavior Processes 20:66–78.

Rosenberg, K. V., and R. J. Cooper. 1990. Approaches to avian diet analysis. Studies in Avian Biology 13:80–90.

Rosenthal, R. 1976. Experimenter effects in behavioral research. Irvington, New York.

Ruggiero, L. F., R. S. Holthausen, B. G. Marcot, K. B. Aubry, J. W. Thomas, and E. C. Meslow. 1988. Ecological dependency: The concept and its implications for research and management. Transactions of the North American Wildlife and Natural Resources Conference 53:115–126.

Rychlik, L., and E. Jancewicz. 2002. Prey size, prey nutrition, and food handling by shrews of different body sizes. Behavioral Ecology 13:216–223.

Sakai, H. F., and B. R. Noon. 1990. Variation in the foraging behaviors of two flycatchers: Associations with stage of breeding cycle. Studies in Avian Biology 13:237–244.

Salewski, V., F. Bairlein, and B. Leisler. 2003. Niche partitioning of two Palearctic passerine migrants with Afrotropical residents in their West African winter quarters. Behavioral Ecology 14:493–502.

Samuel, M. D., and M. R. Fuller. 1994. Wildlife radiotelemetry. Pp. 370–418 *in* T. A. Bookhout, ed. Research and management techniques for wildlife and habitats. 5th edition. The Wildlife Society, Bethesda, MD.

Schleidt, W. M., G. Yakalis, M. Donnelly, and J. McGarry. 1984. A proposal for a standard ethogram, exemplified by an ethogram of the bluebreasted quail (*Coturnix chinensis*). Zeitschrift für Tierpsychologie 64:193–220.

Schmitz, O. J. 1992. Optimal diet selection by white-tailed deer: Balancing reproduction with starvation risk. Evolutionary Ecology 6:125–141.

Schoener, T. W. 1969. Optimal size and specialization in

constant and fluctuating environments: An energy time approach. Brookhaven Symposia in Biology 22:103–114.

Seamans, M. E., and T. A. Sanders. 2014. Mourning dove population status, 2014. US Fish and Wildlife Service, Division of Migratory Bird Management, Population and Habitat Assessment Branch, Laurel, MD.

Severson, K. E., and M. May. 1967. Food preferences of antelope and domestic sheep in Wyoming's Red Desert. Journal of Range Management 20:21–25.

Shettleworth, S. J. 1993. Varieties of learning and memory in animals. Journal of Experimental Psychology: Animal Behavior Processes 19:5–14.

Siegel, S. 1956. Nonparametric statistics for the behavioral sciences. McGraw-Hill, New York.

Siegel, S., and N. J. Castellan. 1988. Nonparametric statistics for the behavioral sciences, 2nd edition. McGraw-Hill, New York.

Slater, P. J. B. 1978. Data collection. Pp. 7–24 *in* P. W. Colgan, ed. Quantitative ethology. John Wiley & Sons, New York.

Snedecor, G. W., and W. G. Cochran. 1980. Statistical methods, 7th edition. Iowa State University Press, Ames.

Sokal, R. R., and F. J. Rohlf. 1995. Biometry, 3rd edition. W. H. Freeman, San Francisco.

Sommer, B., and R. Sommer. 1991. A practical guide to behavioral research: Tools and techniques, 3rd edition. Oxford University Press, New York.

Speakman, J. R., and P. A. Racey. 1987. The energetics of pregnancy and lactation in the brown long-eared bat, *Plecotus auritus*. Pp. 367–395 *in* M. B. Fenton, P. A. Racey, and J. M. V. Rayner, eds. Recent advances in the study of bats. Cambridge University Press, Cambridge.

Stamps, J. A., and V. V. Krishnan. 1994. Territory acquisition in lizards: I. First encounters. Animal Behaviour 47:1375–1385.

Stephens, D. W. 1990. Foraging theory: Up, down, and sideways. Studies in Avian Biology 13:444–454.

Stephens, D. W., and J. R. Krebs. 1986. Foraging theory. Princeton University Press, Princeton, NJ.

Strauss, R. E. 1979. Reliability estimates for Ivlev's electivity index, the forage ratio, and a proposed linear index of food selection. Transactions of the American Fisheries Society 108:344–352.

Sutherland. W. J. 2007. Future directions in disturbance research. Ibis 149(Supplement 1):120–124.

Swanson, G. A., and J. C. Bartonek. 1970. Bias associated with food analysis in gizzards of blue-winged teal. Journal of Wildlife Management 34:739–746.

Swihart, R. K., and N. A. Slade. 1985. Testing for independence of observations in animal movements. Ecology 66:1176–1184.

Tatner, P. 1983. The diet of urban magpies, *Pica pica*. Ibis 125:90–107.

Tener, J. S. 1965. Muskoxen: A biological and taxonomic review. Canadian Wildlife Service Monograph Series 2. Canadian Wildlife Service, Department of Northern Affairs and National Resources, Natural and Historic Resources Branch, Ottawa.

Thill, R. A. 1985. Cattle and deer compatibility on southern forest range. Pp. 159–177 *in* F. H. Baker and R. K. Jones, eds. Proceedings of a conference on multispecies grazing. Winrock International Institute for Agricultural Development, Morrilton, AR.

Tinbergen, N. 1951. The study of instinct. Oxford University Press, Oxford.

Trainer, J. M., D. B. McDonald, and W. A. Learn. 2002. The development of coordinated singing in cooperatively displaying long-tailed manakins. Behavioral Ecology 13:65–69.

Vanderploeg, H. A., and D. Scavia. 1979a. Calculation and use of selectivity coefficients of feeding: Zooplankton grazing. Ecological Modelling 7:135–149.

Vanderploeg, H. A., and D. Scavia. 1979b. Two electivity indices for feeding with special reference to zooplankton grazing. Journal of Fisheries Research Board of Canada 36:362–365.

Van Dyne, G. M. and H. F. Heady. 1965. Botanical composition of sheep and cattle diets on a mature animal range. Hilgardia 36:465–492.

von Frisch, K. 1927. Aus dem Leben der Bienen. Springer-Verlag, Berlin. Translation (1953) by Dora Ilse as The dancing bees: An account of the life and senses of the honey bee. Harvest / HBJ [Harcourt Brace Jovanovich] Book, New York.

Waugh, D. R. 1979. The diet of sand martins in the breeding season. Bird Study 26:123–128.

Waugh, D. R., and C. J. Hails. 1983. Foraging ecology of a tropical aerial feeding bird guild. Ibis 125:200–217.

Weathers, W. W., W. A. Buttemer, A. M. Hayworth, and K. A. Nagy. 1984. An evaluation of time-budget estimates of daily energy expenditures in birds. Auk 101:459–472.

Webb, J. K., and R. Shine. 1993. Prey-size selection, gape limitation and predator vulnerability in Australian blindsnakes (Typhlopidae). Animal Behaviour 45:1117–1126.

Weckerly, F. W. 1994. Selective feeding by black-tailed deer: Forage quality or abundance? Journal of Mammalogy 75:905–913.

Weinberg, S. L., and K. P. Goldberg. 1990. Statistics for the behavioral sciences. Cambridge University Press, Cambridge.

White, G. C., and R. A. Garrott. 1990. Analysis of wildlife radio-tracking data. Academic Press, San Diego.

Wikelski, M., and F. Trillmich. 1994. Foraging strategies of the Galápagos marine iguana (*Amblyrhynchus cristatus*): Adapting behavioral rules to ontogenetic size change. Behaviour 128:255–279.

Williams, P. L. 1990. Use of radiotracking to study foraging in small terrestrial birds. Studies in Avian Biology 13:181–186.

Williams, R. L., A. E. Goodenough, and R. Stafford. 2012. Statistical precision of diet diversity from scat and pellet analysis. Ecological Informatics 7:30–34.

Willms, W., A. McLean, R. Tucker, and R. Ritchey. 1980. Deer and cattle diets on summer range in British Columbia. Journal of Range Management 33:55–59.

Wilmshurst, J. F., J. M. Fryxell, and R. J. Hudson. 1995. Forage quality and patch choice by wapiti (*Cervus elaphus*). Behavioural Ecology 6:209–217.

Wolda, H. 1990. Food availability for an insectivore and how to measure it. Studies in Avian Biology 13:38–43.

Yunger, J. A., P. L. Meserve, and J. R. Gutiérrez. 2002. Small-mammal foraging behavior: Mechanisms for coexistence and implication for population dynamics. Ecological Monographs 72:561–577.

Zar, J. H. 1999. Biostatistical analysis, 4th edition. Prentice Hall, Upper Saddle River, NJ.

5 Modeling Species-Environment Relationships

Introduction

Modeling the relationships between species and environmental conditions that determine faunal distributions is among the more fundamental tools—and challenges—in animal ecology. In this chapter, we explore a wide range of approaches in depicting and evaluating these relationships. We tier off of our basic tenets from Chapter 1 of Morrison et al. (2020), in which habitat is defined strictly as a species-specific concept (Kirk et al. 2018), and depict the vegetation and other physical attributes that influence the presence, relative abundance, and distribution of individual members of a given species. We address species-environment relationships as a step beyond species-habitat relationships by also considering the dynamic and functional dimensions of biotic and abiotic gradients, spatial and temporal changes, and species interactions, all of which affect the presence and distribution of organisms. Such relationships operate at the individual level, with emergent properties at the population and species levels. Millstein (2014) argued that the boundaries of a population define the environment within which selection operates—that is, where individual variation in resource use and survivorship occurs.

We first provide an overview of the categories of species-environment relationships models, which vary according to the outcomes (i.e., response variables) each is intended to provide, and then summarize them by the type of relationship each depicts. We organize these categories to help guide an appropriate selection of a modeling construct, given an assessment objective. We then discuss key considerations in setting modeling objectives; model parameterization, calibration, and validation; analyses of parameter sensitivity and influence; issues of model resolution; and the handling of uncertainties and unknowns in the various modeling constructs. Cautions and caveats are part of the picture. We end with a look to the near future in the evolution of modeling species-environment relationships and identify some needed research projects.

Classification of Species-Environment Relationships Models

Here, we suggest a broad definition of "model," which derives from the Latin *modus*, meaning "manner" or "measure." Model development occurs along a gradient, ranging from simplistic to sophisticated. They can be conceptual, diagrammatic, mathematical, or statistical, as well as having computer-based phases (Hall and Day 1977). In 1 approach, Marcot (2006) suggested using influence diagrams to help structure causal model networks. These diagrams are essentially mind maps, also called "cognitive maps." They can be as simple as a boxes-and-arrows depiction of which variables influence which other ones, or they can also depict degrees of uncertainty or variability (Dodouras and James 2007; Kang et al. 2012). Creating cognitive maps can help a modeler articulate and depict major causal web influences in an ecological system. A next step can be to embellish the cognitive map with logical links (e.g., fuzzy math relationships or conditional probabilities) or statistical relationships (e.g., regression equations). The model can then be made comprehensive—that is, applicable in a variety of structures, such as in a hierarchical Bayes approach, a probability network, a multivariate statistical model, or others constructs.

The arena of modeling species-environment (including species-habitat) relationships has greatly expanded in recent decades, with an increasing array of concepts, algorithms, and tools, as well as access to large datasets of species' locations and remotely sensed environmental information. Here, we provide an overview of this growing field and discuss examples of approaches and tools under 8 categories: modeling occurrence, habitat association, habitat selection, space use, population dynamics, ecological functions, multiple species, and energy flow. Some modeling tools are available as freeware, others are sold commercially, and still others are offered as sets of code that are run under various platforms, such as MATLAB, R, and WinBUGSs. Some have yet to be developed, and we address these in the closing section of the chapter.

First, though, we remind readers to begin any modeling exercise by clearly defining their objective(s) for undertaking such a venture. It is also important to recognize that the goal(s) could be appropriately addressed using a variety of approaches or tools. Remember that all models, to some degree, are inaccurate (Stouffer 2019), and that there is no single definitive "right way" to approach using them in species-environment relationships. The array of models and approaches we discuss here provide a suite of potential response variables that, in themselves, inform potential objectives.

Occurrence Models

Occurrence models include databases of species locations, which can range from aggregates of opportunistic sightings to statistically valid sampling frames from which analytic models can be developed. The nature of the location data largely constrains the types of models that can be applied. Here we place occurrence models into 2 broad groups. Location models represent a wide classification, based on location data alone. Opportunistically collected data are often utilized in these models. Occupancy models, on the other hand, require repeated surveys to estimate the failure to detect a species when it is present. Occupancy models extend beyond locations where an organism's presence is known (based on verified locations) to include areas where that organism is likely to occur but has not yet been detected.

Location Models

We start with perhaps the simplest form of information on species-environment relationships: the locations of known occurrences of individuals. These are often represented in databases compiled from layperson sightings, such as eBird (eBird.org) and Project FeederWatch (feederwatch.org) for birds and iNaturalist (inaturalist.org) for other taxa. Mapping occurrences can provide initial insight into a species' distribution and its general relationships with broad biogeographic conditions, but there are caveats (explored below) on the reliability of species identifications and the precision of locations. We refer to this process as location modeling, although it is more of a simple exercise in mapping point locations in geographic information systems (GIS).

Location models can lend themselves to more exacting evaluations, if they are done with rigor and take advantage of technology. For example, Hefley et al. (2015) compiled opportunistic sightings of whooping cranes (*Grus americana*) from 1988 to 2012. Coupled with expert-elicited information to determine sampling bias, they ascertained key migratory stopover sites for that species. Takeuchi et al. (2012) tested the use of cell phone GPS locations to track raccoon dogs (*Nyctereutes procyonoides*) in Tokyo, Japan, and found that the location reliability varied from 70 percent in a mosaic of vegetation cover to 98 percent in open areas.

Compilations of species locations have been used in multiple resource agencies for years, and many provide important historical data for analyses of changes in geographic ranges and distributions through time. Examples of these databases include compilations by the North Carolina Natural Heritage Program (www.ncnhp.org) and the US Geological Survey's Breeding Bird Survey (www.pwrc.usgs.gov/bbs/).

Non-statistical observations also can have great value, in the form of local ecological knowledge (Bélisle et al. 2018), particularly from long-resident indigenous peoples. For instance, Polfus et al. (2014) compared predictions of woodland caribou (*Rangifer tarandus caribou*) locations and habitat selection (as modeled with resource selection functions) with traditional ecological knowledge garnered from local residents of the Taku River Tlingit First Nation in northern British Columbia, Canada. That study reported comparable results between the quantitative and knowledge-based approaches, with the former predicting more high-quality habitat, and the latter adding valuable information for recovery planning for the species.

Occupancy Models

Occupancy modeling projects the potential distribution of a species, based on a statistical sampling of occurrences that are recorded as detections and non-detections. This process does more than just plot the locations of individuals, as it can account for false absences, the detectability of species, and the effects of environmental covariates on occupancy rates. A major reference source on occupancy modeling was provided by MacKenzie et al. (2018).

Occupancy models have been used with a wide array of taxa over many geographic regions, and statistical modeling, which serves as its basis, has greatly advanced in recent years. Much of occupancy modeling entails the use of PRESENCE and MARK programs, often run in the programming language R. PRESENCE generates estimates of the probability that a site is occupied, and it accounts for potential false absences by the use of repeat surveys to reduce or eliminate such biases. Results can be mapped as depictions of the probability of occurrence of a species and projected change rates in its occurrence, such as the distributional spread of invasive species. Extensions of the model also allow estimations of rates of local colonization and extinction.

The number of permutations in PRESENCE modeling has grown in recent years, accounting for such additional variables as seasonality, false positives (i.e., species' misidentification), breeding status, habitat dynamics, and more (MacKenzie 2006; MacKenzie et al. 2009). Halstead et al. (2018) adopted occupancy modeling to determine the occurrence of amphibians and reptiles by using time to detection as an index of detectability. Furnas and McGrann (2018) used it to track peak vocal activity of passerine birds in northern California. Silvano et al. (2017) employed occupancy analysis to select focal species as indicators of at-risk taxa for use in habitat restoration (Fig. 5.1). Many other examples of adapting this procedure are available in the literature. For example, Baumgardt et al. (2019) took occupancy modeling estimates of rare species, in conjunction with mark-recapture density estimates of common species in rangeland vegetation, and developed competing scenarios for monitoring small mammal populations. In general, because dynamic occupancy modeling accounts for a wide array of covariates, it often outperforms simpler, more static approaches, such as logistic regression, that typically account for fewer covariates, using a binary (i.e., present, absent) response variable.

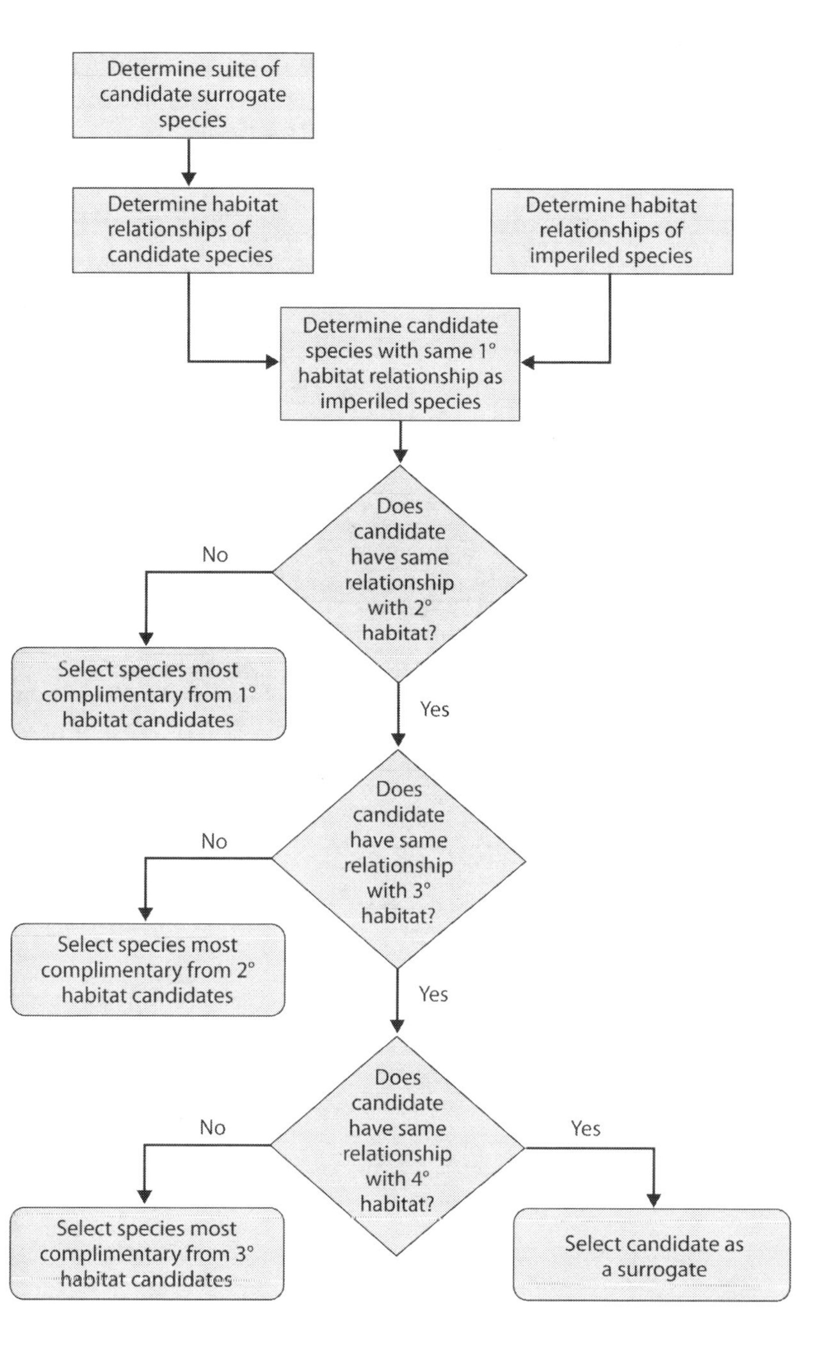

Figure 5.1. A flow diagram showing the procedure for selecting focal species as surrogates for imperiled species. Habitat relationships of candidate and imperiled species are determined using relative sensitivities, calculated from an occupancy analysis. (Reproduced from Silvano et al. (2017), used with the permission of Elsevier)

Habitat Association Models

Both locational and occupancy data can be used to infer habitat associations. For example, in occupancy modeling, detection likelihoods are generally linked to habitat covariates, to decrease detection-generated errors. Here, we concentrate on approaches to explicitly relate species to (i.e., define species in terms of) their habitats. In a broad sense, most of these categories are also referred to as "ecological niche models," although in the classic sense, a species' niche pertains to environmental factors including and beyond those of habitat (see Chapter 1 in Morrison et al. [2020]).

Wildlife-Habitat Relationships Models

The simplest form of habitat association model relates the presence of a species to the occurrence of a specific land cover, or to particular vegetation conditions, as a binary use/nonuse function. The model 1 step above this relates species to environmental conditions on a simple ordinal scale, which formed the basis for the US Forest Service's wildlife-habitat relationships (WHR) program in the 1970s and 1980s (Salwasser et al. 1980). In this program, use levels typically were shown in species-habitat matrices as ordinal categories of no, low, medium, and high use (Fig. 5.2)—or, alternatively, as no, marginal, suitable, and optimum habitat value—albeit often with confusing or scant definitions relating use levels to occurrence frequency, resource use, and population size or trend (Morrison 2001). Much of the information going into WHR species matrices was based on expert knowledge, which may perform well in some instances (Johnson and Gillingham 2004). In our context, however, the combination of ordinal categories, coupled with a lack of transparent relationships between these scores and extant occurrence data, means that most WHR models represent ordinal-scale hybrids that combine occurrence records with opinions. Additionally, only a few studies served to empirically validate and update the baseline species-habitat ratings information (e.g., Raphael and Marcot 1986).

Subsequent programs developed Boolean-based query systems, founded on set theory, from WHR species-habitat matrices (Patton 1992). An example is querying a WHR database to determine which species occur (or do not occur) in a particular vegetation condition, and those that use (or do not use) a specific type of resource for a designated life-history need (e.g., breeding, feeding, resting). Later, Stuber et al. (2017) extended the WHR paradigm by developing a Bayesian method to consider spatial scale in modeling species-habitat relationships. Their case study involved ring-necked pheasants (*Phasianus colchicus*) in Nebraska. Lehmkuhl et al. (2001) used

Ecotype	Whimbrel	Bristle-thighed Curlew	Hudsonian Godwit	Bar-tailed Godwit	Ruddy Turnstone	Black Turnstone	Surfbird	Red Knot	Sanderling	Semipalmated Sandpiper	Western Sandpiper
Alpine Acidic Barrens	0	0	0	0	0	0	3	3	0	0	0
Alpine Acidic Dryas Dwarf Shrub	0	0	0	1	2	0	3	3	0	0	1
Alpine Alkaline Barrens	0	0	0	0	0	0	3	3	0	0	0
Alpine Alkaline Dryas Dwarf Shrub	0	0	0	1	2	0	3	3	0	0	1
Alpine Ericaceous Dwarf Shrub	0	0	0	1	2	0	3	3	0	0	1
Alpine Lake	0	0	0	0	0	0	1	0	0	0	0
Alpine Mafic Barrens	0	0	0	0	0	0	3	3	0	0	0
Alpine Snowfields and Glaciers	0	0	0	0	0	0	0	0	0	0	0

Figure 5.2. A portion of a wildlife-habitat relationships matrix of birds in northwestern Alaska. Numbers denote the ordinal use levels, by individual species, of each land-cover type: 0 = no use, 1 = low use, 2 = medium use, 3 = high use. The full matrix included 162 bird and 39 mammal species in 60 vegetation and other land-cover types ("ecotypes"), and it was used to gauge a difference in habitat use under scenarios of climate change influences on ecotype extents. (Adapted from Marcot et al. (2015), courtesy of SpringerLink)

WHR species-habitat matrix information to evaluate species at risk by using classification and regression trees (discussed below).

Habitat Suitability Index Models

In the early 1980s, the US Fish and Wildlife Service introduced a suite of habitat suitability index (HSI) models, in an attempt to quantify the degree to which species use particular environmental conditions (Schamberger et al. 1982). HSI models typically contained 3 habitat or resource elements, each scaled from 0 to 1 and combined as a harmonic mean, so that the overall HSI would also be scaled from 0 to 1. Examples include HSI models by Schroeder (1982) for yellow warblers (*Setophaga petechia*), by Boyle and Fendley (1987) for bobcats (*Lynx rufus*), and by Rogers and Allen (1987) for black bears (*Ursus americanus*). HSI models have been used in the US Fish and Wildlife Service's Habitat Evaluation Procedures (HEP) to determine the potential impacts of construction and other human activities on fish and wildlife species and their habitats (Wakeley 1988). While HSI scoring was more transparent than WHR, again, there was no direct relationship between the scores and any formal quantitative metric to judge the quality of the index. Importantly, there was no requirement that occurrence data conform to HSI scoring.

Species Distribution Models

A major (and popular) advance in modeling species-habitat relationships came in the 2000s, with a flurry of approaches to species distribution models, or SDMs (Elith and Leathwick 2009). SDMs deviated from previous models with this focus by generally using continuous variables, rather than binary or ordinal relationships. Moreover, because they were directly and statistically based on occurrence data, SDMs were initially developed to map a species' potential presence or probability of occurrence across the landscape. Further applications included modeling the dynamics of range shifts in species (Elith et al. 2010), invasive species' distributions (Robinson et al. 2010), avian reproductive parameters (Brambilla and Ficetola 2012), and many other uses. An SDM modeling package, BIOCLIM, was introduced in Australia in 1984 (Booth 2019). Other programs, such as Genetic Algorithm for Rule Set Production, or GARP (Stockwell and Peters 1999) are also used for SDM modeling. Wintle et al. (2005) addressed the issue of detectability and false absences in occupancy surveys, noting major differences in the effort this can take in noting the presence of diverse owl species in Australia.

Another advance in SDM modeling came with the introduction of a maximum entropy program, or Maxent (Phillips et al. 2004, 2006). It was derived from information theory and used a maximum entropy algorithm (Dudík et al. 2004) to model the geographical distributions of species (Phillips et al. 2004, 2006). Maximum entropy modeling employs presence-only data, along with randomly selected data on background conditions. Several studies have compared the performance of Maxent to GARP and other SDM models, and they generally found a greater concordance of known species distributions with Maxent predictions (Phillips et al. 2006; Ortega-Huerta and Peterson 2008; Costa et al. 2010), although results have been known to vary, depending on the modeling objectives and the role of model bias and uncertainty (Peterson et al. 2007). Fern and Morrison (2017) combined the use of remote sensing data and several methods of species' distribution modeling, including Maxent, to map critical areas for migratory passerine birds in Texas. They concluded that elevation, far more than vegetation, was the relevant factor in identifying such areas. Whereas Maxent's algorithm does generally perform better than its competitors, it is still highly dependent on the occurrence data used to build the model, as well as on the size of the study area relative to the distribution of the species of interest. In many (perhaps most) cases, Maxent has been employed to analyze habitat covariates associated with opportunistically collected occurrence data, which have varying degrees of bias in terms of both when and where information was collected. Further, these data often contain large errors in species identification and the spatial precision of locations—and these mistakes are seldom spatially or temporally uniform. The degree to which such biases and errors affect Maxent depends on many factors. Aubry et al. (2017) found that large spatial biases associated with the collection of high-quality records on the occurrence of fishers (*Pekania* [*Martes*] *pennanti*) were less problematic to habitat modeling than were identification errors associated with much more widely collected but largely anecdotal data. In contrast, Frey et al. (2013), studying white-nosed coatis (*Nasua narica*), found that data quality had little impact on the resulting SDMs. Presumably these differences were associated with the relative rates of misidentification for fishers (high), compared with those for coatis (low, regardless of the source). Given such uncertainties, seeking concordance between SDMs and independent analyses is advisable. At the least, caution should be used when drawing inferences from SDMs when little is known about the species-environment relationships. Williford et al. (2016) used Maxent modeling as an independent line of evidence, comparing it with molecular genetics data to evaluate the phylogeography of the genus *Colinus*. They found that the predicted distributions of the 3 species in this genus of New World quail during the current, past glacial, and interglacial periods were congruent with the phylogeographic patterns suggested by the molecular genetics data. Such hybrid approaches to modeling species distributions, based on independent lines of evidence, clearly have promise for future advances on this front.

Habitat Selection Models

Beyond models of habitat association, some are specifically designed to depict habitat selection. In a sense, the correct definition of "habitat" as a species-specific concept implies—and includes—selection and adaptive advantage (Partridge 1978). Nevertheless, in the literature, the term "habitat selection" is frequently employed in this modeling class to identify environmental and resource conditions chosen by organisms in preference to those assumed to be less used, unused, or avoided. A vast array of statistical techniques (discussed below) have been employed to quantify these relationships, but in all cases—and central to the concept of "selection"—the quantification of the patterns of occurrence, compared with a pattern of expectation, assumes that no selection has occurred. In many cases, the expectation is of random use or random availability across the entire area of analysis. This explicitly assumes that all locations in the defined area are equally available to the organism at all times, and, hence, that all deviations from random use patterns are due to an organism's "selection." As such, choosing the areal extent of the analysis site is of critical importance, as the results of subsequent analyses are dependent on and will change with variations in the dimensions of the chosen locale.

Binary Selection and Classification Models

A general group of statistical modeling approaches is based on presence-absence data, as with general linear models (GLMs). GLM approaches include logistic regression, which is used to help determine the environmental conditions that primarily account for the binary categories of a species' presence or absence. Logistic regression has been used in the ecological literature at least since the 1980s (Brennan et al. 1986, 1987; Hassler et al. 1986), and it has been employed to analyze a wide variety of species-environment relationships, including by Mladenoff et al. (1999) for gray wolves (*Canis lupus*) (Fig. 5.3), by Fecske et al. (2002) for American martens (*Martes americana*), by Browne and Paszkowski (2018) for western toads (*Anaxyrus boreas*), and many more.

Some selection models use classifications of conditions that offer the best distinctions among empirical cases of occupied versus random locations, which is what presence-only models also do. Such approaches include classification trees and regression trees. Classification trees are used to identify which covariates and their values are more successful at

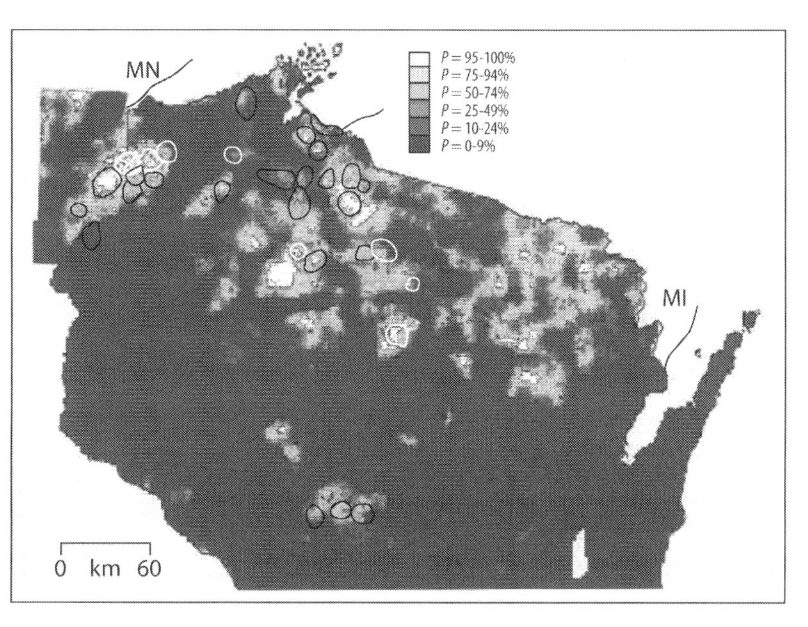

Figure 5.3. A map of gray wolf (*Canis lupus*) habitat in Wisconsin, using a logistic regression model based on road density. *P*-values refer to the probability of occupancy, with *P* >50 percent denoting habitat use. *White boundaries* are wolf pack territories used to derive the model, and *black boundaries* are other wolf pack territories utilized to test the model. (Reproduced from Mladenoff et al. (1999), used with the permission of John Wiley & Sons)

showing contrasts among the cases, whereas regression trees also add information on the values of coefficients for the differentiating covariates. Classification and regression trees are sometimes referred to as "decision trees," in the sense that one can "walk the tree" and determine the category of a new, unknown case by following particular branches, according to the covariates and their values. Other, similar classification algorithms have been around for some time. A popular example is Quinlan's (1986) ID3 classification algorithm and its later variations and improvements (e.g., Utgoff 1989; Quinlan 1993).

Another classification approach that has gained much favor recently is random forests (Cutler et al. 2007). This approach uses more of a machine-learning algorithm to create suites of decision (i.e., classification) trees on randomized subsamples of a dataset, with the results then averaged to a final classification tree (Elith 2019), which helps reduce overfitting and increases the predictive capacity of the final model. Thus random forests modeling improve previous classification and regression tree algorithms and has become popular, thanks to recent increases in computing power. Random forests modeling has been used in a wide array of conservation issues, including determining the influence of productivity, energy, and habitat heterogeneity on breeding bird richness in British Columbia, Canada (Fitterer et al. 2013); the value of topographic parameters in predicting bird species' density over a large geographic scale (Kosicki 2017); and many others (e.g., Elith 2019).

Various tests have demonstrated the value of random forests over other approaches. In a study assessing potential extreme range changes in 220 Australian mammal species, Beaumont et al. (2016) found that random forests models and surface-range envelope models (discussed in the next section) were better predictors of current habitat loss and future range extinction than GLMs and generalized additive models (also discussed below). Random forests models also presaged extreme range shifts more frequently than did the other modeling approaches. The authors noted that the use of random forests and other models had both pros and cons, depending on the modeling objectives and the varying projections of extreme range conditions and trends.

Eskelson et al. (2012) compared the use of random forests to statistical parametric modeling using logistic regression, to estimate snag density by snag decay class. They discovered that random forests more accurately predicted snag densities, whereas a statistical approach provided more-precise estimates (smaller root mean square errors). Lawler et al. (2006) found that random forests models out-predicted other approaches regarding climate-induced range shifts for 100 randomly selected mammal species in the Western Hemisphere. In general, the choice of model can greatly affect the resultant predictions of parameters, such as range change and extinction rates. Therefore, model averaging may be best for forecasting the effects of climate change on species' ranges (Lawler et al. 2006).

Generalized Additive Models

Generalized additive models (GAMs) are essentially linear regression models incorporating non-linear relationships between predictor and response variables. Such non-linear relationships can be expressed through polynomial, spline, or other smoothing functions. An advantage of GAMs over their strictly linear counterparts is that the former can determine more-complex relationships, particularly if a variable or a combination of linear variables interact in non-linear ways, such as with modal, exponential, or other trends. A disadvantage of GAMs lies in clearly interpreting the results, because some GAMs can be complicated, with their associated predictor variables having hazy causal interactions both among themselves and with the response variable. Nevertheless, GAMs are a popular modeling construct. Davey et al. (2012) used GAMs to evaluate how local climate related to bird community diversity and specialization in Britain. Other uses included determining predictive habitat models of freshwater mollusks in northern California (Dunk et al. 2004). The authors chose GAMs instead of GLMs, because almost nothing was known about the mollusks' ecology, so many linear and non-linear functional forms were possible. It is important to at least consider both kinds of functional forms when faced with uncertainty about a species' environmental relationships, while not forgetting that linear forms are more parsimonious, and thus preferable, if they generally describe the same degree of observed variation as non-linear forms would.

Several researchers have compared the performance of GAMs with other approaches, including neural networks, which are models using hidden layers of parameters and various machine-learning algorithms that link predictor and response variables. Dormann et al. (2008) compared the quality and validation outcomes of GLMs, GAMs, and neural network models in predicting climate effects on the distribution of great grey shrikes (*Lanius excubitor*) in Germany. For their particular modeling objective, they found that GLMs and GAMs gave comparable results and were superior to neural networks. García-Callejas and Araújo (2016) evaluated various approaches to predicting species distributions under climate change scenarios, using a surface-range envelope (BIOCLIM), regression (GLMs, GAMs, and multivariate adaptive regression splines, or MARS), and machine-learning algorithms (Maxent, boosted regression trees, and support vector machines, or SVMs). They found the performances to be similar, despite differences in the complexity of the models' computations. The worst models were GLMs and SVMs, the next worst was BIOCLIM, and the rest performed equally well, as measured by area under the curve (AUC) values ≥ 0.6.

Resource Selection Functions

Resource selection functions (RSFs) are a class of habitat selection models that provide estimates of the probability that a given species will use a resource unit (Boyce et al. 2002). As such, RSFs are sometimes viewed as population-scale models. They commonly use GLMs, although other statistical constructs can also be employed. RSFs are crafted from statistical analyses of empirical data on vegetation, land cover, and an organism's selection of these and other conditions as habitat. For instance, Bauder et al. (2018) developed multiscale RSFs for eastern indigo snakes (*Drymarchon couperi*) in Florida, based on various habitat attributes, including undeveloped land cover and land-cover edges, and the negative effects of urban land cover. Johnson et al. (2013) developed RSFs for northern fur seals (*Callorhinus ursinus*) in Alaska, using a point-process modeling approach with telemetry location data. RSFs can also take additional environmental and resource attributes into account, including other aspects of a species' niche, such as the effects of competition with or predation by other species and, particularly, the effects of anthropogenic disturbances. For instance, Scrafford et al. (2017) modeled RSFs of wolverines (*Gulo gulo luscus*) in Alberta, Canada, based on radio-telemetry tracking that showed the effects of logging, roads, seismic lines, and other anthropogenic disturbances. Their results suggested that the wolverines were attracted to active logging areas and intermediate-age cutblocks for movement and foraging opportunities.

RSFs also can be merged with population dynamics modeling. For example, Heinrichs et al. (2017) used RSFs of greater sage-grouse (*Centrocercus urophasianus*) in Wyoming to map seasonal habitat selection, and they incorporated simulations of animal movements to determine population dynamics. Michelot et al. (2019) linked models of individual site selection with population-level RSF models (Fig. 5.4). This is yet another example of merging 2 different modeling approaches into a hybridized format that generated improved inferences of potential linkages between populations and habitat (a goal we set out in Chapter 1 of Morrison et al. [2020]).

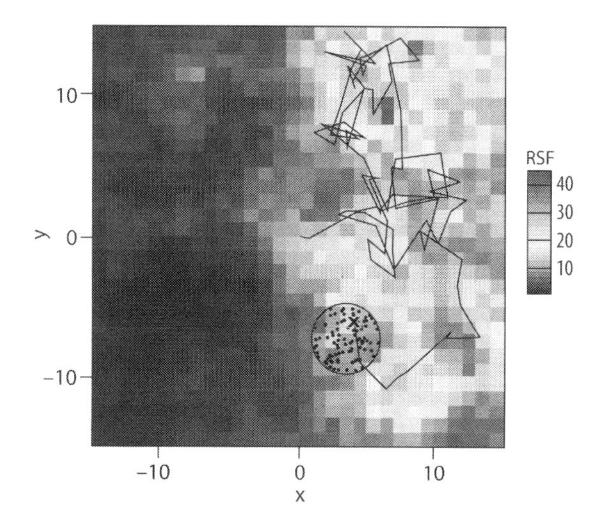

Figure 5.4. An example of combining a map of a resource selection function (RSF) with the simulated movement dynamics (*solid line*) of a hypothetical organism. RSF values ranged from low (10) to high (40). Movements were modeled using a Gibbs sampler algorithm. The *black X* denotes the organism's current location, where it will select its next locale from the *dots within the black circle*, with probabilities related to their RSF values. (Reproduced from Michelot et al. (2019), used with the permission of the Ecological Society of America)

Central-Use Models

In general, central-use models identify organisms' concentrations in locations and frequencies of occurrence, such as sites that are used more heavily by individuals for feeding, resting, and reproduction. The term "core habitat"—usually defined as or assumed to be a geographic area of contiguous resources and environmental conditions supporting particular life-history needs—is sometimes used for such central-use sites (e.g., Johnson and Semlitsch 2003). Publications by various researchers identified core habitat and used it to suggest conservation measures for recovering or sustaining at-risk species, such as Apps and McLellan (2006), for endangered mountain caribou, an ecotype of woodland caribou; Crawford et al. (2016), for mole salamanders (*Ambystoma jeffersonianum*); Dilts et al. (2016), for threatened Mohave ground squirrels (*Xerospermophilus* [*Spermophilus*] *mohavensis*); Kelt et al. (2014), for endangered riparian brush rabbits (*Sylvilagus bachmani riparius*), and Buechley et al. (2018), for vultures in the Middle East and East Africa (Fig. 5.5). Dunk et al. (2019) modeled habitat relationships for northern spotted owls (*Strix occidentalis caurina*) at the core-habitat scale.

Core habitat and central-use locations are identified by utilizing various SDMs, as well as by concentrations of radio-telemetry locations, typically viewed as high-use areas within broader home ranges (Bingham and Noon 1998). As home-range depictions have advanced from delineations of minimum

(a)

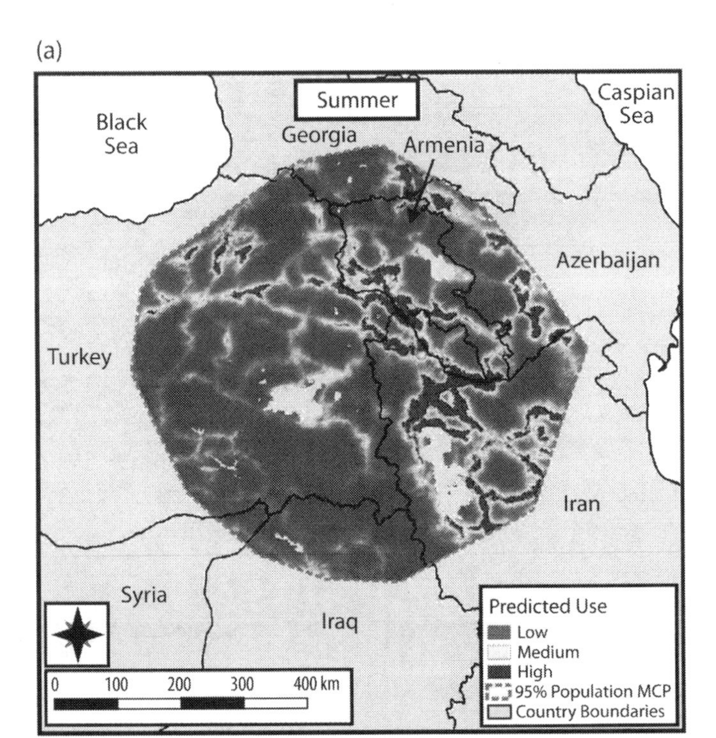

(b)

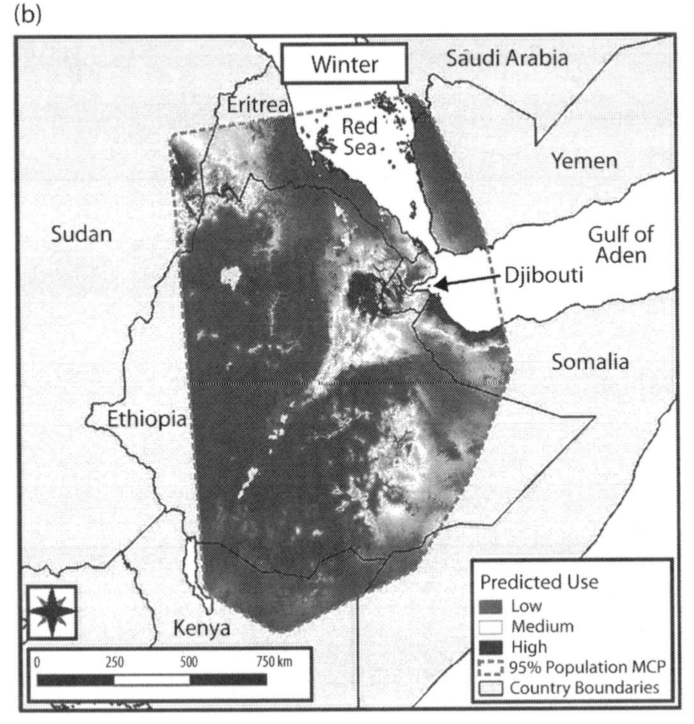

Figure 5.5. Maps of predicted habitat use by Egyptian vultures (*Neophron percnopterus*) in East Africa and the Middle East, during summer (*A*) and winter (*B*), within 95 percent minimum convex polygon (MCP) areas of vulture locations, derived from a dynamic Brownian-bridge movement model. Use levels were deduced from a resource selection model, based on 13 landscape features of human habitation, elevation, vegetation cover, and precipitation. The summer use model was largely influenced by linear anthropogenic features (highways and power lines). (Reproduced from Buechley et al. (2018), used with the permission of Springer)

convex polygons, or MCPs (the area and home-range polygonal shape circumscribed by linearly linking the outermost telemetry locations) to such calculations as harmonic mean and fractal home-range estimators (Samuel and Garton 1987; Loehle 1990), researchers increased their focus on and interest in mapping central-place and core habitats on the individual organism scale. Kenward et al. (2018) created a resource-area-dependence analysis method to examine radio-tracked locations of buzzards (*Buteo buteo*) and 2 species of squirrels (*Sciurus vulgaris* and *S. carolinensis*) in southern England, and to estimate the minimum amount of resources that were required to support individual animals.

Yet the use of radio-telemetry locations to infer resource selection, core habitat, and central use areas can be fraught with various biases, such as inferring selection from use, which can be influenced by the type and degree of local availability of resource conditions. For a central-place forager, such as a bird that must return with resources to its nest, use patterns are constrained by the nest's location, and foraging-site choices are specifically designed to minimize round-trip times. In this case, the expected resource use is not uniform across space, but, rather, can decline with distance from the nest. (See Rosenberg and McKelvey 1999 for an example involving northern spotted owls.) The potential for incorrectly inferring resource selection from utilization led Gautestad and Mysterud (2010) to devise a fractal-based, intra–home range index of habitat suitability that quantified the spatial dispersion of an individual's use of its most-constrained resources, to identify patch preferences.

Climate Envelope Models

The conditions under which a population or species occurs are used to determine an "envelope" of its range of climatic conditions. Climate envelope models, also referred to as "climatic niche models," are generally correlative, derived from some SDM to delineate the geographic extent of a species' distribution and the climatic conditions within that area. These typically include some measurements of temperature and precipitation—which can be determined from field measuring stations—such as means, variations, and extremes of conditions over monthly or annual timeframes. Climate envelope conditions are then compared with those projected under climate-change scenarios, or with conditions in unoccupied locations, to determine the degree to which the climate in that environment may be suitable or unsuitable for a species (Almpanidou et al. 2016).

The Australian Bureau of Rural Sciences developed an online tool, Climatch (Crombie et al. 2008), that can be used to compare climatic conditions in source areas and target areas. Thus a species' climate envelope can be analyzed to determine the degree of match, using Euclidean and closest standard score metrics, based on 8 precipitation and 8 temperature variables (Fig. 5.6). This tool has been used, for example, to determine the climatic conditions favorable for the establishment of exotic vertebrates in Australia and New Zealand (Bomford 2008), and the potential invasiveness of aquarium mollusks (Patoka et al. 2017). Climate envelope modeling can also be used to project shifts in vegetation distribution (Kent et al. 2018), including the response of trees to changing climate conditions (Morgan et al. 2018), which, in turn, may be useful for relating these predictions to wildlife species' habitats.

Climate envelope models are generally useful for comparing current climatic conditions between locations, but, since these are based on spatial and not temporal correlation, they may not perform well for projected changes in conditions and the effects of interspecific competition (Clark et al. 2011). Araújo and Peterson (2012) cautioned that climate envelope modeling should be used only for appropriate queries about climate compatibility, not for analyses of more-complex questions of ecology and evolution. In a test involving models and prediction maps for 15 threatened or endangered species in Florida, Brandt et al. (2017) concluded that statistically based climate envelope modeling is a good first step in determining potential species' responses to climatic conditions

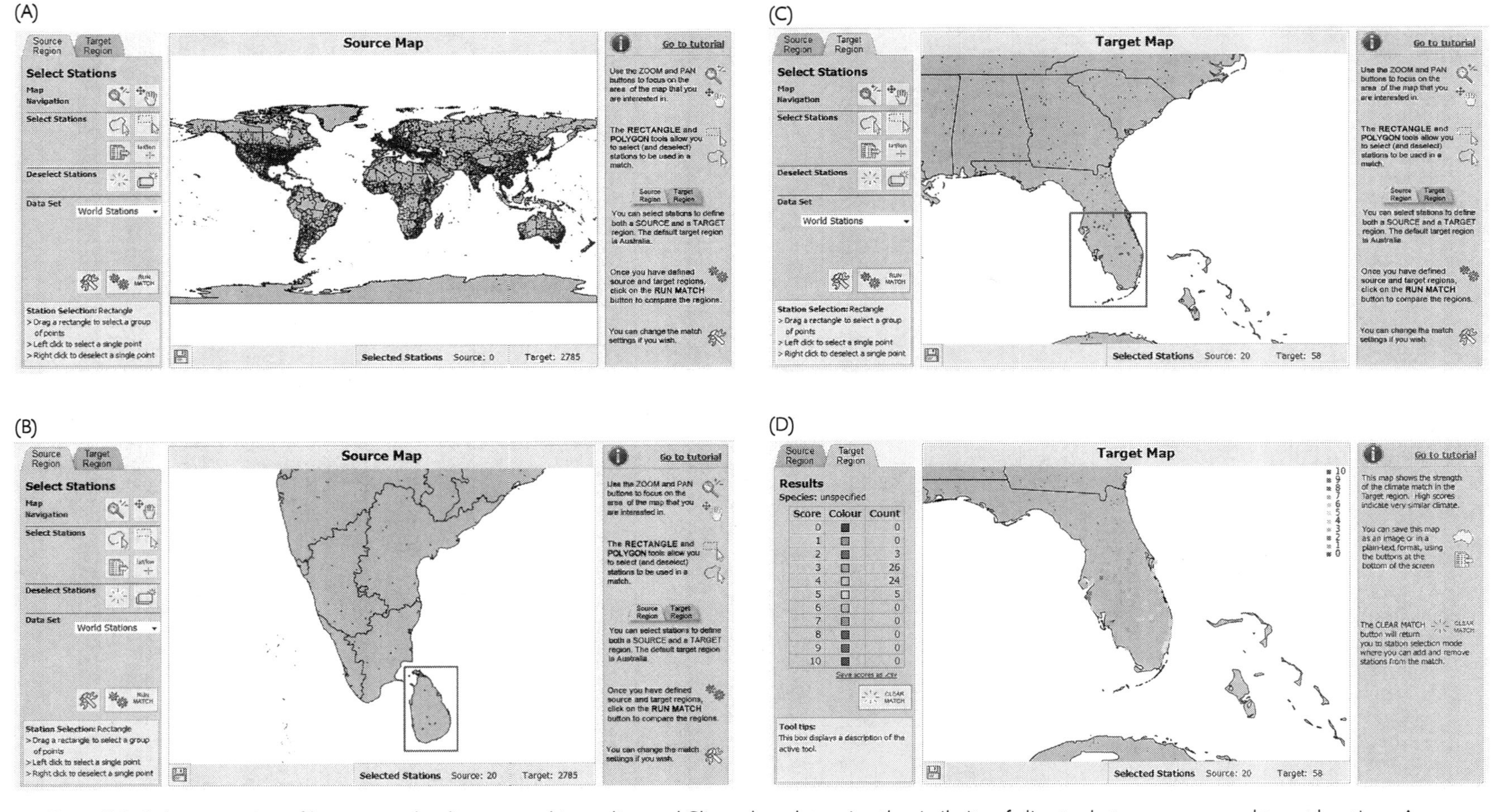

Figure 5.6. A demonstration of how to use the climate-matching online tool Climatch to determine the similarity of climates between source and target locations. An overall source map (*A*) shows meteorological data locations worldwide, with an example selecting Sri Lanka (*B*) as the hypothetical source of a species to be introduced into southern Florida (*C*). The results are shown as the number of locations by climate match score (*D*). This example indicates that there is only a low-to-moderate match between the source and target locales. (Australian Government, Bureau of Rural Sciences, https://climatch.cp1.agriculture.gov.au/climatch.jsp)

and changes, but that expert input would then be useful for identifying further key variables and developing multiple models, to reduce uncertainty and improve predictability (also see Yackulic et al. 2011).

Space-Use Models

Another general class of species-environment relationships modeling is space-use models, which, in various ways, explicitly relate area and the distribution of environmental conditions to occurrence, movement, probability of use, persistence, and other parameters of individuals and populations. An important difference between these models and the habitat selection ones (presented earlier in this chapter) is that the former are frequently used to evaluate communities of plants or animals, whereas the latter have almost exclusively been applied to populations of particular species.

Species-Area Relationships Modeling

Models of species-area relationships have been used for a long time in ecology, and they take several forms. For instance, species-area curves denote the level of species richness (i.e., the number of species) that may be found in various geographic extents of a particular location (e.g., an oceanic island) or environmental condition (e.g., high-elevation alpine). Species-area curves have often been employed in studies of biogeography—for instance, to determine the number of species on islands or in vegetation patches, or various sizes of land-cover conditions (Cowlishaw 1999). Species-area relationships models have been used to evaluate species richness at various scales of site (α), local (β), and regional (γ) diversity. Incidence curves display the probability or frequency of the presence of individual species as a function of habitat area.

Burns (2015) extended the species-area relationships modeling approach to evaluating the occurrence of exotic species on islands off New Zealand. Freeman et al. (2018) employed it to project the effects from changes in matrix lands among coastal forest fragments in South America and Mozambique on native bird species. Similarly, Koh and Ghazoul (2010) used species-area models to gauge the loss of tropical biodiversity resulting changes in landscape matrix lands. Connor et al. (2000) expanded the species-area relationships approach and evaluated the effects of patch or island area on population density per se, examining how densities of individual species are related to area for 287 individual species and 21 faunal groups. The results were mostly positive for insects and birds, but marginal for mammals. Their work suggests a greater conservation value for a few, persistent, large habitat patches over a regional network of mixed small and large patches.

Connectivity Modeling

Connectivity modeling refers to a suite of approaches and programs that evaluate the influence of habitat distribution, isolation, fragmentation, and links, mostly at a landscape scale, on population persistence. In Chapter 1, we stated that the concept of landscape, like habitat, is species specific. Thus habitat connectivity models and analyses must apply to a particular species, because, by the classic definition of the concept of ecological niche, each species uses and selects resources and environmental conditions that, in turn, define its unique niche, habitat, and landscape.

Connectivity modeling can be divided into 2 categories: the spatial pattern of environmental conditions, and the response of organisms to spatial patterns. The former typically focuses on the distribution of vegetation and land cover conditions, using a variety of mathematical measures and indices in GIS that are sensitive to patch size, shape, and the degree of contiguity. Such measures and indices are generally employed to infer connectivity of environments, resources, and habitat for specific species, but they are also used simply as general depictions of geographic patterns of environmental conditions.

A popular program, FRAGSTATS, introduced by McGarigal and Marks (1995), produces a wide array of landscape spatial pattern indices. Among many other examples, it has been used to analyze coastline shapes on the Iberian Peninsula (Chefaoui 2014), and for collaborative forest scenario planning in Michigan (Price et al. 2016). Cushman et al. (2008) reviewed the diversity of landscape structure metrics and identified groupings of metrics with similar attributes. MacLean and Congalton (2015) compared landscape metrics generated by 5 free programs in GIS. They suggested using FRAGSTATS with raster datasets and their PolyFrag program (MacLean and Congalton 2013) with vector datasets. More-recent landscape pattern analysis frameworks include Guidos (Vogt and Riiters 2017), Landscapemetrics (Hesselbarth et al. 2020), and others (Costanza et al. 2019).

Fractal geometry is a type of pattern index. It became popular as a landscape measure (Milne 1988) after the concept was created and introduced by Mandelbrot (1977). A fractal is a measure of partial dimension, such as how the irregularities of a coastline can be depicted somewhere between a 1-dimensional line and a 2-dimensional surface (Pennycuick and Kline 1986; Feng and Liu 2015). Fractal measures have been used, for example, to describe the shape of home ranges (Loehle 1990), river morphology (Stølum 1996), and naked mole-rat (*Heterocephalus glaber*) burrows (Le Comber et al. 2002).

The second category of connectivity modeling directly relates environmental spatial patterns to species-specific utilization and selection. In terms of these responses by organisms, connectivity modeling is generally used to evaluate habitat fragmentation relationships, habitat edge effects, and other aspects of their use and selection of habitat distribution patterns. Some of the landscape indices discussed above have been applied in this way, such as de Beer and van Aarde's (2008) application of FRAGSTATS to a post hoc analysis of habitat use and selection by African elephants (*Loxodonta africana*) in southern Africa woodlands.

The dominant paradigm for movement modeling is landscape resistance, generally applied to raster surfaces. The basic concept is that organisms will travel along paths that are, in some way, easier. What "easier" means is seldom specified, but it might refer to some combination of a track or trail being physically easier to traverse and safer, as well as having more food available, than alternative routes—all of which can be related to an individual's fitness. By far the most common assumption in the literature is that resistance is inversely related to habitat quality—that is, organisms will preferentially choose to travel through high-quality habitat (portions of the landscape with low resistance) and avoid paths that contain poor-quality habitat or completely lack habitat (high resistance). Additionally, these models may suggest that at least some terrestrial organisms will avoid crossing roads (for safety reasons) and urban landscapes. Connectivity tools seek to find the best (i.e., easiest) routes between 2 locations. Virtually all of these approaches entail point-to-point analyses, with specified start and stop locations. In all cases, the tacit assumption is that an animal leaving 1 location has, as a primary goal, getting to a destination.

There are many tools available for such analyses, such as Circuitscape (McRae and Shah 2009), which has been commonly used and serves as a good example of this form of modeling. Circuitscape recasts the landscape as a grid of resistors, connected in a circuit. Each pixel is given a specific resistance value, and certain pixels are additionally designated as terminal locations (i.e., anodes and cathodes). Electrons then flow through the landscape from the anodes to the cathodes, and the paths that they take are connectivity routes. Thus, in Circuitscape, animal movement is modeled as electron movement, going through a landscape modeled as a resistance field. Circuit theory has been used for many wildlife species, including, among others, endangered Mt. Graham ground squirrels (*Tamiasciurus hudsonicus grahamensis*) in Arizona (Merrick and Koprowski 2017) and ovenbirds (*Seiurus aurocapilla*) in Canada (St.-Louis et al. 2014).

Circuit theory itself is part of a broader set of network connectivity models. Other means of analyzing networks include least-cost paths (Pullinger and Johnson 2010), the use of graph theory (Marcot and

Chinn 1982; Cantwell and Forman 1993), and the integration of landscape genetics. Least-cost path analysis assumes a global knowledge of the landscape, so the route that is followed minimizes overall costs. Milanesi et al. (2017) employed this technique to model connectivity among patches of habitat for capercaillie (*Tetrao urogallus*) in Switzerland. Graph theory is a branch of topology that is used to depict and evaluate patterns of how elements are linked within a system. In ecology, it has been employed by Drake et al. (2017) to evaluate habitat linkages of invasive species, by Crist et al. (2017) for greater sage-grouse, and many others. Landscape genetics relates the similarity of genomes in populations of a species to their degree of landscape network connectivity (Manel et al. 2003).

All connectivity modeling is problematic, insofar as landscape resistance values are generally arbitrary and hard to validate. For example, it is difficult to accurately assess the resistance value of a road. There are also additional questions, such as motivation. Do organisms really behave like electrons, having no goals other than to reach a destination? Another relates to knowledge. Do subadults that have never ventured beyond their natal home range really know the entire landscape and how to globally minimize travel costs? Some of the difficulties that arise in landscape connectivity modeling are addressed with individual-based movement simulation modeling, discussed in a later section.

Given the degree of arbitrariness in assigning landscape resistance and navigating organisms through these artificial surfaces, landscape genetics is quite appealing. Gene flow is linked in a concrete way to animal movements, and connectivity models based on gene flow rely directly on pertinent data. There are, however, some concerns over interpreting the results from landscape genetics analyses. Importantly, these include difficulties in determining the temporal depth of the observed patterns, which could easily relate to historical rather than current conditions (Epps and Keyghobadi 2015).

Source-Sink Dynamics

Evaluation of the source-sink dynamics of populations is similar to the use of connectivity analyses. At its simplest, a "population source" is a geographic area in which population growth rates are positive (i.e., the rates of reproduction plus immigration exceed those of mortality plus emigration), and a "population sink" is where these growth rates are negative (vice versa). Source-sink dynamics have been commonly used in modeling predator-prey dynamics (Amezcua and Holyoak 2000). Gilroy and Edwards (2017) pointed out that population sinks can confuse our understanding of species-environment relationships when a species' presence or local abundance is associated with poor-quality habitats. Thus further evidence is needed on demography, dispersal, and genetics to best determine habitat quality and environmental relationships. Kirol et al. (2015) used RSF models of survival, including source-sink dynamics, of greater sage-grouse in Wyoming to map habitat productivity, for use in conservation planning for the species in a region undergoing energy development operations.

Jonzén et al. (2005) recognized that monitoring the dynamics of sinks could provide valuable insight into a species' overall population status, as sinks maybe more sensitive to changes in the reproduction levels of a population. Similarly, in a simulation study of northern spotted owls, Marcot et al. (2013) found that the "floaters" (i.e., non-reproductive, unpaired, and dispersing individuals) in the population were an indicator of that species' overall population status. The floater component became increasingly depleted with greater declines in the entire population.

Some researchers have used source-sink relationships to model the dynamics of ecosystem components other than populations, such as nutrient flow (Gravel et al. 2010). Ager et al. (2012) modeled source-sink dynamics of fire behavior across a fire-prone landscape in Oregon, simulating the initiation and spread of these events. They found that simulated fire behavior was sensitive to the combination of fuel, topography, and weather, beyond the behavioral patterns related to gradients of general fire regimes. These examples do have a bearing on modeling species-environment relationships, insofar as organisms participate in, and respond to, the dynamics of nutrient flow, fire events, and other such disturbances. Identifying sites that function as sources and sinks depends on the scale of the assessment, and evaluations at varying spatial scales can help point out the relative importance of different source and sink sites.

Individual-Based Movement Modeling

Simulations of how organisms disperse, explore, and use the environment are generally referred to as "individual-based models" (IBMs). IBMs are used, in part, to evaluate source-sink dynamics, to determine likely movement corridors or pathways, to evaluate the effects of resistance to movement from low-quality environmental conditions, and to assess the potential effects of management scenarios and environmental changes on population persistence. Early versions of IBMs used cellular automata, where simple rules governed incremental movements across squares and determined whether a cell became unoccupied (Wallentin 2017). Subsequent IBMs have generally employed a more sophisticated, object-oriented or agent-based modeling approach, simulating actual movement patterns of individual organisms across real or devised geographic space. In this way, IBMs differ from connectivity modeling by having patterns of spatial use emerge from the dynamics of the simulation, rather than being defined a priori by a species' environmental relationships.

IBM simulations are typically run for specified time periods, with a number of replicates, from which estimates of quasi-extinction levels can then be calculated. These represent the proportion of replicate runs in which a population dropped below a designated level at a particular point in time. In a sense, IBMs allow organisms to define their landscape as they react to the distributional patterns of habitat patches, edges, and movement filters and blockages, as well as to other organisms, such as competitors, predators, symbionts, and potential mates. IBMs are typically parameterized with an event space that defines annual occurrences, such as movement, exploration, reproduction, survival, and interaction with conspecifics. In some models, this includes encounters with competitors and predators, diseases, and various environmental conditions. The models start with an initial number and location of individuals, and they typically depict age- or stage-structured populations, along with their associated vital rates of survival and reproduction. IBMs incorporate RSFs or other models, to denote how specific environmental conditions provide degrees of habitat quality.

Movement parameters in IBMs are depicted with statistical distributions, equations, or algorithms that denote distance and the direction or angle of movement (e.g., the degree of spatial autocorrelation of movement pathways), and how parameters may vary in different environmental conditions. Movement algorithms include variations on correlated random walk patterns (Bailey et al. 2018), such as the popularly used Lévy walk patterns (named after the French mathematician Paul Lévy), which are characterized by many small steps and a few long ones. For example, Focardi et al. (2009) developed IBMs using Lévy walks to analyze foraging by fallow deer (*Dama dama*).

The HexSim program, developed and distributed for free by the US Environmental Protection Agency, provides a general framework for creating IBMs (Schumaker and Brookes 2018). HexSim works with "hexmaps" that tile the plane with hexagons, among which movements of individuals are simulated (Fig. 5.7). Hexmaps can be developed from GIS maps of environmental conditions (e.g., vegetation, land-cover classes, etc.) and related to use by the organisms in question through RSF-type relationships that are denoted in the model. HexSim models have been developed for a variety of species and conservation issues, including the control of wild dogs (*Canis lupus familiaris*, *Canis lupus dingo*, and hybrids) in Australia (Pacioni et al. 2018), and the response of Agassiz's (or Mohave) desert tortoises (*Gopherus agassi-*

(A)

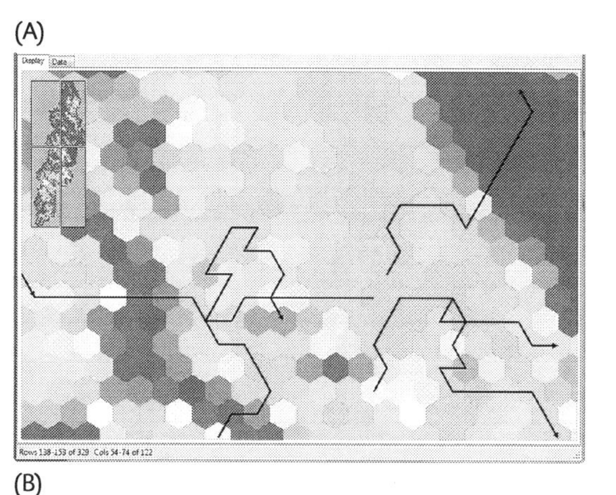

(B)

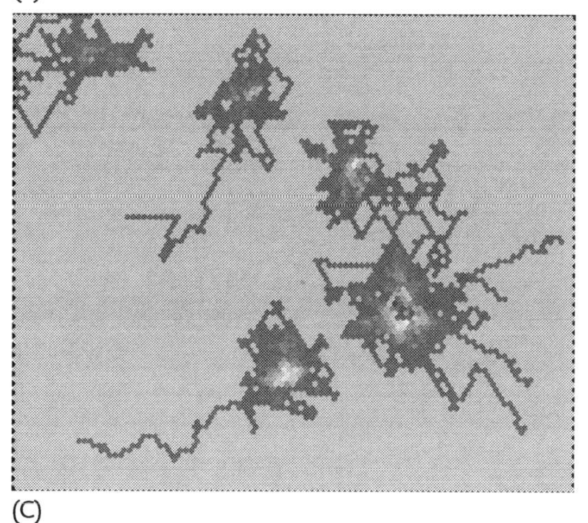

(C)

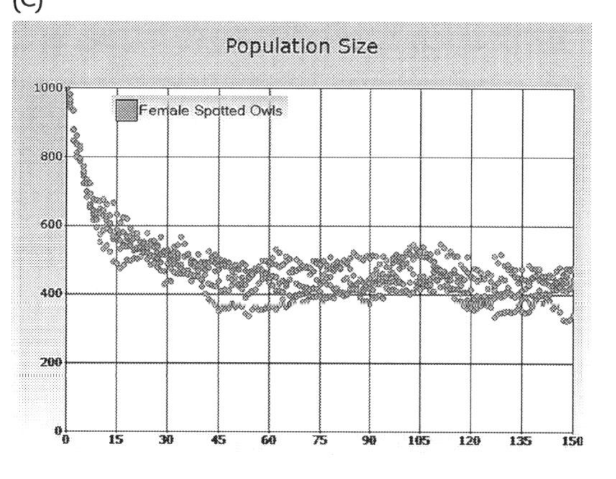

Figure 5.7. An individual-based movement (IBM) model of northern spotted owls (*Strix occidentalis caurina*), developed with the HexSim program, showing an example of the movement dynamics of dispersal and exploration (*A*); a dispersal flux map, illustrating the density of movements along pathways (*B*); and a graph of replicated stochastic model runs (*C*), indicating the population size of female northern spotted owls over time.

zii) to disease and habitat degradation in the Mojave Desert of the southwestern United States. (Tuma et al. 2016). They have also helped identify critical habitat areas for northern spotted owls in the United States (Dunk et al. 2019). As with various other IBMs and connectivity models, HexSim can be used to evaluate source-sink dynamics, to produce "flux maps" of the movement patterns of individuals, to determine the effects on population size and trends from movement filters and barriers, and for other considerations (Schumaker et al. 2014). The problems with IBMs are similar to those of movement models more generally, but with added complexities. In addition to making rules concerning how organisms travel through environments, they also entail many other parameters, such as how organisms interact, how diseases are spread, how and when mating occurs, and what the reproductive and survival consequences of various decisions are. While each of these rules will generally be based on published materials, the accuracy of the behavioral patterns in the overall model is difficult to assess without independent empirical data.

Integrated step selection function (SSF) modeling is 1 form of IBMs. It uses GPS data to generate resource selection functions that are based on known movement patterns of organisms (Thurfjell et al. 2014; Avgar et al. 2016). SSF models have been used to assess movement patterns in pumas (*Puma concolor*) in southern California (Zeller et al. 2016), the effects of roads in impeding the movement of migratory elk (*Cervus elaphus* [*C. e. canadensis*]) in western North America (Prokopenko et al. 2016), and other species and landscape conditions.

Population Dynamics Models

Although some of the model categories discussed above can be used to determine population trends when organisms are affected by environmental conditions, models of population viability and metapopulation dynamics can more explicitly combine traditional population demography with habitat conditions.

Population Viability Analysis Models

Population viability analysis (PVA) has been a mainstay of ecology for several decades, generally entailing calculations of the probability of persistence over specified quasi-extinction levels, or time to extinction (Walsh et al. 1995; See and Holmes 2015). When PVAs are based on particular conditions of habitat and species-environment relationships, they provide a scenario-planning tool for conservation efforts.

A number of modeling programs have been used to conduct PVAs, including VORTEX, RAMAS, and others. Brook et al. (1999) compared 6 PVA modeling tools in analyzing the persistence probability of endangered whooping cranes and found that the tools projected different outcomes, as they were highly sensitive to how variations in the breeding structure were represented in the models. The authors concluded that outcomes of PVA programs should be interpreted with caution, giving due consideration to the underlying biological assumptions of any given model. Similar caveats arose from other such tests (e.g., Lindenmayer et al. 2000). PVA models are of value in suggesting potential conditions for concern, but for most rare or little-known species, habitat influences on their behavior and vital rates—such as on survival, recruitment, Allee effects, density dependency, and interspecific interactions—are poorly understood and could greatly influence the results.

Metapopulation Models

A "metapopulation" is a collection of interacting subpopulations. Thus metapopulation modeling can be used to evaluate how local subpopulations can "rescue" others in the overall population network. Two forms of this are demographic rescue and genetic rescue. Demographic rescue occurs when a declining subpopulation is replenished with numbers of immigrants from another subpopulation to ensure the former's persistence. Genetic rescue is the supplementation of an increasingly homozygous subpopulation with more diverse alleles from immigrating breeding individuals. As a rule of thumb, an order-of-magnitude greater number of immigrating individuals are needed for sustained demographic rescue than for genetic rescue, but this can vary greatly by species, taxon, breeding structure, and other factors. In crafting conservation and recovery plans for at-risk species, both kinds of rescue may be at play in some cases (McRae et al. 2012), and it may be difficult to differentiate their respective contributions, such as Creel (2006) found in conservation efforts for Florida panthers (*Puma concolor concolor*). Providing for either kind of rescue entails understanding species-environment relationships, so individuals can at least disperse through intervening "matrix land" (i.e., non-habitat or poorly suitable) conditions.

Over the decades, many species have been assumed to be structured as metapopulations, and, therefore, modeled using such tools as RAMAS Metapop (e.g., Brook et al. 1999). Nonetheless, field work to determine the specifics of how a population is structured is seldom undertaken, such as by using landscape genetics and determining the degree of interchange in individuals and genes among population centers. For example, golden-cheeked warblers (*Setophaga chrysoparia*) were assumed to occur as a metapopulation, but subsequent work involving genetics revealed that the species was present as a single population (Morrison et al. 2012). As Fronhofer et al. (2012) suggested, metapopulation structures per se may be rarer than is commonly assumed in the ecological literature.

Ecological Functions Models

Another general category of species-environment relations modeling pertains to the ecological functions and roles of organisms in their ecosystem, which may influence the availability of conditions and resources used and selected by other species. In a sense, this is the inverse of competition, which reduces resources. Major ecological functions of species have been depicted as the classic concept of "keystone species," which, if removed from an ecosystem, would have the greatest adverse impact on that system's diversity or productivity, because of their functional influences (Bond 1994). Scherer et al. (2016) argued for a functional-type approach to more effectively model bird responses to climate change in African savannas. And a commonly noted keystone role is primary cavity excavation by woodpeckers (Aubry and Raley 2002; Lorenz et al. 2016).

Ecosystem Engineers

The term "ecosystem engineer" (or "ecological engineer") is used to describe species with behaviors that physically modify major aspects of their environment, as do some keystone species. Examples include the creation of dams and ponds by American beavers (*Castor canadensis*) (Tape et al. 2018); the formation and maintenance of open water and wetlands by wallowing American bison (*Bison bison*) (Nickell et al. 2018); and even footprints of African elephants, which provide microhabitats for aquatic macroinvertebrate communities in Uganda (Remmers et al. 2017). Of concern, however, are the ecosystem engineering behaviors of both floral and faunal invasive species (e.g., Straube et al. 2009; Byers et al. 2012). Some ecosystem engineer species are also termed "zoogeomorphic agents," as they modify terrestrial substrates by digging and burrowing (Mathers et al. 2019).

Key Ecological Functions

A generalized framework for considering the effects of biotic key ecological functions (KEFs) on environmental conditions was provided by Marcot and Vander Hayden (2001) and codified in wildlife-habitat relationship databases. Their classification system of KEFs encompassed 85 categories of biotic functions for 598 species of non-fish vertebrates in Washington and Oregon. That database was applied to several analyses, including the functional roles of arboreal mammals (Marcot and Aubry 2003); the management of decaying wood for wildlife (Marcot 2002); the functional roles of interactions by Pacific salmon (*Oncorhynchus* spp.) with terrestrial wildlife in providing nutrients to riparian and upland ecosystems (Cederholm et al. 2001); and maps of functional redundancy (i.e., the number of wildlife species with similar functional roles) and diversity (i.e., the relative abundance of species with functional roles) of terrestrial fauna communities across the western US–Canada border (Marcot et al. 2006). Each of the examples of ecological functional modeling—including keystones, engineers, and KEFs—highlights the importance of evaluating the biotic roles that *all* organisms play in modifying, enhancing, or worsening environmental conditions used by other species.

Other Models of Species-Environment Relations

We would be remiss in not mentioning a variety of other modeling frameworks that could pertain to, or be informed by, species-environment relationships. These include an assortment of multiple-species models that evaluate species richness and gaps in the conservation coverage of species included in them. Scott et al. (1993) proposed the use of Gap Analysis, and Boykin et al. (2010) and Vimal et al. (2011) evaluated this approach. A number of species-interactions models incorporate various relationships between competition, mutualism, mimicry, and other interactions (e.g., Byrne et al. 2019). The framework of multiple-species models can also encompass food webs, ecosystem trophic structures, predator-prey interactions, and energy flow and transfer rates (e.g., Legović et al. 2010; Baiser et al. 2013).

We have not discussed a number of specific modeling topics and methods using such approaches as data mining, loop analysis, logic modeling, mixed-effects models, *N* mixture models, multimodel inference, and others. Major treatises have been written on most of these topics, and astute readers can probably identify even more modeling categories and frameworks not mentioned here that could impart knowledge from, or be informed by, species-environment relationships.

Deciding on Appropriate Modeling Approaches

This chapter covers a long list of modeling categories, frameworks, and tools pertaining to many aspects of species-environment relationships. How do you select which modeling approach to use? As with any such venture, it is best to begin by clearly stating your objectives, remembering that, in ecology and conservation, "developing a model" is a method, not an objective. Conservation objectives can include a need to determine which portions of a broad geographic area should be utilized to create or maintain specific environmental conditions serving as habitat for a target species (e.g., with landscape design); to conserve biodiversity hotspots; to provide population connectivity via habitat corridors; and so on. Target species can be identified as at-risk and listed (i.e., endangered or threatened) species, species with particular biotic functions of broader ecological value, species of social interest and value, and other categories, as discussed above.

Objectives should also clarify if the goal of an investigation is to evaluate past (i.e., historic), current, or future conditions. If future conditions are indicated, a researcher should further clarify if the purpose of a model is prediction (i.e., assessing possible future outcomes based on initial conditions), forecasting (i.e., assessing the most likely future outcome, based on initial conditions), projection (i.e., assessing possible future outcomes, based on changing future conditions), scenario planning (i.e., assessing the implications of hypothetical alternative management situations), or other needs, such as mitigation (i.e., identifying alternative conditions that could lead to a desired outcome). It is also important to articulate whether the model is intended to provide best-case predictions, projections, and so on; or to identify causal mechanisms; or both (Breiman 2001). The best predictive models, particularly using machine-learning algorithms and constructs such as neural networks, might do well in foretelling outcomes, but act poorly in describing and explaining underlying mechanisms of both individual organisms' behaviors and population responses.

Why is it so important to clearly identify modeling objectives at the outset? The answer has to do with a common tendency to want the resulting model to exceed its original intent, as well as do more than its framework was designed for. We call this tendency "objective creep," as additional (and inappropriate) objectives are piled onto the initial objective for a given modeling structure. For example, an initial objective could be to determine the climate envelope of native wildlife species in their current range, as a way of hypothesizing their potential spread northward (in the Northern Hemisphere) under regional warming, climate change, and habitat shifts. Well and good. But such mechanistic climate niche models typically do not include variations in the ecological responses among individuals, as well as the evolutionary processes of populations and species that collectively influence adaptation through changes in behavior, reproduction, dispersal, and survival (Leroux et al. 2013; Singer et al. 2016). Asking such mechanistic models to project changes in a species' habitat selection behavior under climate change, without accounting for those other factors, would be an example of inappropriate objective creep.

When objectives have been clearly articulated, it should become apparent which type of modeling tool, if appropriate, would serve the task of analysis, as well as whether the results would provide useful information on species' interactions or environmental conditions, or produce accurately mapped outcomes of species' distributions or habitats. It would also be helpful to clarify if the modeling framework focuses on individual organisms (e.g., individual movement models), subpopulations (e.g., metapopulation, rescue, and source-sink models), populations (e.g., PVA models), species-level distributions (e.g., occupancy and RSF models), biotic functions (e.g., KEF models), or other categories. When specifying research objectives and the general form of the desired response, Table 5.1, which summarizes the modeling categories, could

Table 5.1 An overview of categories of species-environment relationships models, their definitions, and how they are depicted or structured

Model category	Definition; structure
	Occurrence Modeling
Location models	Known occurrence of individuals from direct observation or via signs; denotes locations as geographic locators (e.g., UTMs).
Occupancy models	Extrapolation of species occurrence from repeat surveys; denotes occupancy as probabilities of potential occurrence.
	Habitat Association Modeling
Wildlife-habitat relationships models	Expert-summarized depictions of use or ordinal use levels of vegetation and other cover conditions, at species level; denoted in matrices of species by conditions.
Habitat suitability index models	Quantitative index on a normalized (0,1) scale of the degree of habitat suitability, based on a few (typically 3) environmental parameters; produces index calculations that may be mapped.
Species distribution models	Geographic extent of potentially useable, or inferred selection of, environmental conditions providing habitat; variously depicted with a normalized or extended range of values, and may be mapped.
Maximum entropy models	Quantitative index on a normalized (0,1) scale of the degree of suitability of environmental conditions contributing to species' presence; mapped across a specified geographic extent, based on conditions where species' presence is noted.
	Habitat Selection Modeling
Binary selection and classification models	Decision trees displaying environmental variables and their values, most differentiating known cases of species' presence or population density; listed on tree branches in decreasing degrees of differentiating power.
Generalized additive models	Linear regressions using non-linear relationships between predictor and response variables; depicts relationships with a variety of curve-fitting functions.
Resource selection functions	Mostly linear functions or regressions, producing indices or statistical calculations of overall habitat suitability, based on the use and selection of resource conditions; may be mapped.
Central-use models	Concentration of locations of an individual, typically based on radio telemetry, suggesting high-quality or frequently used areas; usually mapped as core habitats.
Climate envelope models	Range of climate conditions (temperature, precipitation) in which a species currently occurs, used to project climate suitability in unoccupied areas; determined as a degree of suitability, and may be mapped.
	Space-Use Modeling
Species-area relationships models	Species richness as a function of the size of a geographic area, and also as an incidence function probability or frequency of occurrence of a species in a given habitat area; shown as cumulative probability distributions.
Connectivity models	Mathematical and topological indices of habitat patch distributional patterns, indicting the degree of patch contiguity or fragmentation; depicted as various index values. Also, a linkage of habitat areas, with corridors or connectors of cover types amenable to movement; shown as a probability or degree of connectivity, and may be mapped.
Source-sink dynamics models	Flux of individuals moving from, or into, specific locations; may be mapped as areas where survival exceeds mortality, or vice versa.
Individual-based movement models	Individual movement dynamics across static or dynamic landscapes for a given timeframe; may be summarized as population size over time, and may be mapped, showing areas of concentration, use, and the flux of movement lanes.
	Population Dynamics Modeling
Population viability analysis models	Projections of population rates of survival and reproduction in response to initial conditions; depicted as an annual rate of change, and as a probability of extinction or quasi-extinction at particular points in time.
Metapopulation models	Degree to which subpopulations exchange individuals (demographic rescue) or genes (genetic rescue); can be combined with mapped models of environmental conditions conducive to movement and dispersal.

(*continued*)

Table 5.1 continued

Model category	Definition; structure
	Ecological Functions Modeling
Ecosystem engineers	Changes in structures or substrates of an environment caused by the mechanical biotic functions of organisms; depicted mostly as qualitative effects on substrates.
Key ecological functions	Biotic roles of organisms affecting resources or environmental conditions for other species; uses wildlife-habitat relationships matrices, denoting the qualitative functional roles of species.
	Multiple-Species Modeling
Species richness and Gap Analysis models	Number of species in a given location; mapped as "gap" locations of higher richness in areas with no protection.
Assemblage, guild models	Number of species of a particular taxonomic or functional group in a given location; can be mapped as above.
Species interactions models	Dynamic influence among species engaged in competition, predation, or symbiotic relationships; typically shows aspatial functional relations.
	Energy-Flow Modeling
Metabolic rate models	Degree to which individuals can meet various levels of metabolic needs; consists of aspatial models of the metabolic rates required for various life-history needs.
Energy transfer models	Energy exchange rates between trophic levels; gives aspatial dynamic functions of transfer rates.

Note: See the chapter text for specific examples

be used as a decision guide to narrow the field of modeling approaches.

Key Modeling Considerations

Beyond the need to clearly specify modeling objectives, as discussed above, there are further key considerations for developing and using species-environment relationships models. They need to be appropriately structured and parameterized, whether they are based on expert knowledge or derived from field data, and whether they use mathematical functions, statistical relations, categorical classes, or ordinal scales. They also need to be appropriately calibrated to the scale at which a given species responds to their environment, as well as to geographic conditions and other factors of particular interest. If possible, models should be evaluated for validity by testing their performance against databases with known parameters, and against other potential situations and outcomes.

Bootstrapping or jackknifing (testing subsets of the same dataset used to create the model) should be done to evaluate a model, unless you have independent test data with which to do a fuller cross-validation. Some modeling structures, such as Maxent, incorporate jackknifing (Shcheglovitova and Anderson 2013) or other means of cross-validation to determine the extent and type of bias that may be present in a parameterized model. Models with high calibration success but low validation success indicate that they are overfit to their datasets and should be applied with great caution if employed outside the realm of the data used for their parameterization. On the other hand, models can be underfit, being far too general, and thus serving as poor tools for prediction and evaluation. Various approaches have been developed to evaluate the degree of model fit, such as the use of null models (e.g., Merckx et al. 2011; Hijmans 2012).

Another consideration is determining the degree of sensitivity and influence in the model's parameters. "Sensitivity analysis" refers to the extent to which the value of a response variable shifts because of incremental changes in the input variables. This is measured as either entropy or variance response for each input variable. "Influence analysis" considers the relative effect on response variables after setting each input variable to its minimum, maximum, and most likely values, and is measured by comparing differences in the resulting values of the response variable (Marcot 2012). In PVAs, the sensitivity structure of Leslie matrices (models of a population's vital rates) is calculated with elasticity analysis (Kajin et al. 2012). Understanding parameter sensitivity and influence can help provide information about the potential control that management activities might have over outcomes. For example, Amstrup et al. (2008) presented Bayesian probability network models, projecting the responses of polar bears (*Ursus maritimus*) to environmental and anthropogenic stressors. They reported that if management efforts were able to stave off all direct anthropogenic stressors (such as hunting, killing for defense of life and property, pollution, and tourism), without changing current trends in sea ice loss from regional climate warming, this would buy only perhaps 20 years' time before the populations would likely greatly decline, whereas initially curtailing the loss of sea ice would allow a long-term persistence of the populations.

A related consideration is understanding the implications of model uncertainties and unknowns (Graham and Kimble 2019), particularly if a model is to be used to help prioritize further surveys, mitigation, and research. For the covariates or inputs that are least well known, sensitivity and influence analyses can be employed to identify which ones might have the greatest effect on model outcomes, and if those effects would qualitatively pertain to important conservation and management decisions. Resource managers might then want to prioritize those variables for further scrutiny. For example, Jay et al. (2011) conducted a set of sensitivity and influence analyses on a network model of Pacific walruses (*Odobenus rosmarus divergens*), which identified key stressors with the greatest degree of uncertainty and those with the greatest effect—information that was later used to guide further research on that species.

It is also important to specify the temporal and spatial resolution of a model, to avoid inappropriate applications. For instance, the polar bear models mentioned above were developed at the scale of 4 major circumpolar ecoregions, and their results should not be applied to the bears' 19 individual circumpolar populations without due caution in considering local conditions, which could vary from the ecoregional ones.

With models used to determine the presence and absence of at-risk species, to prioritize sites for conservation, it may be vital to determine the models' false positive (Type I) error rates (Bingham and Brennan 2004) and false negative (Type II) error rates (Bingham et al. 2007). These 2 types of errors can have drastically different real-world implications. False positives may result in protecting conditions that are unnecessary for the species of interest, thus incurring the opportunity costs of humans' lost access to those resources. Such errors may also threaten the credibility of the model and its use by conservation agencies. False negatives, in contrast, may result in not protecting critical sites or conditions for the relevant species, adding threats to these organisms' recovery and garnering ire from conservation-minded segments of the public.

Another consideration is the appropriate elicitation—and use—of expert knowledge in structuring, parameterizing, improving, and applying species-environment relationships models. This field, known as "expert elicitation," has burgeoned in recent years, with guidelines on rigorous procedures (Burgman 2016; Hemming et al. 2018). Expert elicitation has been used in studies of the effects of climate change on Australia avifauna and marine mammals (Wilcox et al. 2018), the effects of forest management on biodiversity (Filyushkina et al. 2018), and many more.

A number of approaches use machine-learning algorithms with data from specific cases to structure and parameterize those models, such as in studies of animal behavior (Valletta et al. 2017). Such algorithms vary widely in their effectiveness, however, resulting in greater model complexity, greater computational cost, and lack of transparency. Some, such as random forests (discussed above) and expectation maximization (Do and Batzoglou 2008), employ iterative methods (e.g., convergent log-likelihood functions) to find the best model fit, and they can handle missing data through imputation procedures. A user should be aware of the algorithm that was employed, as well as its relevance and appropriateness to the modeling objective at hand.

Caveats and Cautions

Some additional warnings and advice need to be mentioned here regarding specific modeling attributes. The first is the limitations of presence-only models, where defining and determining absence is problematic. A lack of evidence regarding presence is not necessarily evidence of absence, especially in systems undergoing rapid alterations in environmental conditions, such as climate change in far northern or far southern latitudes. A second is that SDMs are generally based on correlative relationships (e.g., climate envelope and climate niche models), which could be misleading if the current distribution of a species belies its longer-term equilibrium. An example of this is the initial flush and, later, further exploration of an invasive species (Chapman et al. 2017).

The correlational nature of most of our knowledge concerning habitat use by various species runs through the vast majority of the modeling methods discussed here, and many of these methods are explicitly used to derive correlational relationships. Such formal relationships are statistically valid only for the sampled populations. Extrapolation from model results and their application to other populations, times, and places that were not analyzed is a non-statistical process. Success in making such transfers critically depends on community-level similarities across space and time. That is, an organism's behavior, habitat use, and land-use preferences are conditioned by the ecosystems in which they exist. Further, we discover these patterns and preferences only if a particular resource is both available and limiting. For example, formal associations with water are more likely to be detected in deserts than in rainforests, but they would not be found in large areas of the Atacama Desert, where surface water is rare. Our failure to detect correlational patterns associated with water in these systems, however, does not mean that water is unimportant. Should climate change lead to a previously wet area becoming arid, the importance of water sources would be more apparent. As we enter a period of rapid ecological change—characterized by both fundamental shifts in climate and a constant barrage of new players, due to the transportation of invasive, exotic species—we can expect the patterns of how organisms use and select resources and environments to change as their ecosystems become increasingly novel (Buma and Wessman 2013). This understanding highlights the need for strong validations of model results. In the strict sense, statistical model *validation* entails testing predictions against independent data, whereas model *evaluation* pertains to other aspects of ensuring a model's credibility, acceptability, calibration, and utility (Marcot et al. 1983).

Fourth, commonly conveyed statistical caveats about modeling, in addition to those discussed above regarding false presence and false absence, pertain to the need to address issues of statistical power and effect size, such as with the use of Cohen's d statistic. It is also important to appropriately interpret resource use as selection and, particularly, non-selection (i.e., "avoidance"), only in instances where a full range of alternative conditions is truly available to individuals. A case in point is with species residing in regions where few alternatives exist, and "use" would suggest a high selection of conditions that would otherwise not indicate deliberate choices, or habitat factors that may even be selected against in the presence other conditions. The models for species that have had their ranges significantly reduced may be particularly vulnerable to such circumstances. Fifth, another caution (alluded to further above) is to not draw unfounded conclusions about environmental conditions conferring adaptive advantages, as well as about those under which a species made selections and evolved its characteristics and adaptations. This caution pertains to many modeling frameworks, including RSFs and climate envelopes.

Sixth, cast a wary eye on data sources derived from non-statistically structured sampling procedures. We are speaking here of crowd-sourced information and sightings data from social media and online sites. Although popular websites, such as eBird and the like, might provide useful information for evaluating broad-scale patterns and trends (e.g., Martin 2018), these databases can suffer from false positives (i.e., incorrect identifications), false negatives (i.e., a lack of reported sightings does not necessarily mean absence of a species), bias (i.e., more information gathered on popular or charismatic species), and false indications of abundance hot spots (Johnston et al. 2015). Similarly, data from ever-expanding citizen-science programs need to be used with caution, to ensure rigorous and vetted data-collection methods (e.g., Vermeiren et al. 2016; Steenweg et al. 2017).

Lastly, a modeler should be wary of excessive emphasis on the importance of "habitat fragmentation" (Wilcove 1987; Fahrig 2002), particularly when using measures of habitat distribution patterns that are not applied to a particular species (e.g., Liu et al. 2018). Miller et al. (2019), however, presented a counter-example of how multiscale correlates of habitat fragmentation can be applied to analyses of long-term trends for northern bobwhites (*Colinus virginianus*), a species that is declining over a broad geographic region. Remember the emphasis in Chapter 1, which underlined that "habitat" is essentially a species-specific concept.

What's Next for Modeling Species-Environment Relationships?

From the materials reviewed in this chapter, we can identify some key modeling topics that need further research. Just a few are listed below, and we invite readers—particularly engaged students—to identify others:

- more-rigorous comparisons of alternative approaches, especially SDMs, to provide clearer guidance on which ones to use in particular situations and for specific modeling objectives;
- more work in integrating the evolutionary basis and adaptive causes and consequences of habitat use and selection, particularly in environments undergoing rapid change;
- in classification-type models, better analysis methods, as well as clearer implications and interpretations of model errors (Type I and Type II), bias, overfitting, and underfitting;
- further development of standardized model frameworks, to better amalgamate genomics (especially the behavioral expression of genetics) and analyses of local adaptations with SDMs, RSFs, and other modeling constructs;
- the development of integrated or hybridized approaches to modeling and assimilating habitat-population linkages, as a promising area for future research—for example, combining occupancy and mark-recapture modeling approaches on the same site, to use them in developing monitoring scenarios for both rare and common species, and linking SDMs with phylogenetic analyses based on molecular genetic data, to provide independent lines of evidence that can strengthen inferences made from a single approach;
- greater examination of many new technological field techniques and data sources that provide data to build species-environment relationships models (e.g., the use of drones, automatic recording units, geotags, inertial sensors, and others), to assess their reliability, utility, and appropriate application; and
- more development in the area of modeling key ecological (biotic) functions, particularly in field studies aimed at quantifying the rates and amounts of functional effects by organisms. This area holds promise for guiding management activities in maintaining or restoring ecosystem functionality, particularly if the models incorporate genomics.

Summary

The field of species-environment relationships modeling has never been richer, or more apt to borrow so many tools and methods from developments in mathematics, physics, and computer science. We all need to keep alert, however, to ensure that the exponential increases in computational prowess and the capacity of our modeling systems do not outstrip our understanding and interpretation of basic natural history. In the end, it may be a race between ensuring the ecological validity and credibility of the many new and enhanced modeling frameworks, and keeping pace with the ways in which our natural world is being homogenized and changed.

LITERATURE CITED

Ager, A. A., N. M. Vaillant, M. A. Finney, and H. K. Preisler. 2012. Analyzing wildfire exposure and source-sink relationships on a fire prone forest landscape. Forest Ecology and Management 267:271–283.

Almpanidou, V., G. Schofield, A. S. Kallimanis, O. Türkozan, G. C. Hays, and A. D. Mazaris. 2016. Using climatic suitability thresholds to identify past, present and future population viability. Ecological Indicators 71:551–556.

Amezcua, A. B., and M. Holyoak. 2000. Empirical evidence for predator-prey source-sink dynamics. Ecology 81(11):3087–3098.

Amstrup, S. C., B. G. Marcot, and D. C. Douglas. 2008. A Bayesian network modeling approach to forecasting the 21st century worldwide status of polar bears. Pp. 213–268 *in* E. T. DeWeaver, C. M. Bitz, and L.-B. Tremblay, eds. Arctic sea ice decline: Observations, projections, mechanisms, and implications. Geophysical Monograph 180. American Geophysical Union, Washington, DC.

Apps, C. D., and B. N. McLellan. 2006. Factors influencing the dispersion and fragmentation of endangered mountain caribou populations. Biological Conservation 130(1):84–97.

Araújo, M. B., and A. T. Peterson. 2012. Uses and misuses of bioclimatic envelope modeling. Ecology 93(7):1527–1539.

Aubry, K. B., and C. M. Raley. 2002. The pileated woodpecker as a keystone habitat modifier in the Pacific Northwest. Pp. 257–274 *in* W. F. Laudenslayer Jr., P. J. Shea, B. E. Valentine, C. P. Weatherspoon, and T. E. Lisle, eds. Proceedings of the Symposium on the Ecology and Management of Dead Wood in Western Forests, 2–4 November 1999, Reno, Nevada. General Technical Report PSW-181. US Department of Agriculture, Forest Service, Pacific Southwest Research Station, Albany, CA.

Aubry, K. B., C. M. Raley, and K. S. McKelvey. 2017. The importance of data quality for generating reliable distribution models for rare, elusive, and cryptic species. PLoS ONE 12(6):e0179152.

Avgar, T., J. R. Potts, M. A. Lewis, and M. S. Boyce. 2016. Integrated step selection analysis: Bridging the gap between resource selection and animal movement. Methods in Ecology and Evolution 7:619–630.

Bailey, J. D., J. Wallis, and E. A. Codling. 2018. Navigational efficiency in a biased and correlated random walk model of individual animal movement. Ecology 99(1):217–223.

Baiser, B., N. Whitaker, and A. M. Ellison. 2013. Modeling foundation species in food webs. Ecosphere 4:1–14. http://dx.doi.org/10.1890/ES13-00265.1/.

Bauder, J. M., D. R. Breininger, M. R. Bolt, M. L. Legare, C. L. Jenkins, B. B. Rothermel, and K. McGarigal. 2018. Multi-level, multi-scale habitat selection by a wide-ranging, federally threatened snake. Landscape Ecology 33(5):743–763.

Baumgardt, J. A., M. L. Morrison, L. A. Brennan, B. L. Pierce, and T. A. Campbell. 2019. Development of multi-species, long-term monitoring programs for resource management. Rangeland Ecology and Management 72:168–181.

Beaumont, L. J., E. Graham, D. E. Duursma, P. D. Wilson, A. Cabrelli, J. B. Baumgartner, W. Hallgren, M. Esperón-Rodríguez, D. A. Nipperess, D. L. Warren, S. W. Laffan, and J. VanDerWal. 2016. Which species distribution models are more (or less) likely to project broad-scale, climate-induced shifts in species ranges? Ecological Modelling 342:135–146.

Bélisle, A. C., H. Asselin, P. LeBlanc, and S. Gauthier. 2018. Local knowledge in ecological modeling. Ecology and Society 23(2):art. 14.

Bingham, B. B., and B. R. Noon. 1998. The use of core areas in comprehensive mitigation strategies. Conservation Biology 12(1):241–243.

Bingham, R. L., and L. A. Brennan. 2004. Comparison of Type I error rate for statistical analyses of resource selection. Journal of Wildlife Management 68:206–212.

Bingham, R. L., L. A. Brennan, and B. M. Ballard. 2007. Misclassified resource selection: Compositional analysis and unused habitat. Journal of Wildlife Management 71:1369–1374.

Bomford, M. 2008. Risk assessment models for establishment of exotic vertebrates in Australia and New Zealand. Invasive Animals Cooperative Research Centre, Canberra.

Bond, W. J. 1994. Keystone species. Pp. 237–253 *in* E. D. Schulze and H. A. Mooney, eds. Biodiversity and ecosystem function. Springer-Verlag, Berlin.

Booth, T. H. 2019. BIOCLIM: The pioneering SDM package. Bulletin of the Ecological Society of America 100(1):e01475.

Boyce, M. S., P. R. Vernier, S. E. Nielsen, and F. K. A. Schmiegelow. 2002. Evaluating resource selection functions. Ecological Modelling 157:281–300.

Boykin, K. G., B. C. Thompson, and S. Propeck-Gray. 2010. Accuracy of gap analysis habitat models in predicting physical features for wildlife-habitat associations in the southwest U.S. Ecological Modelling 221(23):2769–2775.

Boyle, K. A., and T. T. Fendley. 1987. Habitat suitability index models: Bobcat. Biological Report FWS/OBS-82/10.147. US Department of the Interior, Fish and Wildlife Service, Research and Development, Washington, DC.

Brambilla, M., and G. F. Ficetola. 2012. Species distribution models as a tool to estimate reproductive parameters: A case study with a passerine bird species. Journal of Animal Ecology 81:781–787.

Brandt, L. A., A. M. Benscoter, R. Harvey, C. Speroterra, D. Bucklin, S. S. Romañach, J. I. Watling, and F. J. Mazzotti. 2017. Comparison of climate envelope models developed using expert-selected variables versus statistical selection. Ecological Modelling 345:10–20.

Breiman, L. 2001. Statistical modeling: The two cultures. Statistical Science 16(3):199–231.

Brennan, L. A., W. M. Block, and R. J. Gutiérrez. 1986. The use of multivariate statistics for developing habitat suitability index models. Pp. 177–182 *in* J. Verner, M. L. Morrison, and C. J. Ralph, eds. Wildlife 2000: Modeling habitat relationships of terrestrial vertebrates. University of Wisconsin Press, Madison.

Brennan, L. A., W. M. Block, and R. J. Gutiérrez. 1987. Habitat use by mountain quail in northern California. Condor 89:66–74.

Brook, B. W., J. R. Cannon, R. C. Lacy, C. Mirande, and R. Frankham. 1999. A comparison of the population viability analysis packages GAPPS, INMAT, RAMAS and VORTEX for the whooping crane (*Grus americana*). Animal Conservation 2:23–31.

Browne, C. L., and C. A. Paszkowski. 2018. Microhabitat selection by western toads (*Anaxyrus boreas*). Herpetological Conservation and Biology 13(2):317–330.

Buechley, E. R., M. J. McGrady, E. Çoban, and Ç. H. Şekercioğlu. 2018. Satellite tracking a wide-ranging endangered vulture species to target conservation actions in the Middle East and East Africa. Biodiversity and Conservation 27:2293–2310.

Buma, B., and C. A. Wessman. 2013. Forest resilience, climate change, and opportunities for adaptation: A specific case of a general problem. Forest Ecology and Management 306:216–225.

Burgman, M. A. 2016. Trusting judgements: How to get the best out of experts. Cambridge University Press, Cambridge.

Burns, K. C. 2015. A theory of island biogeography for exotic species. American Naturalist 186(4):441–451.

Byers, J. E., P. E. Gribben, C. Yeager, and E. E. Sotka. 2012. Impacts of an abundant introduced ecosystem engineer within mudflats of the southeastern US coast. Biological Invasions 14(12):2587–2600.

Byrne, M. E., A. E. Holland, K. L. Turner, A. L. Bryan, and J. C. Beasley. 2019. Using multiple data sources to investigate foraging niche partitioning in sympatric obligate avian scavengers. Ecosphere 10(1):e02548.

Cantwell, M. D., and R. T. T. Forman. 1993. Landscape graphs: Ecological modeling with graph theory to detect configurations common to diverse landscapes. Landscape Ecology 8(4):239–255.

Cederholm, C. J., D. H. Johnson, R. E. Bilby, L. G. Dominguez, A. M. Garrett, W. H. Graeber, E. L. Greda, M. D. Kunze, B. G. Marcot, J. F. Palmisano, R. W. Plotnikoff, W. G. Pearcy, C. A. Simenstad, and P. C. Trotter. 2001. Pacific salmon and wildlife: Ecological contexts, relationships, and implications for management. Pp. 628–685 *in* D. H. Johnson and T. A. O'Neill, eds. Wildlife-habitat relationships in Oregon and Washington. Oregon State University Press, Corvallis.

Chapman, D. S., R. Scalone, E. Štefanić, and J. M. Bullock. 2017. Mechanistic species distribution modeling reveals a niche shift during invasion. Ecology 98(6):1671–1680.

Chefaoui, R. M. 2014. Landscape metrics as indicators of coastal morphology: A multi-scale approach. Ecological Indicators 45:139–147.

Clark, J. S., D. M. Bell, M. H. Hersh, and L. Nichols. 2011. Climate change vulnerability of forest biodiversity: Climate and competition tracking of demographic rates. Global Change Biology 17:1834–1849.

Connor, E. F., A. C. Courtney, and J. M. Yoder. 2000. Individuals-area relationships: The relationship between animal population density and area. Ecology 81(3):734–748.

Costa, G. C., C. Nogueira, R. B. Machado, and G. R. Colli. 2010. Sampling bias and the use of ecological niche modeling in conservation planning: A field evaluation in a biodiversity hotspot. Biodiversity and Conservation 19(3):883–899.

Costanza, J. K., K. Riitters, P. Vogt, and J. Wickham. 2019. Describing and analyzing landscape patterns: Where are we now, and where are we going? Landscape Ecology 34(9):2049–2055.

Cowlishaw, G. 1999. Predicting the pattern of decline of African primate diversity: An extinction debt from historical deforestation. Conservation Biology 13(5):1183–1193.

Crawford, J. A., W. E. Peterman, A. R. Kuhns, and L. S. Eggert. 2016. Towards a better mechanistic understanding of edge effects. Landscape Ecology 31(10):2231–2244.

Creel, S. 2006. Recovery of the Florida panther: Genetic rescue, demographic rescue, or both? Response to Pimm et al. (2006). Animal Conservation 9(2):125–126.

Crist, M. R., S. T. Knick, and S. E. Hanser. 2017. Range-wide connectivity of priority areas for greater sage-grouse: Implications for long-term conservation from graph theory. Condor 119(1):44–57.

Crombie, J., L. Brown, J. Lizzio, and G. Hood. 2008. Climatch user manual. Australian Government, Bureau of Rural Sciences, Canberra.

Cushman, S. A., K. McGarigal, and M. C. Neel. 2008. Parsimony in landscape metrics: Strength, universality, and consistency. Ecological Indicators 8(5):691–703.

Cutler, D. R., T. C. Edwards Jr., K. H. Beard, A. Cutler, K. T. Hess, J. Gibson, and J. J. Lawler. 2007. Random forests for classification in ecology. Ecology 88(11):2783–2792.

Davey, C. M., D. E. Chamberlain, S. E. Newson, D. G. Noble, and A. Johnston. 2012. Rise of the generalists: Evidence for climate driven homogenization in avian communities. Global Ecology and Biogeography 21(5):568–578.

de Beer, Y., and R. J. van Aarde. 2008. Do landscape heterogeneity and water distribution explain aspects of elephant home range in southern Africa's arid savannas? Journal of Arid Environments 72(11):2017–2025.

Dilts, T. E., P. J. Weisberg, P. Leitner, M. D. Matocq, R. D. Inman, K. E. Nussear, and T. C. Esque. 2016. Multiscale connectivity and graph theory highlight critical areas for conservation under climate change. Ecological Applications 26(4):1223–1237.

Do, C. B., and S. Batzoglou. 2008. What is the expectation maximization algorithm? Nature Biotechnology 26:897–899.

Dodouras, S., and P. James. 2007. Fuzzy cognitive mapping to appraise complex situations. Journal of Environmental Planning and Management 50(6):823–852.

Dormann, C. F., O. Purschke, J. R. G. Márquez, S. Lautenbach, and B. Schröder. 2008. Components of uncertainty in species distribution analysis: A case study of the great grey shrike. Ecology 89(12):3371–3386.

Drake, J. C., K. L. Griffis-Kyle, and N. E. McIntyre. 2017. Graph theory as an invasive species management tool: Case study in the Sonoran Desert. Landscape Ecology 32(8):1739–1752.

Dudík, M., S. J. Phillips, and R. E. Schapire. 2004. Performance guarantees for regularized maximum entropy density estimation. Pp. 472–486 *in* J. Shawe-Taylor and Y. Singer, eds. Proceedings of the Seventeenth Annual Conference on Computational Learning Theory. Springer, Berlin.

Dunk, J. R., B. Woodbridge, N. Schumaker, E. M. Glenn, B. White, D. W. LaPlante, R. G. Anthony, R. J. Davis, K. M. Dugger, K. Halupka, P. Henson, B. G. Marcot, M. Merola-Zwartjes, B. R. Noon, M. G. Raphael, J. Caicco, D. L. Hansen, M. J. Mazurek, and J. Thrailkill. 2019. Conservation planning for species recovery under the Endangered Species Act: A case study with the northern spotted owl. PLoS ONE 14(1):e0210643.

Dunk, J. R., W. J. Zielinski, and H. K. Preisler. 2004. Predicting the occurrence of rare mollusks in northern California forests. Ecological Applications 14(3):713–729.

Elith, J. 2019. Machine learning, random forests and boosted regression trees. Pp. 281–297 *in* L. A. Brennan, A. N. Tri, and B. G. Marcot, eds. Quantitative analyses in wildlife science. Johns Hopkins University Press, Baltimore.

Elith, J., M. Kearney, and S. Phillips. 2010. The art of modelling range-shifting species. Methods in Ecology and Evolution 1:330–342.

Elith, J., and J. R. Leathwick. 2009. Species distribution models: Ecological explanation and prediction across space and time. Annual Review of Ecology, Evolution, and Systematics 40:677–697.

Epps, C. W., and N. Keyghobadi. 2015. Landscape genetics in a changing world: Disentangling historical and contemporary influences and inferring change. Molecular Ecology 24(24):6021–6040.

Eskelson, B. N. I., H. Temesgen, and J. C. Hagar. 2015. A comparison of selected parametric and imputation methods for estimating snag density and snag quality attributes. Forest Ecology and Management 272:26–34.

Fahrig, L. 2002. Effect of habitat fragmentation on the extinction threshold: A synthesis. Ecological Applications 12(2):346–353.

Fecske, D. M., J. A. Jenks, and V. J. Smith. 2002. Field evaluation of a habitat-relation model for the American marten. Wildlife Society Bulletin 30(3):775–782.

Feng, Y., and Y. Liu. 2015. Fractal dimension as an indicator for quantifying the effects of changing spatial scales on landscape metrics. Ecological Indicators 53:18–27.

Fern, R. R., and M. L. Morrison. 2017. Mapping critical areas for migratory songbirds using a fusion of remote sensing and distributional modeling techniques. Ecological Informatics 42:55–60.

Filyushkina, A., N. Strange, M. Löf, E. E. Ezebilo, and M. Boman. 2018. Applying the Delphi method to assess impacts of forest management on biodiversity and habitat preservation. Forest Ecology and Management 409:179–189.

Fitterer, J. L., T. A. Nelson, N. C. Coops, M. A. Wulder, and N. A. Mahony. 2013. Exploring the ecological processes driving geographical patterns of breeding bird richness in British Columbia, Canada. Ecological Applications 23(4):888–903.

Focardi, S., P. Montanaro, and E. Pecchioli. 2009. Adaptive Lévy walks in foraging fallow deer. PLoS ONE 4(8):e6587, doi:10.1371/journal.pone.0006587.

Freeman, M. T., P. I. Olivier, and R. J. van Aarde. 2018. Matrix transformation alters species-area relationships in fragmented coastal forests. Landscape Ecology 33(2):307–322.

Frey, J. K., J. C. Lewis, R. K. Guy, and J. N. Stuart. 2013. Use of anecdotal occurrence data in species distribution models: An example based on the white-nosed coati (*Nasua narica*) in the American Southwest. Animals 3(2):327–348.

Fronhofer, E. A., A. Kubisch, F. M. Hilker, T. Hovestadt, and H. J. Poethke. 2012. Why are metapopulations so rare? Ecology 93(8):1967–1978.

Furnas, B. J., and M. C. McGrann. 2018. Using occupancy modeling to monitor dates of peak vocal activity for passerines in California. Condor 120:188–200.

García-Callejas, D., and M. B. Araújo. 2016. The effects of model and data complexity on predictions from species distributions models. Ecological Modelling 326:4–12.

Gautestad, A. O., and I. Mysterud. 2010. Spatial memory, habitat auto-facilitation and the emergence of fractal home range patterns. Ecological Modelling 221(23):2741–2750.

Gilroy, J. J., and D. P. Edwards. 2017. Source-sink dynamics: A neglected problem for landscape-scale biodiversity conservation in the tropics. Current Landscape Ecology Reports 2:51–60.

Graham, J., and M. Kimble. 2019. Visualizing uncertainty in habitat suitability models with the hyper-envelope modeling interface, version 2. Ecology and Evolution 9(1):251–264.

Gravel, D., F. Guichard, M. Loreau, and N. Mouquet. 2010. Source and sink dynamics in meta-ecosystems. Ecology 91(7):2172–2184.

Hall, C. A. S., and J. W. Day. 1977. Systems and models: Terms and basic principles. Pp. 6–36 *in* C. A. S. Hall and J. W. Day, eds. Ecosystem modeling in theory and practice. Wiley Interscience, New York.

Halstead, B. J., P. M. Kleeman, and J. P. Rose. 2018. Time-to-detection occupancy modeling: An efficient method for analyzing the occurrence of amphibians and reptiles. Journal of Herpetology 52(4):416–425.

Hassler, C. C., S. A. Sinclair, and E. Kallio. 1986. Logistic regression: A potentially useful tool for researchers. Forest Products Journal 36:16–18.

Hefley, T. J., D. M. Baasch, A. J. Tyre, and E. E. Blankenship. 2015. Use of opportunistic sightings and expert knowledge to predict and compare whooping crane stopover habitat. Conservation Biology 29(5):1337–1346.

Heinrichs, J. A., C. L. Aldridge, M. S. O'Donnell, and N. H. Schumaker. 2017. Using dynamic population simulations to extend resource selection analyses and prioritize habitats for conservation. Ecological Modelling 359:449–459.

Hemming, V., M. A. Burgman, A. M. Hanea, M. F. McBride, and B. C. Wintle. 2018. A practical guide to structured expert elicitation using the IDEA protocol. Methods in Ecology and Evolution 9(1):169–180.

Hesselbarth, M. H. K., M. Sciaini, K. A. With, K. Wiegand, and J. Nowosad. 2020. Landscapemetrics: An open-source R tool to calculate landscape metrics. Ecography 42(10):1648–1657.

Hijmans, R. J. 2012. Cross-validation of species distribution models: Removing spatial sorting bias and calibration with a null model. Ecology 93(3):679–688.

Jay, C. V., B. G. Marcot, and D. C. Douglas. 2011. Projected status of the Pacific walrus (*Odobenus rosmarus divergens*) in the twenty-first century. Polar Biology 34(7):1065–1084.

Johnson, C. J., and M. P. Gillingham. 2004. Mapping uncertainty: Sensitivity of wildlife habitat ratings to expert opinion. Journal of Applied Ecology 41:1032–1041.

Johnson, D. S., M. B. Hooten, and C. E. Kuhn. 2013. Estimating animal resource selection from telemetry data using point process models. Journal of Animal Ecology 82(6):1155–1164.

Johnson, J. R., and R. D. Semlitsch. 2003. Defining core habitat of local populations of the gray treefrog (*Hyla versicolor*) based on choice of oviposition site. Oecologia 137(2):205–210.

Johnston, A., D. Fink, M. D. Reynolds, W. M. Hochachka, B. L. Sullivan, N. E. Bruns, E. Hallstein, M. S. Merrifield, S. Matsumoto, and S. Kelling. 2015. Abundance models improve spatial and temporal prioritization of conservation resources. Ecological Applications 25(7):1749–1756.

Jonzén, N., J. R. Rhodes, and H. P. Possingham. 2005. Trend detection in source-sink systems: When should sink habitats be monitored? Ecological Applications 15(1):326–334.

Kajin, M., P. J. A. L. Almeida, M. V. Vieira, and R. Cerqueira. 2012. The state of the art of population projection models: From the Leslie matrix to evolutionary demography. Oecologia Australis 16(1):13–22.

Kang, B., Y. Deng, R. Sadiq, and S. Mahadevan. 2012. Evidential cognitive maps. Knowledge-Based Systems 35:77–86.

Kelt, D. A., P. A. Kelly, S. E. Phillips, and D. F. Williams. 2014. Home range size and habitat selection of reintroduced *Sylvilagus bachmani riparius*. Journal of Mammalogy 95(3):516–524.

Kent, A., T. D. Drezner, and R. Bello. 2018. Climate warming and the arrival of potentially invasive species into boreal forest and tundra in the Hudson Bay Lowlands, Canada. Polar Biology 41(10):2007–2022.

Kenward, R. E., E. M. Arraut, P. A. Robertson, S. S. Walls, N. M. Casey, and N. J. Aebischer. 2018. Resource-area-dependence analysis: Inferring animal resource needs from home-range and mapping data. PLoS ONE 13(10):e0206354.

Kirk, D. A., A. C. Park, A. C. Smith, B. J. Howes, B. K. Prouse, N. G. Kyssa, E. N. Fairhurst, and K. A. Prior. 2018. Our use, misuse, and abandonment of a concept: Whither habitat? Ecology and Evolution 8:4197–4208.

Kirol, C. P., J. L. Beck, S. V. Huzurbazar, M. J. Holloran, and S. N. Miller. 2015. Identifying greater sage-grouse source and sink habitats for conservation planning in an energy development landscape. Ecological Applications 25:968–990

Koh, L. P., and J. Ghazoul. 2010. A matrix-calibrated species-area model for predicting biodiversity losses due to land-use change. Conservation Biology 24(4):994–1001.

Kosicki, J. Z. 2017. Should topographic metrics be considered when predicting species density of birds on a large geographical scale? A case of random forest approach. Ecological Modelling 349:76–85.

Lawler, J. J., D. S. White, R. P. Neilson, and A. R. Blaustein. 2006. Producing climate-induced range shifts: Model differences and model reliability. Global Change Biology 12(8):1568–1584.

Le Comber, S. C., A. C. Spinks, N. C. Bennett, J. U. M. Jarvis, and C. G. Faulkes. 2002. Fractal dimension of African mole-rat burrows. Canadian Journal of Zoology 80(3):436–441.

Legović, T., J. Klanjšček, and S. Geček. 2010. Maximum sustainable yield and species extinction in ecosystems. Ecological Modelling 221(12):1571–1576.

Lehmkuhl, J. F., B. G. Marcot, and T. Quinn. 2001. Characterizing species at risk. Pp. 474–500 *in* D. H. Johnson and T. A. O'Neill, eds. Wildlife-habitat relationships in Oregon and Washington. Oregon State University Press, Corvallis.

Leroux, S. J., M. Larrivée, V. Boucher-Lalone, A. Hurford, J. Zuloaga, J. T. Kerr, and F. Lutscher. 2013. Mechanistic models for the spatial spread of species under climate change. Ecological Applications 23(4):815–828.

Lindenmayer, D. B., R. C. Lacy, and M. L. Pope. 2000. Testing a simulation model for population viability analysis. Ecological Applications 10(2):580–597.

Liu, J., M. Wilson, G. Hu, J. Liu, J. Wu, and M. Yu. 2018. How does habitat fragmentation affect the biodiversity and ecosystem functioning relationship? Landscape Ecology 33(3):341–352.

Loehle, C. 1990. Home range: A fractal approach. Landscape Ecology 5(1):39–52.

Lorenz, T. J., K. T. Vierling, J. M. Kozma, and J. E. Millard. 2016. Foraging plasticity by a keystone excavator, the white-headed woodpecker, in managed forests: Are there consequences for productivity? Forest Ecology and Management 363:110–119.

MacKenzie, D. I. 2006. Modeling the probability of resource use: The effect of, and dealing with, detecting a species imperfectly. Journal of Wildlife Management 70(2):367–374.

MacKenzie, D. I., J. D. Nichols, J. A. Royle, K. H. Pollock, L. L. Bailey, and J. E. Hines. 2018. Occupancy estimation and modeling: Inferring patterns and dynamics of species occurrence, 2nd edition. Academic Press, London.

MacKenzie, D. I., J. D. Nichols, M. E. Seamans, and R. J. Gutiérrez. 2009. Modeling species occurrence dynamics with multiple states and imperfect detection. Ecology 90(3):823–835.

MacLean, M. G., and R. G. Congalton. 2013. PolyFrag: A vector-based program for computing landscape metrics. GIScience & Remote Sensing 50(6):591–603.

MacLean, M. G., and R. G. Congalton. 2015. A comparison of landscape fragmentation analysis programs for identifying possible invasive plant species locations in forest edge. Landscape Ecology 30(7):1241–1256.

Mandelbrot, B. B. 1977. The fractal geometry of nature. W. H. Freeman, San Francisco.

Manel, S., M. K. Schwartz, G. Luikart, and P. Taberlet. 2003. Landscape genetics: Combining landscape ecology and population genetics. Trends in Ecology & Evolution 18(4):189–197.

Marcot, B. G. 2002. An ecological functional basis for managing decaying wood for wildlife. Pp. 895–910 *in* W. F. Laudenslayer Jr., P. J. Shea, B. E. Valentine, C. P. Weatherspoon, and T. E. Lisle, eds. Proceedings of the Symposium on the Ecology and Management of Dead Wood in Western Forests, 2–4 November 1999, Reno, Nevada. General Technical Report PSW-181. US Department of Agriculture, Forest Service, Pacific Southwest Research Station, Albany, CA.

Marcot, B. G. 2006. Habitat modeling for biodiversity conservation. Northwestern Naturalist 87(1):56–65.

Marcot, B. G. 2012. Metrics for evaluating performance and uncertainty of Bayesian network models. Ecological Modelling 230:50–62.

Marcot, B. G., and K. B. Aubry. 2003. The functional diversity of mammals in coniferous forests of western North America. Pp. 631–664 *in* C. J. Zabel and R. G. Anthony, eds. Mammal community dynamics: Management and conservation in the coniferous forests of western North America. Cambridge University Press, Cambridge.

Marcot, B. G., and P. Z. Chinn. 1982. Use of graph theory measures for assessing diversity of wildlife habitat. Pp. 69–70 *in* R. Lamberson, ed. Mathematical models of renewable resources: Proceedings of the First Pacific Coast Conference on Mathematical Models of Renewable Resources. Humboldt State University, Arcata, CA.

Marcot, B. G., M. T. Jorgenson, J. Lawler, C. M. Handel, and A. R. DeGange. 2015. Projected changes in wildlife habitats in Arctic natural areas of northwest Alaska. Climatic Change 130(2):145–154.

Marcot, B. G., T. A. O'Neill, J. B. Nyberg, A. MacKinnon, P. J. Paquet, and D. H. Johnson. 2006. Analyzing key ecological functions for transboundary subbasin assessments. Pp. 37–50 *in* C. W. Slaughter and N. Berg, eds. Watersheds across boundaries: Science, sustainability, security; Proceedings of the Ninth Biennial Watershed Management Council Conference, November 3–7, 2002, Stevenson, Washington. Center for Water Resources Report No. 107. University of California, Riverside.

Marcot, B. G., M. G. Raphael, and K. H. Berry. 1983. Monitoring wildlife habitat and validation of wildlife-habitat relationships models. Transactions of the North American Wildlife and Natural Resources Conference 48:315–329.

Marcot, B. G., M. G. Raphael, N. H. Schumaker, and B. Galleher. 2013. How big and how close? Habitat patch size and spacing to conserve a threatened species. Natural Resource Modeling 26(2):194–214.

Marcot, B. G., and M. Vander Heyden. 2001. Key ecological functions of wildlife species. Pp. 168–186 *in* D. H. Johnson and T. A. O'Neill, eds. Wildlife-habitat relationships in Oregon and Washington. Oregon State University Press, Corvallis.

Martin, C. A. 2018. Bird seasonal beta-diversity in the contiguous USA. Journal of Ornithology 159(2):565–569.

Mathers, K. L., S. P. Rice, and P. J. Wood. 2019. Predator, prey, and substrate interactions: The role of faunal activity and substrate characteristics. Ecosphere 10(1):e02545.

McGarigal, K., and B. J. Marks. 1995. FRAGSTATS: Spatial pattern analysis program for quantifying landscape structure. General Technical Report PNW-351. US Department of Agriculture, Forest Service, Pacific Northwest Research Station, Portland, OR.

McRae, B. H., S. A. Hall, P. Beier, and D. M. Theobald. 2012. Where to restore ecological connectivity? Detecting barriers and quantifying restoration benefits. PLoS ONE 7(12):e52604.

McRae, B. H., and V. B. Shah. 2009. Circuitscape user guide. University of California, Santa Barbara. http://www.circuitscape.org.

Merckx, B., M. Steyaert, A. Vanreusel, M. Vincx, and J. Vanaverbeke. 2011. Null models reveal preferential sampling, spatial autocorrelation and overfitting in habitat suitability modelling. Ecological Modelling 222(3):588–597.

Merrick, M. J., and J. L. Koprowski. 2017. Circuit theory to estimate natal dispersal routes and functional landscape connectivity for an endangered small mammal. Landscape Ecology 32(6):1163–1179.

Michelot, T., P. G. Blackwell, and J. Matthiopoulos. 2019. Linking resource selection and step selection models for habitat preferences in animals. Ecology 100(1):e02452.

Milanesi, P., R. Holderegger, K. Bollmann, F. Gugerli, and F. Zellweger. 2017. Three-dimensional habitat structure and landscape genetics: A step forward in estimating functional connectivity. Ecology 98(2):393–402.

Miller, K. S., L. A. Brennan, H. L. Perotto-Baldivieso, F. Hernández, E. D. Grahmann, A. Z. Okay, X. B. Wu, M. J. Peterson, H. Hannusch, J. Mata, J. Robles, and T. Shedd. 2019. Correlates of habitat fragmentation and northern bobwhite abundance in the Gulf Prairie Landscape Conservation Cooperative. Journal of Fish and Wildlife Management 10(1):3–17. https://fwspubs.org/doi/pdf/10.3996/112017-JFWM-094/.

Millstein, R. L. 2014. How the concept of "population" resolves concepts of "environment." Philosophy of Science 81(5):741–755.

Milne, B. T. 1988. Measuring the fractal geometry of landscapes. Applied Mathematics and Computation 27:67–79.

Mladenoff, D. J., T. A. Sickley, and A. P. Wydeven. 1999. Predicting gray wolf landscape recolonization: Logistic regression models vs. new field data. Ecological Applications 9(1):37–44.

Morgan, J. W., J. D. Vincent, and J. S. Camac. 2018. Upper range limit establishment after wildfire of an obligate-seeding montane forest tree fails to keep pace with 20th century warming. Journal of Plant Ecology 11(2):200–207.

Morrison, M. L. 2001. A proposed research emphasis to overcome the limits of wildlife-habitat relationships studies. Journal of Wildlife Management 65(4):613–623.

Morrison, M. L., L. A. Brennan, B. G. Marcot, W. M. Block, and K. S. McKelvey. 2020. Foundations for advancing animal ecology. Johns Hopkins University Press, Baltimore.

Morrison, M. L., B. A. Collier, H. A. Mathewson, J. E. Groce, and R. N. Wilkins. 2012. The prevailing paradigm as a hindrance to conservation. Wildlife Society Bulletin 36(3):408–414.

Nickell, Z., S. Varriano, E. Plemmons, and M. D. Moran. 2018. Ecosystem engineering by bison (*Bison bison*) wallowing increases arthropod community heterogeneity in space and time. Ecosphere 9(9):e02436.

Ortega-Huerta, M. A., and A. T. Peterson. 2008. Modeling ecological niches and predicting geographic distributions: A test of six presence-only methods. Revista Mexicana de Biodiversidad 79:205–216.

Pacioni, C., M. S. Kennedy, O. Berry, D. Stephens, and N. H. Schumaker. 2018. Spatially explicit model for assessing wild dog control strategies in Western Australia. Ecological Modelling 368:246–256.

Partridge, L. 1978. Habitat selection. Pp. 351–376 *in* J. R. Krebs and N. B. Davies, eds. Behavioural ecology: An evolutionary approach. Sinauer Associates, Sunderland, MA.

Patoka, J., O. Kopecký, V. Vrabec, and L. Kalous. 2017. Aquarium molluscs as a case study in risk assessment of incidental freshwater fauna. Biological Invasions 19(7):2039–2046.

Patton, D. R. 1992. Wildlife habitat relationships in forested ecosystems. Timber Press, Portland OR.

Pennycuick, C. J., and N. C. Kline. 1986. Units of measurement for fractal extent, applied to the coastal distribution of bald eagle nests in the Aleutian Islands, Alaska. Oecologia 68:254–258.

Peterson, A. T., M. Papes, and M. Eaton. 2007. Transferability and model evaluation in ecological niche modeling: A comparison of GARP and Maxent. Ecography 30:550–560.

Phillips, S. J., R. P. Anderson, and R. E. Schapire. 2006. Maximum entropy modeling of species geographic distributions. Ecological Modelling 190:231–259.

Phillips, S. J., R. E. Schapire, and M. Dudík. 2004. A maximum entropy approach to species distribution modeling. Pp. 655–662 *in* R. Greiner and D. Schuurmans, eds. Proceedings of the Twenty-First International Conference on Machine Learning. Association for Computing Machinery, New York.

Polfus, J. L., K. Heinemeyer, and M. Hebblewhite. 2014. Comparing traditional ecological knowledge and Western science woodland caribou habitat models. Journal of Wildlife Management 78(1):112–121.

Price, J. M., J. Silbernagel, K. Nixon, A. Swearingen, R. Swaty, and N. Miller. 2016. Collaborative scenario modeling reveals potential advantages of blending strategies to achieve conservation goals in a working forest landscape. Landscape Ecology 31(5):1093–1115.

Prokopenko, C. M., M. S. Boyce, and T. Avgar. 2016. Characterizing wildlife behavioural responses to roads using integrated step selection analysis. Journal of Applied Ecology 54(2):470–479.

Pullinger, M. G., and C. J. Johnson. 2010. Maintaining or restoring connectivity of modified landscapes: Evaluating the least-cost path model with multiple sources of ecological information. Landscape Ecology 25(10):1547–1560.

Quinlan, J. R. 1986. Induction of decision trees. Machine Learning 1:81–106

Quinlan, J. R. 1993. C4.5: Programs for machine learning. Morgan Kaufmann, San Mateo, CA.

Raphael, M. G., and B. G. Marcot. 1986. Validation of a wildlife-habitat-relationships model: Vertebrates in a Douglas-fir sere. Pp. 129–138 *in* J. Verner, M. L. Morrison, and C. J. Ralph, eds. Wildlife 2000: Modeling habitat relationships of terrestrial vertebrates. University of Wisconsin Press, Madison.

Remmers, W., J. Gameiro, I. Schaberl, and V. Clausnitzer. 2017. Elephant (*Loxodonta africana*) footprints as habitat for aquatic macroinvertebrate communities in Kibale National Park, south-west Uganda. African Journal of Ecology 55(3):342–351.

Robinson, T. P., R. D. van Klinken, and G. Metternicht. 2010. Comparison of alternative strategies for invasive species distribution modeling. Ecological Modelling 221(19):2261–2269.

Rogers, L. L., and A. W. Allen. 1987. Habitat suitability index models: Black bear, Upper Great Lakes Region. Biological Report FWS/OBS-82/10.144. US Department of the Interior, Fish and Wildlife Service, Research and Development, Washington, DC.

Rosenberg, D. K., and K. S. McKelvey. 1999. Estimation of habitat selection for central-place foraging animals. Journal of Wildlife Management 63(3):1028–1038.

Salwasser, H., J. C. Capp, H. Black, and J. F. Hurley. 1980. The California Wildlife Habitat Relationships Program: An overview. Pp. 369–378 *in* R. M. DeGraff and N. G. Tilghman, comps. Workshop proceedings: Management of western forests and grasslands for nongame birds. General Technical Report INT-86. US Department of Agriculture, Forest Service, Intermountain Forest and Range Experiment Station, Ogden, UT.

Samuel, M. D., and E. O. Garton. 1987. Incorporating activity time in harmonic home range analysis. Journal of Wildlife Management 51:254–257.

Schamberger, M., A. H. Farmer, and J. W. Terrell. 1982. Habitat suitability index models: Introduction. Biological Report FWS/OBS-82/10. US Department of the Interior, Fish and Wildlife Service, Division of Ecological Services, Office of Biological Services, Washington, DC.

Scherer, C., F. Jeltsch, V. Grimm, and N. Blaum. 2016. Merging trait-based and individual-based modelling: An animal functional type approach to explore the responses of birds to climatic and land use changes in semi-arid African savannas. Ecological Modelling 326:75–89.

Schroeder, R. L. 1982. Habitat suitability index models: Yellow warbler. Biological Report FWS/OBS-82/10.147. US Department of the Interior, Fish and Wildlife Service, Western Energy and Land Use Team, Washington, DC.

Schumaker, N. H., and A. Brookes. 2018. HexSim: A modeling environment for ecology and conservation. Landscape Ecology 33(2):197–211.

Schumaker, N. H., A. Brookes, J. R. Dunk, B. Woodbridge, J. A. Heinrichs, J. J. Lawler, C. Carroll, and D. LaPlante. 2014. Mapping sources, sinks, and connectivity using a simulation model of northern spotted owls. Landscape Ecology 29:579–592.

Scott, J. M., F. Davis, B. Csuti, R. Noss, B. Butterfield, C. Groves, H. Anderson, S. Caicco, F. D'Erchia, T. C. Edwards Jr., J. Ulliman, and R. G. Wright. 1993. Gap Analysis: A geographic approach to protection of biological diversity. Wildlife Monographs 123:1–41.

Scrafford, M. A., T. Avgar, B. Abercrombie, J. Tigner, and M. S. Boyce. 2017. Wolverine habitat selection in response to anthropogenic disturbance in the western Canadian boreal forest. Forest Ecology and Management 395:27–36.

See, K. E., and E. E. Holmes. 2015. Reducing bias and improving precision in species extinction forecasts. Ecological Applications 25(4):1157–1165.

Shcheglovitova, M., and R. P. Anderson. 2013. Estimating optimal complexity for ecological niche models: A jackknife approach for species with small sample sizes. Ecological Modelling 269:9–17.

Silvano, A. L., C. Guyer, T. D. Steury, and J. B. Grand. 2017. Selecting focal species as surrogates for imperiled species using relative sensitivities derived from occupancy analysis. Ecological Indicators 73:302–311.

Singer, A., K. Johst, T. Banitz, M. S. Fowler, J. Groeneveld, A. G. Gutiérrez, F. Hartig, R. M. Krug, M. Liess, G. Matlack, K. M. Meyer, G. Pe'er, V. Radchuk, A.-J. Voinopol-Sassu, and J. M. J. Travis. 2016. Community dynamics under environmental change: How can next generation mechanistic models improve projections of species distributions? Ecological Modelling 326:63–74.

Steenweg, R., M. Hebblewhite, R. Kays, J. Ahumada, J. T. Fisher, C. Burton, S. E. Townsend, C. Carbone, J. M. Rowcliffe, J. Whittington, J. Brodie, J. A. Royle, A. Switalski, A. P. Clevenger, N. Heim, and L. N. Rich. 2017. Scaling-up camera traps: Monitoring the planet's biodiversity with networks of remote sensors. Frontiers in Ecology and Evolution 15(1):26–34.

St.-Louis, V., J. D. Forester, D. Pelletier, M. Bélisle, A. Desrochers, B. Rayfield, M. A. Wulder, and J. A. Cardille. 2014. Circuit theory emphasizes the importance of edge-crossing decisions in dispersal-scale movements of a forest passerine. Landscape Ecology 29(5):831–841.

Stockwell, D. R. B., and D. G. Peters. 1999. The GARP modelling system: Problems and solutions to automated spatial prediction. International Journal of Geographic Information Systems 13:143–158.

Stølum, H.-H. 1996. River meandering as a self-organizing process. Science 271(5256):1710–1713.

Stouffer, D. B. 2019. All ecological models are wrong, but some are useful. Journal of Animal Ecology 88(2):192–195.

Straube, D., E. A. Johnson, D. Parkinson, S. Scheu, and N. Eisenhauer. 2009. Nonlinearity of effects of invasive ecosystem engineers on abiotic soil properties and soil biota. Oikos 118(6):885–896.

Stuber, E. F., L. F. Gruber, and J. J. Fontaine. 2017. A Bayesian method for assessing multi-scale species-habitat relationships. Landscape Ecology 32(12):2365–2381.

Takeuchi, T., R. Matsuki, and M. Nashimoto. 2012. GPS cell phone tracking in the Greater Tokyo area: A field test on raccoon dogs. Urban Ecosystems 15(1):181–193.

Tape, K. D., B. M. Jones, C. D. Arp, I. Nitze, and G. Grosse. 2018. Tundra be dammed: Beaver colonization of the Arctic. Global Change Biology 24(10):4478–4488.

Thurfjell, H., S. Ciuti, and M. S. Boyce. 2014. Applications of step-selection functions in ecology and conservation. Movement Ecology 2:art. 4.

Tuma, M. W., C. Millington, N. Schumaker, and P. Burnett. 2016. Modeling Agassiz's desert tortoise population response to anthropogenic stressors. Journal of Wildlife Management 80(3):414–429.

Utgoff, P. E. 1989. Incremental induction of decision trees. Machine Learning 4(2):161–186.

Valletta, J. J., C. Torney, M. Kings, A. Thornton, and J. Madden. 2017. Applications of machine learning in animal behavioural studies. Animal Behaviour 124:203–220.

Vermeiren, P., C. Munoz, M. Zimmer, and M. Scheaves. 2016. Hierarchical toolbox: Ensuring scientific accuracy of citizen science for tropical coastal ecosystems. Ecological Indicators 66:242–250.

Vimal, R., A. S. L. Rodrigues, R. Mathevet, and J. D. Thompson. 2011. The sensitivity of gap analysis to conservation targets. Biodiversity and Conservation 20(3):531–543.

Vogt, P., and K. Riiters. 2017. GuidosToolbox: Universal digital image object analysis. European Journal of Remote Sensing 50(1):352–361.

Wakeley, J. S. 1988. A method to create simplified versions of existing habitat suitability index (HSI) models. Environmental Management 12(1):79–83.

Wallentin, G. 2017. Spatial simulation: A spatial perspective on individual-based ecology—a review. Ecological Modelling 350:30–41.

Walsh, P. D., H. R. Akcakaya, M. Burgman, and A. H. Harcourt. 1995. PVA in theory and practice. Conservation Biology 9(4):704–708.

Wilcove, D. S. 1987. From fragmentation to extinction. Natural Areas Journal 7:23–29.

Wilcox, C., A. J. Hobday, and L. E. Chambers. 2018. Using expert elicitation to rank ecological indicators for detecting climate impacts on Australian seabirds and pinnipeds. Ecological Indicators 95:637–644.

Williford, D., R. W. DeYoung, R. L. Honeycutt, L. A. Brennan, and F. Hernández. 2016. Phylogeography of the bobwhite (*Colinus*) quails. Wildlife Monographs 193:1–49.

Wintle, B. A., R. P. Kavanagh, M. A. McCarthy, and M. A. Burgman. 2005. Estimating and dealing with detectability in occupancy surveys for forest owls and arboreal marsupials. Journal of Wildlife Management 69(3):905–917.

Yackulic, C. B., S. Blake, S. Deem, M. Kock, and M. Uriarte. 2011. One size does not fit all: Flexible models are required to understand animal movement across scales. Journal of Animal Ecology 80(5):1088–1096.

Zeller, K. A., K. McGarigal, S. A. Cushman, P. Beier, T. W. Vickers, and W. M. Boyce. 2016. Using step and path selection functions for estimating resistance to movement: Pumas as a case study. Landscape Ecology 31(6):1319–1335.

6 Where We Go from Here

New Imperatives

Under normal conditions the research scientist is not an innovator but a solver of puzzles, and the puzzles upon which he concentrates are just those which he believes can be both stated and solved within the existing scientific tradition.

T. S. Kuhn (1977:234)

Introduction

Although the eminent scientist and philosopher Thomas Kuhn was making a commentary on science in general, his remarks about a research scientist being "not an innovator but a solver of puzzles" applies equally well to the field of animal ecology. Kuhn was best known for his criticism of key positivist doctrines, and he initiated a new style in the philosophy of science that was closely linked with the actual history of science. As such, Kuhn (1996) noted that the development of science was marked by periods of slow progress, punctuated by revisionary revolutions when a prevailing paradigm was broken. He defined a "scientific paradigm" as what was to be observed, the types of questions that were to be asked, how these questions were to be structured, and how the results were to be interpreted. In other words, there was a set of questions to be asked, and a standard way to pursue research on them.

Adhering to the prevailing paradigm is comfortable because, by definition, other scientists are doing the same thing. The ultimate risk of failure—namely, not being able to get published—is minimized. Scientists seldom subject a favored hypothesis to the often unrealistically high standards they set for the work of others (Hull 1988:348). Prior beliefs are not likely to be swayed by data that are inconsistent with them. In addition, Hull (1988:113) cautioned against overstating the differences between competing conceptual systems (e.g., for our case here, "community" and "metacommunity"). In relation to our emphasis on individuals and biological populations, this quote by Hull (1988:113) seems appropriate: "Research groups as social groups are as difficult to treat accurately as are biological populations."

In his critique of how we approach wildlife-habitat studies, Morrison (2012) discussed how we were (and still are) "stuck in a rut." We identify a convenient study area, sample a range of vegetation and other environmental parameters, apply a series of increasingly sophisticated statistical analyses, compare our results with other studies, and justify publication by extrapolating findings to some unspecified larger area. In our publications, we provide suggestions for additional research, as well as a list of usually rather vague recommendations for the management and, ultimately, conservation of the species being studied. As was previously noted (Morrison 2012), we are probably overstating the need to have a paradigm shift per se, because we are really just calling for a different and clearer focus on how we pursue our studies of animals and the environment. If there is a paradigm that needs changing, it is along the lines of breaking through the "we cannot identify biological populations and related entities because it is too difficult" mantra that is expressed so often by critics of our recommended approach.

Similar to what Bennett (1987) stated over 30 years ago regarding the field of ecological physiology, we tend to gather more data on phenomena that are already generally understood, but we do so either with different species or with the same species at different times and places. That is, animal ecologists often tend to be risk adverse when it comes to devising and applying new approaches to field research, preferentially opting for traditional approaches that will pass muster with journals' editorial and publication processes. In research-oriented universities and other research organizations, it makes sense to be risk adverse if, for example, you are an assistant professor seeking tenure. But after the initial career hurdles are successfully crossed, scientists are generally expected to conduct studies that advance knowledge, rather than just accumulating more papers on similar topics. Unfortunately, most funding organizations are likewise risk adverse, because they want something definitive—not speculative or theoretical—to show for their support.

Solving the boundary issue delineating biological populations is difficult, because it takes time and effort, which means it costs a lot of money and may not lend itself to the typical short-term studies required for academic degrees. As such, unless substantially more financial support becomes available for investigations in animal ecology, organizations instead would need to focus their research funds on fewer studies. Although it is easy to pontificate on how other people should spend their money, if the goal is to advance the kind of knowledge that leads to the persistence of animal populations, then we should probably worry about those populations in the first place. Academic institutions may need to craft new and clever ways in which degree candidates can contribute to longer-term studies in a productive yet timely manner. And journal editors should be as willing to publish papers on what did not happen or go as planned (i.e., the misnamed "negative results") as they are for those that produce "positive results"—although both should contain statistically significant findings, with appropriate study designs and sufficient sampling intensities. Otherwise, someone will just come along and do the same thing without knowing it was tried before.

We have been witnessing (and will continue to see) a rapid increase in technological achievements, including the miniaturization of static and, especially, mobile monitoring devices; in advances to our ability to study genetics and physiology; in faster computers that allow us to construct more-sophisticated population models; in far more detailed remote sensing imagery; and so on. Yet we are applying these enhanced techniques without similarly improving the way we think about organisms in space and time. For example, being able to obtain remote sensing imagery at a high resolution is not likely to tell us much more than what we can already gather from on-the-ground observations (albeit faster and at less cost). Imagery is easy to obtain; you do not even need to leave your office.

Humans are exerting both direct and indirect pressures on the land base and its inhabitants, which bring along with them an increase in stressors and extinctions, as well as worries about human-induced alterations in the environment. These dislocations present ecologists with several closely related challenges, including the need to conduct studies that address land-use problems in relation to anthropogenic changes across multiple spatial and temporal scales. Such events will be increasingly represented by shifts in species distributions and environmental conditions never before seen, and they will be occurring at an ever-quickening pace. The great challenge here is to delineate populations and their dynamics against a rapidly shifting backdrop. Further, we must be able to keep abreast of the latest advances in computer software, modeling capabilities, and analytical tools, as well as new methods in metagenetics, physiology, and remote sensing. These requirements necessitate a higher level of education and experience than previously seen in almost any field of ecology.

In this chapter we discuss several problems that will confront ecologists—and scientists in general—in the coming years, as well as the types of information needed to adequately address them. In earlier chapters, we have already made the case for a need to focus on understanding what drives animal persistence, so we will not revisit that theme in any specificity here. (See also Morrison et al. 2020 for a thorough theoretical development of this concept.) Rather, we will focus on the appropriate scale for studies, some suggested areas of emphasis for research, the need to better tailor our research to meet practical applications, and the training and education required at all levels of the research-to-implementation process.

Scale of Conservation

Beginning in the 1970s and accelerating into the 1980s and beyond, the historical concentration on single-season, site-specific studies conducted at the microhabitat scale began to give way to multiseason, multisite investigations that examined ecological relationships at broader spatial scales. These shifts were driven by both the ecological and practical needs of resource managers. Ecologically, scientists began to better understand the hierarchical nature of habitat selection, as developed by Johnson (1980) and Hutto (1985). We also started to see the mistakes that had been made in previous studies of habitat, including mismatching spatial scales within an investigation (Wiens 1989:227–233; Scott et al. 2002). Managers were increasingly asking for research that incorporated multiple species over broader spatial scales than had historically been employed (Scott et al. 2002). Pressures from an ever-expanding human population were driving the need to view land and wildlife from a more focused perspective.

The move toward investigations that considered multiple and broader spatial scales also reflected the natural progression of our knowledge. The study of the ecology of animals properly begins with species-specific, detailed investigations of life-history parameters. As our scientific knowledge about organisms grows, so does our ability to explore and understand the interactions between them and the host of factors influencing their survival and persistence. Stauffer (2002) presented a concise history of habitat studies in animal ecology, and there have been an increasing number of papers at what has been considered to be the "landscape" scale since the 1980s. The burgeoning interest in multiple-species research over broader spatial scales, along with a growing realization of the adverse impacts human activities have on the natural world, led, in part, to the formation of organizations such as the Society for Conservation Biology and the initiation of more-targeted journals, exemplified by *Conservation Biology* and *Landscape Ecology*.

Regardless of our interests and the need for such a focus, has our knowledge of spatial relationships been adequate to meet the conservation challenge? Clearly, additional attention should be paid to disturbance dynamics and habitat configuration, in order to augment autoecological studies in such a broad-scale context. Based on the concepts and approaches we developed throughout this book, it should be evident that we are not promoting "broad scale" per se. Rather, we should shape our investigations by applying appropriate boundaries for the question(s) being asked.

Advancing Habitat Research

In an earlier book that served as a partial foundation for what we have presented in this volume, Morrison et al. (2006) outlined some thoughts on how we could advance the study of wildlife-habitat relationships. Although we draw on that material again as a foundation, because it represented a synthesis of the ideas of other scientists, here we expand upon those concepts to increase their relevance to our current focus on biological populations.

The distinction between extensive and intensive

variables made by Slobodkin (1992), the perception of them by individual organisms, and the resulting impact on habitat selection theory are all directly related to this focus. "Intensive variables" could be directly sensed by an individual organism, whereas "extensive variables" could only be perceived by considering an overview of an entire system. By "system," the author was referring to the spatial extents in which individual animals were located, such as a research site or a broader ecosystem. As an example, he said that the extensive variable "population size" meant less to an individual organism within that population than the intensive variables that were directly associated with food, such as local population density, which may affect intraspecific competition. This extensive-intensive dichotomy is similar to the construct developed by Johnson (1980), where investigations of habitat use and studies of feeding belonged to different orders of complexity. Food is at a lower order than (i.e., is a subset of) habitat, and thus is a relatively more intensive and proximate variable. Slobodkin (1992) concluded that separating intensive and extensive variables in investigations of habitat selection allowed us to better explain the proximate mechanisms driving the behavior of individual organisms and the ultimate context in which such behaviors arise and function. This reinforces our emphasis on identifying and working within biological populations as a basis for research on habitat use. If we assume that selection and drift occur at the population level, we cannot really draw any meaningful conclusions regarding animal ecology if we base our work on an arbitrary set of conspecifics as study units (sensu Millstein 2009).

Thus the broad spatial-scale (i.e., macrohabitat, or landscape) patterns we describe are only the accumulated picture of many individual organisms selecting resources at finer scales. As such, these patterns do not mean that each animal recognizes the distribution of resources or their constraints at population scales, or, if so, that these aspects are of primary importance to that individual. Natural selection operates on variations in fitness among individuals, and the accumulated picture of this action is expressed in the resultant population parameters. Linking studies of habitat use to the Causal Interactionist Population Concept, or CIPC (Millstein 2014; see also Morrison et al. 2020:Chapter 1), thus makes sense in an evolutionary context, and it provides a powerful way to compare different investigations within and between temporal and spatial scales. That is, rather than trying to assess and find explanations for dissimilarities between studies, conducted in divergent places and times, that were based on some unknown portion of a biological population, placing our research in a known population context allows us to make more meaningful conclusions on why we see the similarities and differences that we do. Meta-analyses do not provide a work-around for this population issue.

The methodology employed in applied studies of habitat ecology is based on a natural history approach, where the focus is on a description of *what* is being used or done. Such research stands as the foundation upon which additional knowledge is built. While this is essential, it is not sufficient if we hope to gain knowledge about why organisms are doing what we observe, as well as how those behaviors lead to persistence through time. Investigations of habitat use do not imply any active selection process or represent consequences for survival and fitness (e.g., Morse 1980:89–90; Hutto 1985; Hall et al. 1997). Asking what habitat features a species uses is neither a "how" nor a "why" question (Gavin 1991).

Most habitat studies usually equate "selection" with the non-random use of some item, often vegetation (see reviews by Johnson 1980; Thomas and Taylor 1990; Alldredge and Ratti 1992). Thus animals can be non-selective under such designs, whereas statistical evidence of avoidance may really be just an artifact of the investigation, or even a remnant of short-lived or quirky environmental conditions. A case in point is that studies of mammal species in the Alexander Archipelago of southeastern Alaska may show a statistically significant selection for old-growth forests (e.g., Person et al. 1996), but this pattern may not reflect past conditions, because the landscape there now mostly consists of either old, uncut forests or recently logged areas in their early successional stages.

A more evolutionary approach is to regard habitat selection as constantly occurring, because it is a process (Hall et al. 1997). Its fundamental expression is at the scale of the individual; its patterns can shift over time and space in response to different degrees of available conditions (such as in the southeastern Alaska example); and it should be viewed in terms of intensive variables. This framework separates the application of extensive studies of population phenomena (e.g., density-dependency, animal-vegetation relationships) from intensive explorations of resource acquisition. CIPC characterizes populations in ecological and evolutionary contexts, in order to delineate the environment in which the species are evolving.

Translating Ecological Research into Management

The plea to make ecological research more accessible and useable by resource managers has a long history. For example, Risser (1993) called for theoretical ecologists to accept responsibility for explaining how such research can be utilized by managers, stating that it was inappropriate to blame or criticize these individuals for not finding the information or for finding it but then using it incorrectly. He urged journals such as *Ecological Applications* to include a closing section in each article that described results in ways that could be helpful to resource managers. Such sections have long been a standard feature in many applied ecological journals (most notably, the *Journal of Wildlife Management* and the *Wildlife Society Bulletin*). Unfortunately, many journals (including *Ecological Applications*) still have no dedicated section on the application of research results, although, to its credit, the Ecological Society of America, which publishes *Ecological Applications*, also issues a more manager-friendly journal, *Frontiers in Ecology and the Environment*, and The Wildlife Society prints *The Wildlife Professional*. In contrast, other scientific journals are striving to make research results more accessible to managers, including *Restoration Ecology*, which highlights specific "Implications for Practice" immediately following the abstract of an article. Another is the *Journal of Applied Ecology*, which has a brief "Synthesis and Applications" bullet point as part of each abstract. *The Condor: Ornithological Applications*, however, is surprisingly inconsistent in its use of a specific "Implications" section. More often than not, the "Management Implications" or "Implications for Practice" portions of articles in natural resource journals are just a continuation or extension of the "Discussion" section, rather than 1 or 2 short paragraphs about what the information in the paper, and its inferences, can mean to managers (Guthery 2011; Brennan et al. 2013).

It seems to us that, as scientists, we can substantially advance conservation by giving more attention to how our research results can be, or cannot be, used in terms of real-world situations. Although such work may have no direct utility per se, this certainly does not mean that it should not have been conducted. Here we are simply calling for more attention to be paid and focus directed toward what applications are possible, based on the type of investigation. When a researcher is reviewing the literature, and then deciding on what methods to use and which variables to measure, thought should be given to how the study's results might be specifically employed in management contexts. For example, it is standard practice for investigators to divide vegetation measurements into extremely fine categories, which might make good sense for the ecological aspects of that research. But they seldom ask, "How could the results of such methods and analyses be translated into management practices?" Clearly, a manager cannot reduce foliage volume in the 2–4 m height category without that being an extremely expensive procedure. Yet scientists could analyze their findings in a way that would lead to their practical utilization in management efforts. They could also provide comments on the potential value of doing so.

There is no reason for not including a few additional analyses that could translate to more useful management applications, in addition to those that would be impractical (e.g., based on the best frequentist *P*-value or Bayesian credible interval). We need not be totally consumed by generating the best R^2 and *P*-values (Anderson et al. 2001).

Alternatively, we also think that it is the clear responsibility of the wildlife biology and wildlife management communities to become well versed in the theoretical underpinnings of the ecological systems in which they work, as well as to keep current on new research results. This does not mean that we need to be theoreticians per se and have to read reams of the latest journal articles. Rather, it is our responsibility to at least possess a basic understanding of the theories and concepts on which our management decisions are or could be made.

Education and Training for the Future

"Our current understanding amounts to bungalows. To continue on the present course will only lead to better bungalows . . . Instead, it is within our capacity to take a revolutionary course and develop understanding to the state of Taj Mahals" (Romesburg 1991:744). The author used this analogy to introduce his argument about why we are failing to advance our understanding of ecological relationships and wildlife management as rapidly as we could. Moreover, he argued strongly that we are falling short in educating our students in a manner that will lead to such advancements in the future (see also Knight 1993; Romesburg 1993). Romesburg is not alone, nor was he the first to raise serious questions regarding the rigor with which we educate our wildlife students (e.g., Clark 1986; Gavin 1989; Hunter 1989). As noted by Kroll (2009) in his review, however, although we should always strive for rigor, rigor per se has not necessarily translated into better management skills.

As technology improves, we are forced to fill our students' minds with even more facts. Romesburg (1991) emphasized that separate undergraduate and graduate courses in research philosophy are vital in teaching students how to learn to think, and he provided specific topics that such courses should contain. While they do show how the process of science proceeds, he also argued persuasively that we should use a broader set of disciplines to teach students how to invent new ideas. Courses of this sort would introduce them to the diverse frameworks of theories from many disciplines and demonstrate how such theories can be applied to the natural resources. In addition, we should educate students on how to identify potentially significant problems (Keppie 1990; Romesberg 1991).

To expand on these suggestions, we also see a role for critical thinking, including how it needs to address the topics of uncertainty and variability. For instance, in traditional frequentist statistical approaches, if a study fails to reject a null hypothesis, it does not mean that the null hypothesis is proven, or even that it is necessarily correct. On the other hand, rejecting a null hypothesis does not prove that an alternate hypothesis is valid, particularly when multiple alternate hypotheses are plausible. Further, students—and professionals—need to understand the concepts and roles of hidden and latent variables, those pesky causal factors that are not studied (or even observable) but which could be the underlying reasons for observed patterns of habitat use and selection. In addition, there are constructs in the world of Bayesian statistics, such as Bayesian networks, that explicitly model and denote the interplay of uncertainties in observed conditions and outcomes. Ultimately, understanding what is well known, and what is uncertain, should inform management decisions by being placed in a risk assessment and risk management framework. As Marcot et al. (2016) declared in the title of their article, "Uncertainty is information, too."

Programs in animal ecology would serve their students well if they required training in multiple areas, prior to completing a graduate program. Although we recognize that individuals do need flexibility in their choice of coursework, all students should have a common core to serve as a foundation for more discipline-specific courses. To this end, the following are suggested topics, and many of them could be folded into just a few courses:

- history of science and the historical and current links (or gaps) between disciplines
- general philosophy of knowledge, theory, and creativity
- research philosophy in the sciences
- the ways in which theories and hypotheses in science are developed
- theory and methods of study design, including impact assessment
- statistics, up to and including non-parametric and multivariate analyses, as well as Bayesian methods
- an understanding of mathematical and computer modeling

In addition, some knowledge of how people, in general, perceive and act on contradictory or uncertain information, and how they integrate new findings into various environmental philosophical perspectives, would go far in helping researchers and managers alike to think beyond preset or traditional viewpoints. An entire discipline has been developed in recent decades that addresses much of what we advocate, under the rubric of "conservation psychology" (Saunders 2003; Clayton and Myers 2009).

Kroll (2009) provided 4 priorities that he thought would improve students' preparations to be effective wildlife professionals. Although he acknowledged that researchers and managers have different professional roles, which require specific skills to meet these challenges, he noted that success in wildlife management and conservation required interactions between these 2 occupations. He posited that (1) all wildlife programs should offer a mandatory planning course, which integrates the disciplines that participate in resource planning activities; (2) wildlife programs should use the requirements for The Wildlife Society's Certified Wildlife Biologist (CWB) Program as a template for better integration in wildlife programs; (3) professional internships should be mandatory for students who are working toward a master's or doctoral degree; and (4) departments could be more selective in their admissions processes, as a way to identify students who have the potential to be effective wildlife scientists and managers. Although these suggestions were tailored to a wildlife curriculum, his comments apply equally well to any program that focuses on animal ecology, regardless of its moniker.

There has been a steady stream of books designed to bridge the gap between scientists and resource managers, from such classics as *Game Management* (Leopold 1933) and *Wildlife Biology* (Dasmann 1964) to more-current works. These include *Introduction to Wildlife Management: The Basics* (Krausman 2001) and *Wildlife Ecology, Conservation, and Management* (Fryxell et al. 2014), which are useful at the undergraduate level; and *Wildlife Habitat Relationships: Concepts and Applications* (Morrison et al. 2006) and *Wildlife Science: Connecting Research with Management* (Sands et al. 2012), which are applicable at the graduate level. Recognizing the continuing difficultly in translating animal ecology research to management applications, *Applied Wildlife Habitat Management* (Lopez et al. 2017) was written to appeal to and ultimately link undergraduate education and professionals in the field. Similarly, *Restoring Wildlife: Ecological Concepts and Practical Applications* (Morrison 2009) attempted to directly bring advanced concepts in animal ecology to the field of ecological restoration.

Moreover, education and training must continue after a formal university education ends. Fortunately, the rapid increase in online distance courses allows working professionals to access quality continuing education from virtually any geographic location. Attendance at professional meetings also exposes individuals to new people and new ideas. Such meetings usually have numerous workshops and interactive field outings that help keep participants current on methods, analyses, and other innovations. An increasing number of journals are publishing open-source material, accessible to all, which can be part of one's ongoing learning toolkit. All such training is ultimately the responsibility of each individual, but it can certainly be promoted within every relevant organization, be it governmental or private (including for-profits and nonprofits). Individual supervisors, as well as organizational policies, can create an atmosphere where staying current, creatively thinking about new ways to approach old problems, establishing programs for continuing education, encouraging membership and participation in professional societies, and the like could lead to new responsibilities, job advancement, and, most importantly, sound resource management as a legacy for future generations. The continual advances in technology, and the sheer volume of scientific research, require all of us in the field of animal ecology to stay current and properly interpret what all of this information means.

Postlude

We began writing this book, as well as Morrison et al. (2020), *Foundations for Advancing Animal Ecology*, with an acknowledgment that we have learned—and continue to learn—a lot about where different species occur, and how organisms interact, and we have published innumerable descriptive papers on habitat and resource use that were taking place at various locales and times. But we caution that these studies alone, primarily because of the way in which they are being conducted, do not provide a framework for taking the next step necessary to substantially advance our understanding of animal ecology, including the conservation goal of long-term animal persistence. Thus we have taken the approach that research in animal ecology should begin with a focus on the behaviors and characteristics of individual organisms, which are then put into the context of how they are organized into collections—that is, breeding pairs, leks, herds, flocks, and, of course, populations. Only by tackling the problem of determining the physical boundaries of these groupings can we hope to truly understand the factors that lead to their survival and, ultimately, their persistence. Artificial and often arbitrary labels—such as study populations, metapopulations, communities, and metacommunities—do not provide a clear understanding of intra- and interspecific interactions (e.g., competition, habitat selection, mate choice, food webs). We have focused instead on the issue of resolving the boundary problem, so relevant concepts, including landscapes and fragmentation, can be examined from a solid foundation.

It is our thesis that as scientists, we must acknowledge the problems we face in the field of animal ecology and not sweep them under the rug, simply because they are difficult to solve. A necessary first step for every investigation—without exception—is to discuss what you know and do not know, and what you can solve and cannot solve, regarding the structure of the biological population you are attempting to study, given the practical constraints of available time and resources. Not knowing the boundaries of a population (or some justifiable segment thereof) does not imply that your findings should not be published. What it does mean is that we, as researchers, are being fully transparent, in order to place our results in a justifiable context. Your manuscript might be submitted to a top-tier journal, but will it be published and, more importantly, accessible to interested people? What we are calling for is really nothing more than what is required in a properly designed study of any type: clearly explaining your sampling frame in relation to your target population (e.g., Morrison et al. 2008). We must move beyond using the terms "population" and "habitat" in vague and inappropriate ways. We must place the segment of interest in the context of what we know and do not know about the overall biological population (Brennan and Block 2018) and refocus our attention on habitat—that is, on those conditions pertinent to particular organisms and species. And then let the buyer (i.e., reader) beware.

We also emphasized that our human perspective

of the natural world is important, because if the ultimate utility of ecology is conservation, then we must be able to translate our work into actions that can and will be implemented by resource managers, as well as be understandable to the public, policy makers, and legislators. Much of the research needed to accomplish these goals will be far more difficult to conduct than what we have been doing. Investigations—scaled to the species of interest—will usually require sampling over larger spatial extents and longer periods of time. They will often involve working across multiple land ownerships, and interacting with individuals having contrary views and expectations, so compromises will have to be made. But in the end, we will accumulate knowledge that allows us to start generalizing in a much more robust manner than is currently possible. Despite this progress, our research will undoubtedly generate yet another call to break through 1 more seemingly insurmountable impediment to advancing what we know about animal ecology. There is, indeed, still much work to be done.

LITERATURE CITED

Alldredge, J. R., and J. T. Ratti. 1992. Further comparison of some statistical techniques for analysis of resource selection. Journal of Wildlife Management 56:1–9.

Anderson, D. R., W. A. Link, D. H. Johnson, and K. P. Burnham. 2001. Suggestions for presenting the results of data analyses. Journal of Wildlife Management 65:373–378.

Bennett, A. F. 1987. The accomplishments of ecological physiology. Pp. 1–8 *in* M. E. Fedes, A. F. Bennett, W. W. Burggren, and R. B. Huey, eds. New directions in ecological physiology. Cambridge University Press, Cambridge.

Brennan, L. A., and W. M. Block. 2018. Population ecology. Pp. 685–710 in M. L. Morrison, A. D. Rodewald, G. Voelker, M. R. Colón, and J. F. Prather, eds. Ornithology: Foundations, analysis, and application. Johns Hopkins University Press, Baltimore.

Brennan, L. A., J. Wallace, and T. E. Boal. 2013. Implications, management. Wildlife Society Bulletin 37–247.

Clark, R. W. 1986. Case studies in wildlife policy education. Renewable Resources Journal 4:11–16.

Clayton, S., and G. Myers. 2009. Conservation psychology: Understanding and promoting human care for nature. Wiley-Blackwell, Chichester, UK.

Dasmann, R. F. 1964. Wildlife biology. John Wiley & Sons, New York.

Fryxell, J. M., A. R. E. Sinclair, and G. Caughley. 2014. Wildlife ecology, conservation, and management, 3rd edition. John Wiley & Sons, West Sussex, UK.

Gavin, T. A. 1989. What's wrong with the questions we ask in wildlife research? Wildlife Society Bulletin 17:345–350.

Gavin, T. A. 1991. Why ask "why": The importance of evolutionary biology in wildlife science. Journal of Wildlife Management 55:760–766.

Guthery, F. S., 2011. Opinions on management implications. Wildlife Society Bulletin 35:519–522.

Hall, L. S., P. R. Krausman, and M. L. Morrison. 1997. The habitat concept and a plea for standard terminology. Wildlife Society Bulletin 25:173–182.

Hull, D. L. 1988. Science as a process: An evolutionary account of the social and conceptual development of science. University of Chicago Press, Chicago.

Hunter, M. L. 1989. Aardvarks and Arcadia: Two principles of wildlife research. Wildlife Society Bulletin 17:350–351.

Hutto, R. L. 1985. Habitat selection by nonbreeding, migratory land birds. Pp. 455–476 *in* M. L. Cody, ed. Habitat selection in birds. Academic Press, San Diego.

Johnson, D. H. 1980. The comparison of usage and availability measurements for evaluating resource preference. Ecology 61:65–71.

Keppie, D. M. 1990. To improve graduate student research in wildlife education. Wildlife Society Bulletin 18:453–458.

Knight, R. L. 1993. On improving the natural resources and environmental sciences: A comment. Journal of Wildlife Management 57:182–183.

Krausman, P. R. 2001. Introduction to wildlife management: The basics. Pearson, London.

Kroll, A. J. 2009. Integrating professional skills in wildlife student education. Wildlife Society Bulletin 71:226–230.

Kuhn, T. S. 1977. The essential tension: Selected studies in scientific tradition and change. University of Chicago Press, Chicago.

Kuhn, T. S. 1996. The structure of scientific revolutions, 3rd edition. University of Chicago Press, Chicago.

Leopold, A. 1933. Game management. C. Scribner's Sons, New York.

Lopez, R. R., I. D. Parker, and M. L. Morrison. 2017. Applied wildlife habitat management. Texas A&M University Press, College Station.

Marcot, B. G., M. P. Thompson, T. W. Bonnot, and F. R. Thompson. 2016. Uncertainty is information, too. Wildlife Professional 10(1):30–33.

Millstein, R. L. 2009. Populations as individuals. Biological Theory 4:267–73.

Millstein, R. L. 2014. How the concept of "population" resolves concepts of "environment." Philosophy of Science 81:741–755.

Morrison, M. L. 2009. Restoring wildlife: Ecological concepts and practical applications. Island Press, Washington, DC.

Morrison, M. L. 2012. The habitat sampling and analysis paradigm has limited value in animal conservation: A prequel. Journal of Wildlife Management 76:438–450.

Morrison, M. L., W. M. Block, M. D. Strickland, B. A. Collier, and M. J. Peterson. 2008. Wildlife study design, 2nd edition. Springer-Verlag, New York.

Morrison, M. L., L. A. Brennan, B. G. Marcot, W. M. Block, and K. S. McKelvey. 2020. Foundations for Advancing Animal Ecology. Johns Hopkins University Press, Baltimore.

Morrison, M. L., B. G. Marcot, and R. W. Mannan. 2006. Wildlife-habitat relationships: Concepts and applications, 3rd edition. Island Press, Washington, DC.

Morse, D. H. 1980. Behavioral mechanisms in ecology. Harvard University Press, Cambridge, MA.

Person, D. K., M. Kirchhoff, V. van Ballenberghe, G. C. Iverson, and E. Grossman. 1996. The Alexander Archipelago wolf: A conservation assessment. General Technical Report PNW-384. US Department of Agriculture, Forest Service, Pacific Northwest Research Station, Portland, OR.

Risser, P. G. 1993. Making ecological information practical for resource managers. Ecological Applications 3:37–38.

Romesburg, H. C. 1991. On improving the natural resources and environmental sciences. Journal of Wildlife Management 55:744–756.

Romesburg, H. C. 1993. On improving the natural resources and environmental sciences: A reply. Journal of Wildlife Management 57:184–189.

Sands, J. P., S. J. DeMaso, M. J. Schnupp, and L. A. Brennan, eds. 2012. Wildlife science: Connecting research with management. CRC Press / Taylor & Francis Group, Boca Raton, FL.

Saunders, C. D. 2003. The emerging field of conservation psychology. Human Ecology Review 10(2):137–149.

Scott, J. M., P. J. Heglund, M. L. Morrison, J. B. Haufler, M. G. Raphael, W. A. Wall, and F. B. Samson. 2002. Introduction. Pp. 1–5 *in* J. M. Scott, P. J. Heglund, M. L. Morrison, J. B. Haufler, M. G. Raphael, W. A. Wall, and F. B. Samson, eds. 2002. Predicting species occurrences: Issues of accuracy and scale. Island Press, Washington, DC.

Slobodkin, L. B. 1992. A summary of the special feature and comments on its theoretical context and importance. Ecology 73:1564–1566.

Stauffer, D. F. 2002. Linking populations and habitats: Where have we been? Where are we going? Pp. 53–61 *in* J. M. Scott, P. J. Heglund, M. L. Morrison, J. B. Haufler, M. G. Raphael, W. A. Wall, and F. B. Samson, eds. Predicting species occurrences: Issues of accuracy and scale. Island Press, Washington, DC.

Thomas, D. L., and E. J. Taylor. 1990. Study designs and tests for comparing resource use and availability. Journal of Wildlife Management 54:322–330.

Wiens, J. A. 1989. The ecology of bird communities. Vol. 2, Processes and variations. Cambridge University Press, Cambridge.

Index

Page numbers followed by b, f, and t indicate boxes, figures, and tables, respectively.